Multilayered Aquifer Systems

Civil and Environmental Engineering

A Series of Reference Books and Textbooks

Editor

Michael D. Meyer

Department of Civil and Environmental Engineering
Georgia Institute of Technology
Atlanta, Georgia

Additional Volumes in Production

Multilayered Aquifer Systems

Fundamentals and Applications

Alexander H.-D. Cheng
University of Delaware
Newark, Delaware

CRC Press
Taylor & Francis Group
Boca Raton London New York

CRC Press is an imprint of the
Taylor & Francis Group, an **informa** business

CRC Press
Taylor & Francis Group
6000 Broken Sound Parkway NW, Suite 300
Boca Raton, FL 33487-2742

First issued in paperback 2019

© 2000 by Taylor & Francis Group, LLC
CRC Press is an imprint of Taylor & Francis Group, an Informa business

No claim to original U.S. Government works

ISBN-13: 978-0-8247-9875-8 (hbk)
ISBN-13: 978-0-367-39835-4 (pbk)

**Visit the Taylor & Francis Web site at
http://www.taylorandfrancis.com**

**and the CRC Press Web site at
http://www.crcpress.com**

To
Daisy, Jackie, and Julia

Preface

The non-equilibrium pumping well solution for a single-layer aquifer was presented by C.V. Theis in 1935. The solution came as a recognition of an analogy between heat flow and porous medium flow, much as Forchheimer understood it some fifty years ago for the steady state case. Since the 1930s, the Theis theory has been widely accepted in field practice.

Jacob in the 1940s, and Jacob and Hantush in the 1950s, extended Theis' work to the leaky aquifer theory, further considering the flow exchange of an aquifer with its adjacent layers. Many other refinements, such as delayed yield in phreatic aquifers, partially penetrating wells, and well storage followed. However, the formalism of a multilayered aquifer system, taking into account storage as well as hydraulic conductivity for all layers, was not introduced until 1969 by the simultaneous efforts of Neuman and Witherspoon, and Herrera.

Since the 1970s, part of the leaky aquifer theory of Hantush and Jacob has become standard instructional material. The multilayered aquifer theory, on the other hand, has not gained appropriate coverage in groundwater textbooks. The problem with its adoption appears to stem from the mathematical difficulty in its presentation.

During the last several decades, the increase of human water consumption has accelerated the exploitation of groundwater from deeper and deeper aquifers. As there exist interactions among aquifers and aquitards, these adjacent layers must be studied together as a system. As a consequence, the multilayered aquifer theory has increasingly found more applications. To help groundwater professionals practicing the theory, and students entering the profession, a systematic treatment of this material, in a textbook style presentation, has become necessary.

The first part of this book, Chapters 1 through 8, is aimed at providing a pedagogical presentation of the multilayered aquifer theory. It takes a unified approach that builds up from the classical Theis theory step by step to the multilayer theory. A key to the simplified mathematics that makes the unified presentation possible is the use of an automated numerical Laplace inversion procedure.

To enable readers to apply the theory, two types of tools are constructed—one based on the long-standing engineering computation tool, *Fortran* programs, and the other a new breed, macro packages for *Mathematica*™, a computer algebra software. Demonstration of the applications is found throughout the book.

In the second part of the book, Chapters 9 to 12, a few unconventional subjects are introduced. In view of their applications in aquifer management, two chapters, one on sensitivity analysis and the other on stochastic analysis, are provided. These are followed by a chapter on parameter determination, using a computer automated, nonlinear least square approach, in place of the traditional type-curve matching approach. Finally, a chapter on fractured aquifers is presented, due to the increasing exploitation of groundwater in these non-traditional aquifers.

This book is primarily designed for practicing professionals as a self-study and reference book. It contains ideas that are important and are increasingly being applied, but are not covered in a traditional groundwater textbook. It can also be used as a textbook for a graduate course. This book, however, should not be the first book on groundwater that students are exposed to. Students need to gain some qualitative knowledge on groundwater hydrology, for which the current book is not designed. However, it can be used as a supplement book in a first course.

In the first part of the book, efforts are made to keep the level of mathematics low, suitable for an upperclass engineering student. Prerequisites include calculus and ordinary differential equations. Readers also need to be familiar with the notations of partial differential equations, but not with the methods for solving them.

I am indebted to a number of people who have helped in the development of this book. I first learned about porous medium flow from a course taken with Professor Wilfred Brutsaert of Cornell University. Professor Jacob Bear's books provided the basic and advanced knowledge needed to undertake this project. I first became aware of the multilayered aquifer theory in a seminar given by Professor Ismael Herrera, an originator of the theory. I have since become acquainted with Professor Herrera. My subsequent investigation of the multilayered aquifer theory led to the completion of a Master thesis by Mr. Kwotsong Ou and a Doctoral dissertation by Dr. Olushola Morohunfola. Part of their work is incorporated herein. Another former student, Dr. Paston Sidauruk, contributed to the *Mathematica* macro packages and assisted in the preparation of the Parameter Determination chapter. Their contributions are gratefully acknowledged.

A large part of the book was written while I was on a sabbatical leave at the National Taiwan University. I would like to express my sincere gratitude for the hospitality of the Department of Civil Engineering at NTU, for the financial support of the National Science Council, Taiwan, and particularly for the sponsorship of my host Professor Der-Liang (Frank) Young. Several colleagues, Professors Ismael Herrera and Pual Imhoff and Dr. Yongke Mu, kindly reviewed the draft of the book, and provided valuable comments and corrections. The permission given by the authors of the book *Numerical Recipes* to reprint three *Fortran* subroutines in Appendix B is deeply appreciated. I also thank B.J. Clark at Marcel Dekker, Inc. for his patient monitoring of my progress for two years.

Finally, my hearty gratitude goes to my wife Daisy, and daughters Jacqueline and Julia, for enduring the separation of my sabbatical leave, and the long hours that I spent working on this book.

Alexander H.-D. Cheng

Contents

Chapter 1

INTRODUCTION

1.1 Groundwater Resources

Groundwater is a major source of water supply in many parts of the world. It supports domestic consumption, irrigation, and industrial processing. The use of groundwater has been rising steadily in the last several decades. It has been exploited to sustain a growing population and economy. The loss of surface water to pollution has further increased the stress on groundwater extraction. By now, as much as one third of the world's drinking water is derived from groundwater.

Generally speaking, there is an ample amount of freshwater on earth for human consumption. However, its spatial and temporal distribution is uneven. There are places with too much water, and places with too little. There are seasons that are too wet or too dry. Year-to-year fluctuations can cause disasters like flooding or drought. Although overall there is an ample amount, water may not appear at the place where we want it, at the time that we need it, and in the quality that we desire. Hence engineering interventions that include storage and conveyance of water are needed.

To spatially redistribute water, hydraulic structures such as pipelines and open channels have been constructed, but the cost can be quite high. To regulate the time fluctuation of surface water, storage facilities such as reservoirs have been built. Reservoirs inundate land areas. In populated regions, land is difficult to acquire due to the high cost, or it may be unavailable due to unfavorable topography and geology. For existing reservoirs, the loss of water to evaporation and seepage may be significant. The safety of hydraulic structures, such as dams and spillways, during catastrophic events is of great concern. The surge of dam-break flood waves has damaged property and claimed lives. In the last few decades, scientists have realized that the interception and impoundment of a large quantity of water can have detrimental effects on the ecology of a watershed. Several large dam projects were labeled environmental disasters for this reason. As a result of this, international financing or assistance for large dam projects has virtually ceased.

Aquifers, on the other hand, can be used as storage facilities and even as conveyance systems. Excess surface water can be recharged into groundwater reservoirs for future extraction. This type of reservoir does not require any surface area. There is, however, a need for protecting recharge regions, which can be achieved by restricting types of land use. By studying groundwater movement, recharge can be conducted in remote regions while extraction takes place at downstream locations. In semi-arid regions, more aggressive recharge measures, such as using reclaimed wastewater, have been exercised.

Some interesting statistics on groundwater quantity are examined below. It is estimated that of all the water on earth, 99.4% (1.35×10^9 km^3) is surface water. Thus, groundwater occurs only as 0.6% (8.47×10^6 km^3) of that total. The amount of water retained in the atmosphere is negligible (0.001% or 1.3×10^4 km^3). At first glance, it appears that surface water is most abundant. However, most of the surface water is in the form of saltwater in oceans and inland seas (97.2%). Fresh surface water accounts for only 2.1% of the total volume of water.

To have a clearer picture of the freshwater distribution, saltwater is excluded in the accounting from this point on. The balance between surface and groundwater now becomes 77.6% to 22.4%. It still seems that the volume of surface freshwater is three times greater than that of groundwater. However, a further breakdown of freshwater reveals that most of the surface freshwater is in the form of ice that is locked in ice-caps and glaciers in polar regions (77.3% of the total volume of freshwater). As a result, the fresh surface water resources accessible for human consumption are water in lakes (0.3%) and streams (0.003%). This is dwarfed by the amount of shallow (less than 800 m deep) and deep (more than 800 m) groundwater, each of which represents 11.2% of the total freshwater. Generally speaking, shallow groundwater is relatively easy to extract using conventional water well technologies. Deeper formations are harder to access. However, to meet the rapid increase in water demands, advanced technologies developed in the petroleum industry have been used to reach deeper groundwater.

Clearly, surface water is more accessible and is the first to be utilized. But it is also more susceptible to contamination. Past mistakes stemming from ignorance and apathy have polluted large quantities of surface water. Therefore, the deterioration of water quality has effectively diminished its quantity. For several decades, the burden of meeting increases in water demands has been largely transferred to groundwater which, as demonstrated earlier, is relatively abundant.

Despite its abundance, unregulated extraction of groundwater can easily impose significant stresses, causing localized problems. Actually, the above discussion of the quantity of groundwater in terms of its availability is somewhat misleading. Except on rare occasions, it is generally a poor practice to "mine" groundwater, as in the sense of mining minerals or fossil fuels. In mining, a resource is removed without being replaced. Groundwater should be replaced. An acceptable practice is to estimate a "safe yield" based on the average annual recharge amount. Short-term deviation from the safe yield is acceptable, but in the long run, the quantity extracted should not exceed the recharge. Excessive extraction causes the decline of the water table. Old wells may run dry as the water table falls below the screened sections. New wells then need to be drilled deeper to reach the water table, so the cost of lifting water increases. Also, the reduction of pore pressure head in a confined aquifer causes land subsidence. Cities like Venice and Mexico City are well known for their subsidence due to excessive withdrawal of groundwater. The growth of population and the increase in recreational activities in coastal regions has significantly increased groundwater extraction. The reduced freshwater outflow to the sea in coastal aquifers has caused saltwater encroachment.[14] Chloride concentration increases in freshwater wells until they must be abandoned.

Although better protected than surface water, groundwater can also be contaminated. Once contaminated—because of its subsurface, hidden, and inaccessible nature—detection and remediation are more difficult. A stockpile of industrial waste leaking a toxic substance, entering the topsoil through vadose zone, reaching the saturated groundwater, transported by advection and dispersion, can stay undetected until its trace emerges from a water well somewhere downstream. This process can take months, years, or even decades, depending on the type of soil, groundwater gradient, the distance traveled, etc. Once detected, its removal or remediation is difficult, ineffective, or even impossible with present technologies. Because soil particles can sorb contaminants onto its surface, simply pumping out contaminated groundwater to treat it is not sufficient. As a result, clean water flowing into the region will soon become contaminated due to the release of sorbed substances. To solve this problem, repeated flushing is necessary. A constant percentage of the remaining substance is generally removed with each flushing. However, the amount removed diminishes with each subsequent application. Thus, the process becomes more and more ineffective and expensive.

Compounded by the increase in water demands, the shortage of new surface water sources, and the loss of existing surface water to pollution, society is becoming more and more reliant on groundwater resources. A major challenge facing the water resources and hydrogeology professions is the safe extraction of groundwater at a sustainable rate and the protection of its quality.

1.2 Historical Notes

Historical evidence shows that human utilization of groundwater can be traced back to almost four thousand years. For example, Joseph's well in Egypt, which dated from the 17th century B.C., was excavated nearly 300 feet through rocks.[104] The qanāt system in Persia, the earliest of which are three thousand years old, is considered one of the greatest achievements in utilizing groundwater in ancient times.[16] A qanāt is a horizontal well dug into a hill to reach an aquifer. Water is carried over a long distance through a slightly inclined tunnel, which is serviced by vertical shafts. Originating in Armenia, spreading to Persia, and in later centuries to North Africa and Spain, many qanāt systems are still in use today.[109]

In modern history, the foundation of the mechanics of groundwater flow was laid by Henry Philibert Gaspard Darcy (1803-1858).[16] In a study of purification of water by filtration through sand beds for the city of Dijon, France, Darcy suggested that the discharge of the filter flow is directly proportional to the head difference in a relationship now known as Darcy's law:[39]

$$Q = \frac{KA(H+L)}{L} \tag{1.1}$$

In the above, L and A are respectively the thickness and the cross-sectional area of the filtration bed, Q is the discharge, H is the depth of impounded water above the bed, and K is a constant known as hydraulic conductivity. Darcy also correctly reasoned that an artesian aquifer (an aquifer confined at top and bottom by impermeable layers) was analogous to a pipe connecting two reservoirs at different elevations. Artesian wells (self-flowing wells) were similar to standing pipes through which water rose up under pressure.

Darcy's work on groundwater was soon extended by another Frenchman, Arsène Jules Emile Juvenal Dupuit (1804-1866).[16] In his 1863 treatise,[45] Dupuit derived the equation for axial flow toward a pumping well in an unconfined aquifer:

$$Q_w = \frac{\pi K(H^2 - h^2)}{\ln R/r} \tag{1.2}$$

in which Q_w is the well discharge, r is the radial distance from the well, h is the height of water table at r, R is the radius of influence, and H is the height of the water table beyond the zone of influence. He also presented a similar equation for an artesian (confined) aquifer

$$Q_w = \frac{2\pi Kb(H - h)}{\ln R/r} \tag{1.3}$$

where b is the thickness of the aquifer. Dupuit did not provide an explanation of the choice of the value of R.

The pioneering work of Darcy and Dupuit in France was later taken up by Germans and Austrians.[45] Adolph Thiem[115] (1836-1908), a German engineer, re-derived Eqs. (1.2) and (1.3). He conducted extensive field observations to provide empirical formulae for supporting the use of a radius of influence R. These two equations, (1.2) and (1.3), are now known as the Thiem equilibrium equations.

Stimulated by the mathematical development in heat flow, Philip Forchheimer (1852-1933) combined Darcy's law and Dupuit approximation to derive equations for confined and unconfined aquifers, in the form of the Laplace equation:[48]

$$\frac{\partial^2 h}{\partial x^2} + \frac{\partial^2 h}{\partial y^2} = 0 \qquad \text{for confined aquifer} \qquad (1.4)$$

$$\frac{\partial^2 h^2}{\partial x^2} + \frac{\partial^2 h^2}{\partial y^2} = 0 \qquad \text{for unconfined aquifer} \qquad (1.5)$$

where h is the piezometric head. Forchheimer was among the first to apply advanced mathematics, such as complex variable, conformal mapping, and potential theory to solve these equations for groundwater applications. He was also the first to introduce the method of images to groundwater flow.

In the 20th century, Charles V. Theis (1900-1987) is one of the most prominent names associated with the development of groundwater flow theory. While working for the U.S. Geological Survey in the 1930s, Theis became aware of the limitation of the Thiem equilibrium equation for the determination of hydraulic conductivity using drawdown data obtained from pumping tests. Following his keen physical insight, Theis recognized that there existed an analogy between transient groundwater flow and heat flow, similar to Forchheimer's recognition some 50 years ago, for the steady state case. Theis described the problem to his friend Lubin, an engineering professor at the University of Cincinnati, in these words:[122]

The flow of groundwater has many analogies to the flow of heat by conduction. We have exact analogies in ground water theory for thermal gradient, thermal conductivity, and specific heat. I think a close approach to the solution of some of our problems is probably already worked out in the theory of heat conduction. Is this problem in radial flow worked out? Given a plate of given constant thickness and with constant thermal characteristics ...

Consulting a mathematical text,[19] Lubin provided the solution. Theis published a paper[114] in 1935 which contained the following equation*:

$$s = \frac{Q_w}{2\pi T} W(u) \tag{1.6}$$

where $W(u)$ is the well function, $u = r^2 S/4Tt$, s is the drawdown, T is the transmissivity, S is the storativity, r is the radial distance from the well, and t is the time. This equation, known as the Theis equation or the non-equilibrium equation, marked a new era of groundwater aquifer and water well theory.

Theis also suggested an inverse procedure known as the type-curve matching method.[72, 121] Based on the observed water table drawdown in a well due to step pumping, hydrogeologists can plot the drawdown data versus time on a log-log paper. The data is compared with a theoretical curve, known as the type curve, to seek a match point. From this procedure, aquifer transmissivity and storativity can be determined. This kind of graphical procedure for parameter determination has since found many applications in groundwater as well as in underground oil and gas reservoirs.

Taking into account aquifer interaction with its adjacent layers, C. E. Jacob in 1946 first introduced the concept of a leaky aquifer,[74] where water from one aquifer "leaks" into another. His solution was initially limited to the steady state. Mahdi S. Hantush, together with Jacob, provided the non-equilibrium leaky aquifer solution[57] in 1955, written as

$$s = \frac{Q_w}{2\pi T} W(u, \frac{r}{\lambda}) \tag{1.7}$$

where λ is the leakage factor. Hantush continued the refinement of the theory. Other researchers also modified the solution for various

---

*Lubin was invited as a co-author, but declined, citing that *"I would not want to appear as a co-author, first because my part in it was very small, second because from the standpoint of mathematics the work is not of fundamental importance, i.e. to mathematicians the mathematical part is not significant."*

physical considerations, such as partially penetrating wells, wellbore storage, delayed yield in phreatic aquifers, etc.

However, it was not until in the late 1960s and early 1970s that the formalism of multiple-layered, aquifer-aquitard systems was established through the efforts of Shlomo P. Neuman and Ismael Herrera. Two papers, one by Neuman and Witherspoon[88] and the other by Herrera and Figueroa,[63] in the same issue of *Water Resources Research* in 1969, introduced the two-aquifer-one-aquitard system, which was later generalized to multilayered systems.

Another pioneering work worthy of mentioning within the scope of the book is the double porosity model by Grigory I. Barenblatt. In view of the delayed response in fissured reservoirs, Barenblatt[6] proposed in 1960 that for a naturally fractured rock formation there exist two porosity systems with interchanging flow between them. This double porosity theory has found important applications in hydrocarbon producing reservoirs.

Further developments of these models will be described in the appropriate chapters to follow.

1.3 Scope and Organization

The objective of this book is to examine the concepts behind single-layered and multiple-layered aquifer theories, the rationale for adopting such a physical view, the mathematical models that represents the physical system as well as their solutions in various applications, and the physical understanding uncovered from these solutions.

From the view of groundwater mechanics, geological formations can be considered as a single porous continuum with heterogeneous properties. It seems that three-dimensional governing equations and three-dimensional spatial discretization can be indiscriminately applied. However, with the exception of modeling at the local scale, it is hardly the case.

Groundwater systems at the regional scale are conceptualized by geologists as consisting of interacting hydrostratigraphic units of aquifers, confining beds, leaky layers, recharge regions, etc. Information needed in the modeling is found in geologic maps, cross sections, and well logs, which are typically organized and represented in such a fashion. For hydraulic purposes, several formations may be lumped into one hydrostratigraphic unit, and one geologic formation may be subdivided into several units, according to the similarity and dissimilarity in their storage and hydraulic conveyance properties. The resulting units typically have large contrasts in these properties such that the flow pattern is strongly affected by them.

The *aquifer theory* considers each of these hydrostratigraphic units as a hydraulic entity. Dependent on the dominant feature of the unit, certain simplifying assumptions are imposed. The simplifications typically involve the reduction in spatial dimensions or in restriction in flow directions. By considering the interactions among them, these units can be assembled into a groundwater system. The resultant mathematical model for such a system is not a single continuum in a fully three-dimensional space, but is made of pieces that are adapted to the individual hydrostratigraphic units. Its geometry is often referred to as *quasi-three-dimensional*. The rationalization and derivation of such models are the focus of this book.

The aquifer theory was pioneered by Theis in the 1930s, and extended by Jacob and Hantush in the 1940s to 1960s. The more general multilayered aquifer theory was developed in the late 1960s and the early 1970s by the separate efforts of Neuman and Witherspoon, and Herrera. Despite its importance, the multilayered aquifer theory has not found a prominent place in groundwater textbooks. Part of the reason may be attributed to the relatively complex mathematics involved.

The book is intended to provide a clear and systematic derivation of the governing equations of the multilayered aquifer theory. Analytical solutions of pumping wells are presented and applied. The mathematical difficulty is significantly reduced by the use of the Laplace transform. The normally cumbersome Laplace inversion is circumvented by the adoption of an automated and robust numerical inverse algorithm. These analytical solutions enable us to uncover some fundamental insights into the physical system. Examples are provided to illustrate the use of these basic solutions in practical situations.

The organization of the book is as follows. Chapter 1 gives an introduction and a historical perspective. Chapter 2 provides a concise presentation of the fundamental theories of groundwater flow. Chapter 3 investigates the single-layer, non-leaky aquifer solutions. The presentation differs somewhat from the conventional approach by the introduction of Green's function, the Laplace transform, and the numerical Laplace inversion. Both the *Fortran* and the *Mathematica*™ tools are constructed for the numerical evaluation of the solutions. Chapter 4 sets up the fundamental assumptions of leaky aquifer theory and explores the classical work of Jacob and Hantush. In Chapter 5, the solutions developed in the two preceding chapters are applied in various scenarios.

Chapter 6 gives a step-by-step derivation of the governing equations of multilayered aquifer theory. Chapter 7 is devoted to the pumping well solutions of multilayered aquifer systems. The applications of these solutions are found in Chapter 8. Starting with Chapter 9, several non-traditional topics are introduced, led by a sensitivity analysis. Chapter 10 discusses the determination of aquifer parameters. Chapter 11 delves into the uncertainty issue by providing a stochastic analysis. Finally, Chapter 12 presents the double porosity model of Barenblatt, corresponding to fractured aquifers, and its generalization to the multiple porosity model.

Chapter 2

FUNDAMENTALS

2.1 Darcy's Law

Darcy presented his law of seepage as an empirical law by observing flow through filtration bed. Although we can derive the equation via a number of more rigorous continuum mechanical or statistical approaches by averaging Navier-Stokes equations over a *Representative Elementary Volume* (REV),[11] we shall retain Darcy's intuitive approach for simplicity of presentation.

We first introduce a *piezometric head h* defined as

$$h = \frac{p}{\gamma} + z \qquad (2.1)$$

where p is the pressure, γ is the specific weight of water, and z is the elevation above the datum. The concept of *head* as an energy has long been established by hydraulic engineers. It is rooted in *Bernoulli's law* in fluid mechanics, in which the *total head* is given by

$$h_T = \frac{p}{\gamma} + z + \frac{v^2}{2g} \qquad (2.2)$$

where v is the flow velocity, and g is the gravity acceleration. In Eq. (2.2), we can identify terms of kinetic energy (velocity head), $v^2/2g$, potential energy (elevation head), z, and work done by pressure (pressure head), p/γ. These quantities have the dimension of length because they are *energy per unit weight* of the substance.

A *piezometer* is a tube inserted into the flow with its opening not facing the flow, see Figure 2.1. It measures the total head without

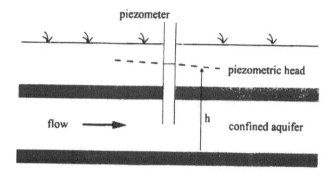

FIGURE 2.1. Illustration of piezometric head.

the velocity head. This leads to the reference of the piezometric head, Eq. (2.1). A water well is naturally a piezometer.

A key assumption of groundwater flow is that the piezometric head is regarded as the total head (energy) of the flow. The presence of kinetic energy is neglected due to the smallness of seepage velocity. This assumption conveniently leads to a linear theory as implied by Darcy's law. Conversely, it places a restriction on the application of Darcy's law. When the velocity gets large, as it can happen in flow in coarse sand and gravel, the theory needs to be modified and becomes nonlinear. This part of groundwater flow theory, however, is beyond the present scope, hence is not elaborated.

From Darcy's law, we define a *specific discharge* as

$$q = \frac{Q}{A} \qquad (2.3)$$

where Q is the fluid discharged from a cross-section of porous medium, and A is the cross-sectional area. Although q has the dimension of velocity, the term "specific discharge" is used to distinguish it from the *average velocity*, given by

$$v = \frac{Q}{A\phi} = \frac{q}{\phi} \qquad (2.4)$$

where ϕ is the *porosity*, defined as the ratio of the void volume in a porous medium to the total volume (solid and void). In fact, ϕ should be referred to as the *effective porosity*, as only the voids that contribute to the flow channels are counted. The average velocity is more closely related to the velocity of flow in the pores. It is the quantity that should be referred to if solute transport in porous media is considered. However, for water supply purposes, the use of specific discharge is preferred as it requires neither the introduction of the porosity concept nor its measurement.

Darcy recognized that the specific discharge q is proportional to the energy available to drive the flow and is inversely proportional to the distance that the fluid needs to travel. This is analogous to pipe flow in which energy is dissipated over the distance to overcome frictional loss resulting from fluid viscosity. When the magnitude of q is small (to justify the use of piezometric rather than total head), the following linear relation is established:

$$q = -K \frac{\Delta h}{\Delta L} \qquad (2.5)$$

in which L is a measure of length along the flow path, Δ denotes the difference of a quantity between two points, consistently applied for h and L, and K is a proportionality constant known as *hydraulic conductivity*. The negative sign shows that flow moves in the direction of decreasing head. From Eq. (2.5) it is clear that K has the dimension of velocity.

Example: *Consider a sand filtering bed with thickness $b = 1$ m and hydraulic conductivity 100 m/day. Water is ponded on top of the bed to a depth of $d = 2$ m, see Figure 2.2. The bottom of the bed is supported by a perforated slab and is exposed to the atmosphere. What is the specific discharge? How much water is filtered per square meter of the bed in one day?*

The piezometric head is 3 m at the top of the sand filter, following the definition Eq. (2.1) (2 m from the hydrostatic pressure of ponded water, and 1 m from elevation), and zero at the bottom. Hence

$$q = -100 \, \frac{\text{m}}{\text{day}} \times \frac{(3\,\text{m} - 0\,\text{m})}{1\,\text{m}} = -300 \, \frac{\text{m}}{\text{day}}$$

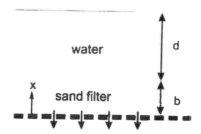

FIGURE 2.2. Seepage through a sand filter bed.

Total water volume filtered per square meter in one day is

$$\text{Volume} = 300\,\frac{\text{m}}{\text{day}} \times 1\,\text{m}^2 \times 1\,\text{day} = 300\,\text{m}^3$$

The original Darcy's law considered only water as the fluid; hence, the properties of water were incorporated into the definition of hydraulic conductivity. If different fluids are considered, the conductibility of the porous medium will change. It is desirable to be able to isolate the fluid properties from that of the porous solid in the definition of K. One way to derive the relation is to build a "conceptual model". For example, the porous medium flow may be considered as a laminar flow inside a group of straight tubes, as in the *capillary tube model*.[11] However, we shall take a different approach based on the reasoning of dimensional analysis.

It is obvious that the fluid dynamic viscosity μ plays an important role in determining the value of hydraulic conductivity. If the pore water in the porous medium is replaced by a more viscous liquid, the conductance will obviously decrease. Hence viscosity should be explicitly modeled. Other important factors are some bulk parameters that represent the microscopic geometry of the porous medium. These properties can only have the dimension of length, or are dimensionless. The first choice seems to be a typical pore size, or a typical grain size, denoted as d. The combination of these two parameters, μ and d, however cannot create the dimension of hydraulic conductivity. Some quantity is still missing, which requires further physical insight.

Going back to Darcy's law Eq. (2.5), we observe that the "head" h is not a fundamental quantity. It is the energy per unit *weight* of fluid. Its definition is tied to earth's gravitational field. A correction factor, the specific weight γ, should be used to convert it to γh, the energy per unit *volume* of substance. The definition of K hence should include a factor γ.

Judging from the dimension of these quantities, we form the following relation

$$K \sim \frac{\gamma d^2}{\mu} = \frac{\rho g d^2}{\mu} \tag{2.6}$$

where ρ is the fluid density and g is the gravitational acceleration. Introducing a proportionality constant C, we can write the equality

$$K = \frac{C \rho g d^2}{\mu} = \frac{\rho g k}{\mu} \tag{2.7}$$

In the above we have defined a quantity known as the *intrinsic permeability*, as

$$k = C d^2 \tag{2.8}$$

The intrinsic permeability k has the dimension of length square and is a property of the porous medium alone. C is also a geometric factor that can be further related to the porosity, pore shape, specific surface, tortuosity, pore size distribution, etc.[107]

Another way to write Eq. (2.5) follows the petroleum engineering convention:

$$q = -\frac{k}{\mu} \frac{\Delta p}{\Delta L} = -\kappa \frac{\Delta p}{\Delta L} \tag{2.9}$$

in which p takes the meaning of *dynamic pressure*, namely the perturbation of pressure from its hydrostatic value, and

$$\kappa = k/\mu \tag{2.10}$$

is known as the *mobility coefficient*.

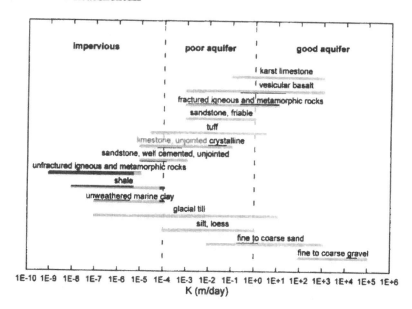

FIGURE 2.3. Typical values of hydraulic conductivity for various geological materials. (*Source*: After Driscoll.[44])

To have a broader application, the one-dimensional Darcy's law must be generalized. First, the *difference equation* (2.5) should be transformed into a *differential equation*:

$$q = -K \frac{dh}{dx} \qquad (2.11)$$

Furthermore, in three spatial dimensions, Darcy's law is extended to

$$
\begin{aligned}
q_x &= -K \frac{\partial h}{\partial x} \\
q_y &= -K \frac{\partial h}{\partial y} \\
q_z &= -K \frac{\partial h}{\partial z} \qquad (2.12)
\end{aligned}
$$

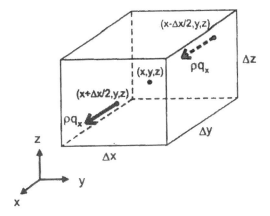

FIGURE 2.4. Derivation of fluid mass conservation.

which, in vector form, becomes

$$
\begin{aligned}
\mathbf{q} &= q_x\,\mathbf{i} + q_y\,\mathbf{j} + q_z\,\mathbf{k} \\
&= -K\left(\frac{\partial h}{\partial x}\,\mathbf{i} + \frac{\partial h}{\partial y}\,\mathbf{j} + \frac{\partial h}{\partial z}\,\mathbf{k}\right) \\
&= -K\,\nabla h
\end{aligned}
\tag{2.13}
$$

where ∇ is the *gradient operator*.

In Figure 2.3 the typical ranges of hydraulic conductivity for various geological formations are illustrated.

2.2   Continuity Equation

Although not explicitly shown, Darcy's law has taken into consideration the momentum and the energy conservation. In this section, the mass conservation is further examined.

Given an infinitesimal cubic volume of a porous medium centered at (x, y, z), with side lengths Δx, Δy, and Δz (see Figure 2.4), the conservation of fluid mass can be stated as

fluid mass influx across the surfaces

−fluid mass outflux across the surfaces

= rate of change of fluid mass in porous frame   (2.14)

First, we check the net flux in the x-direction, across the two shaded surfaces shown in Figure 2.4, located at $x - \Delta x/2$ and $x + \Delta x/2$ respectively:

net flux in x-direction =

$$\left[\rho q_x \Big|_{(x-\frac{\Delta x}{2}, y, z)} - \rho q_x \Big|_{(x+\frac{\Delta x}{2}, y, z)} \right] \Delta y \, \Delta z \qquad (2.15)$$

We can take Taylor series expansion of these terms about the center of the box, (x, y, z), such that

$$\rho q_x \Big|_{(x-\frac{\Delta x}{2}, y, z)} = \rho q_x \Big|_{(x,y,z)} - \frac{\partial \rho q_x}{\partial x} \Big|_{(x,y,z)} \frac{\Delta x}{2} + \cdots \qquad (2.16)$$

$$\rho q_x \Big|_{(x+\frac{\Delta x}{2}, y, z)} = \rho q_x \Big|_{(x,y,z)} + \frac{\partial \rho q_x}{\partial x} \Big|_{(x,y,z)} \frac{\Delta x}{2} + \cdots \qquad (2.17)$$

where we have neglected the higher order terms as they vanish in the process of taking the limit $\Delta x \to 0$. Subtracting these two equations we obtain from Eq. (2.15)

$$\text{net flux in } x\text{-direction} = -\frac{\partial \rho q_x}{\partial x} \Delta x \Delta y \, \Delta z \qquad (2.18)$$

Similar equations can be written for the net flux in the y- and z-direction. Summing up the net flux in all three directions, we can equate it to the fluid mass change in the cubic volume, following Eq. (2.14):

$$-\left(\frac{\partial \rho q_x}{\partial x} + \frac{\partial \rho q_y}{\partial y} + \frac{\partial \rho q_z}{\partial z} \right) \Delta x \Delta y \, \Delta z = \frac{\rho \Delta V_f}{\Delta t} \qquad (2.19)$$

where ΔV_f is the volume of fluid gained in the frame.

The change of fluid mass in a porous frame is normally accompanied by a change in pore pressure, when water is compressed into or expands out of a fixed amount of pore space. However, this is not the only mechanism through which water can enter or leave a frame. Compressibility exists not only for the fluid, but also for the solid frame. Consider, for example, an unconsolidated soil sample composed of nearly incompressible solid grains. The grains are stacked in a random arrangement forming a skeleton containing a significant proportion of interstitial space. When such a sample is compressed, its bulk volume readily decreases. The compressibility obviously does not come from solid grains, as they are incompressible. All deformation must come from the change in pore space, which in turn causes a change in fluid volume, as the pore space is entirely occupied by the fluid. This can take place without a fluid pressure change.

Taking into account both mechanisms, we separate the fluid volume change ΔV_f into two parts, a part due to the fluid compression by fluid pressure, $\Delta V_f^{(1)}$, and the other caused by the frame (pore space) deformation, $\Delta V_f^{(2)}$, such that

$$\Delta V_f = \Delta V_f^{(1)} + \Delta V_f^{(2)} \tag{2.20}$$

In the following, these two effects are separately examined under the condition of a change in aquifer piezometric head due to water extraction or injection.

Based on the definition of piezometric head, Eq. (2.1), a change in piezometric head Δh means a change in fluid pressure of the amount $\Delta p = \gamma \Delta h$, as the elevation does not change. Given a fluid compressibility C_f, the amount of fluid compressed into the frame due to a positive pressure change is

$$\Delta V_f^{(1)} = C_f \Delta p V_f = C_f \Delta p \phi V \tag{2.21}$$

where ϕ is the porosity, V_f is the fluid volume, and V is the total volume (solid and fluid) of the sample.

In a confined aquifer compressed under a constant *overburden pressure*, i.e. the weight of the material above it, the load is shared by the solid skeleton and the fluid. A decrease in pore pressure reduces the fluid's share, causing a load transfer to the solid. The frame is then further compressed and a corresponding amount of fluid is expelled.

A mechanical law of this phenomenon can be established using Terzaghi's *effective stress* concept.[113] Assuming that the pore water and the solid grains are nearly incompressible as compared to the compressibility of the solid frame, Terzaghi decomposed the total overburden compressive stress, σ_T, into a pressure part, p, and an effective stress part, σ_e, such that

$$\sigma_T = \sigma_e + p \tag{2.22}$$

Terzaghi deduced that the deformation of the solid frame is proportional to the effective stress, not the total stress.

Defining a solid frame compressibility C_s, the change of solid frame volume is

$$-\Delta V = C_s \Delta \sigma_e V \qquad (2.23)$$

where a positive ΔV defines expansion. For a constant overburden, an increase in pore pressure Δp causes a decrease in effective stress of the amount $\Delta \sigma_e = -\Delta p$. Furthermore, as the solid grains are incompressible, the entire change of volume in Eq. (2.23) is derived from pore space change. Since the fluid resides in the pore space, an equal amount of fluid must be expelled or absorbed corresponding to the variation of pore space. Hence it can be concluded that $\Delta V = \Delta V_f^{(2)}$. Using these relations in Eq. (2.23) we obtain

$$\Delta V_f^{(2)} = C_s \Delta p V \qquad (2.24)$$

Finally, corresponding to a piezometric head change Δh, the total fluid volume change ΔV_f is the sum of the two parts, Eqs. (2.21) and (2.24):

$$\Delta V_f = (C_s + \phi C_f)\Delta p V = (C_s + \phi C_f)\gamma \Delta h V \qquad (2.25)$$

This is the *constitutive law* that relates the fluid volume change inside the frame to a piezometric head change.

In field practices, it is normally difficult to separate aquifer responses into a fluid portion and a solid frame portion. Therefore a lumped property, known as *specific storage* is adopted. A specific storage (or *specific storativity*) is defined as the volume of fluid released from a unit volume of porous medium following a unit decline of head,

$$S_s = \frac{\Delta V_f}{V \Delta h} \qquad (2.26)$$

which has the dimension of $[L^{-1}]$. Based on Eq. (2.25), S_s is apparently a composite property of solid frame and fluid

$$S_s = \gamma(C_s + \phi C_f) \qquad (2.27)$$

TABLE 2.1. Typical specific storage of various geological materials. (Source: Converted from Domenico & Schwartz[43] and Detournay & Cheng.[41])

Aquifer Material	C_s (m²/N)	ϕ	S_s (1/m)
Plastic clay	1×10^{-6}	0.5	9.8×10^{-3}
Stiff clay	2×10^{-7}	0.4	2.0×10^{-3}
Medium-hard clay	1×10^{-7}	0.4	9.8×10^{-4}
Loose sand	8×10^{-8}	0.4	7.9×10^{-4}
Dense sand	2×10^{-8}	0.3	2.0×10^{-4}
Dense sandy gravel	8×10^{-9}	0.3	8.0×10^{-5}
Fissured jointed rock	5×10^{-10}	0.1	5.4×10^{-6}
Boise sandstone	1×10^{-10}	0.26	2.2×10^{-6}
Weber sandstone	3×10^{-11}	0.06	5.9×10^{-7}

In Table 2.1 we summarize some typical values of vertical frame compressibility of aquifer materials. Referring to the compressibility of water $C_f = 5 \times 10^{-10}$ (m²/N), the specific storage is calculated from Eq. (2.27) and presented in Table 2.1. Figure 2.5 shows the observed range of specific storage for some geological materials.

Example: *A hydraulically isolated sandstone aquifer has an area of 100 km², a thickness of 50 m, and a porosity of 0.2. It was estimated that the volume of water extracted from this aquifer over a period of time was 1×10^6 m³. The average drop of piezometric head in this aquifer during this period was 20 m. Estimate the specific storage and also the compressibility of the sandstone.*

In 1928, Meinzer[82] first realized that the solid matrix compressibility is an important mechanism for releasing water from a confined aquifer. Prior to his time, water compressibility was thought to be the dominating factor. We can estimate the effect of water compressibility as follows.

The observed 20 m head drop is equivalent to a pressure drop of

$$\Delta p = \gamma \Delta h = 9800 \, \text{N/m}^3 \times 20 \, \text{m} = 1.96 \times 10^5 \, \text{N/m}^2$$

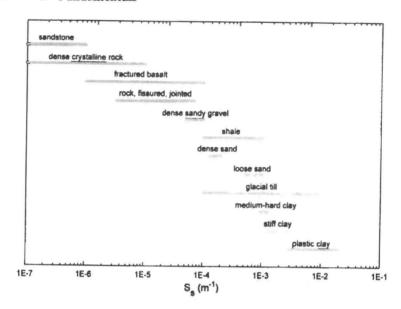

FIGURE 2.5. Range of specific storage for various geological materials. (*Source*: After Domenico,[42] Mercer, et al.,[83] and Leap.[78])

By the compressibility of water, $C_f = 5 \times 10^{-10}$ (m^2/N), water will expand by the amount

$$C_f \Delta p = \left(5 \times 10^{-10}\, \mathrm{m^2/N}\right) \times \left(1.96 \times 10^5\, \mathrm{N/m^2}\right) = 9.8 \times 10^{-5}$$

per unit volume. Since water occupies 20% of aquifer volume, the total water released from water expansion is

$$
\begin{aligned}
\Delta V &= \left(9.8 \times 10^{-5}\right) \times \left(1 \times 10^8\, \mathrm{m^2}\right) \times 50\,\mathrm{m} \times 0.2 \\
&= 9.8 \times 10^{-4}\, \mathrm{m^3}
\end{aligned}
$$

This is about 10% of total water extracted. 90% of it is missing! Based on a similar estimate, Meinzer deduced that the aquifer matrix compressibility plays an important part in the specific storage of aquifer.

For the present problem, the specific storage is calculated using the definition:

$$S_s = \frac{\text{volume of fluid released}}{\text{volume of aquifer} \times \text{head drop}}$$

$$= \frac{1 \times 10^6 \, \text{m}^3}{(1 \times 10^8 \, \text{m}^2) \times 50 \, \text{m} \times 20 \, \text{m}} = 1 \times 10^{-5} \, \text{m}^{-1}$$

Based on Eq. (2.27), we calculate $C_s = 9.2 \times 10^{-10} \, \text{m}^2/\text{N}$.

We observe from Table 2.1 that for materials ranging from plastic clay to dense sandy gravel, the contribution of water compressibility to specific storage is less than 2%.

To complete the derivation, we substitute Eq. (2.26) into Eq. (2.19), take the limit of Δh, $\Delta t \to 0$, and produce the *continuity equation*

$$S_s \frac{\partial h}{\partial t} = - \left(\frac{\partial q_x}{\partial x} + \frac{\partial q_y}{\partial y} + \frac{\partial q_z}{\partial z} \right) \tag{2.28}$$

in which $V = \Delta x \Delta y \Delta z$ has been used. We also note that the spatial variation of density, $\partial \rho / \partial x$, $\partial \rho / \partial y$, etc., which can arise from the different degree of water compression at various locations, has been neglected in the above derivation. Its magnitude is small due to the small deformation involved. The approximation is necessary in order to achieve a linear mathematical theory.

2.3   Heterogeneity and Anisotropy

A porous medium is said to be *homogeneous* if its hydraulic conductivity K and specific storage S_s are the same everywhere. If not, it is *inhomogeneous*, or *heterogeneous*. Heterogeneous properties can be expressed as $K = K(x, y, z)$ and $S_s = S_s(x, y, z)$ to emphasize their dependence on spatial coordinates. If a porous medium can be divided into zones and within each zone K and S_s are constant, the medium is called *piece-wise homogeneous*.

In Sec. 2.1 the porous medium is assumed to be *isotropic*, namely that the material's ability to conduct water at a given point is the same in all directions. However, hydraulic conductivity (but not specific storage) can have directional properties. A number of possibilities exist that can create preferential flow channels in porous media:

- Flat shaped sedimentary particles have the tendency to lie on their flat sides during sedimentation and consolidation processes.

- Geological formations are often layered. If several layers are lumped together to form a hydrostratigraphic unit, its equivalent hydraulic conductivity is larger parallel to the layers than perpendicular to them.

- In carbonate rocks, fluid can dissolve rock into channels parallel to flow directions.

- Tectonic movements induce shear in rocks creating fissures aligned in the preferential stress directions.

- Anisotropic *in situ* stresses can deform pores into shapes that favor directional permeability.

Under these circumstances the porous medium is *anisotropic* and Darcy's law is extended to

$$
\begin{aligned}
q_x &= -K_{xx}\frac{\partial h}{\partial x} - K_{xy}\frac{\partial h}{\partial y} - K_{xz}\frac{\partial h}{\partial z} \\
q_y &= -K_{yx}\frac{\partial h}{\partial x} - K_{yy}\frac{\partial h}{\partial y} - K_{yz}\frac{\partial h}{\partial z} \\
q_z &= -K_{zx}\frac{\partial h}{\partial x} - K_{zy}\frac{\partial h}{\partial y} - K_{zz}\frac{\partial h}{\partial z}
\end{aligned}
\tag{2.29}
$$

where K_{xx}, K_{xy}, etc. are directional hydraulic conductivities, which can be put into a matrix form:

$$
\mathbf{K} = \left[\begin{array}{ccc} K_{xx} & K_{xy} & K_{xz} \\ K_{yx} & K_{yy} & K_{yz} \\ K_{zx} & K_{zy} & K_{zz} \end{array} \right]
\tag{2.30}
$$

One of the implications of Eq. (2.29) is that the flow in a certain direction is driven not only by the head gradient in the same direction, but also by gradients in the perpendicular directions.

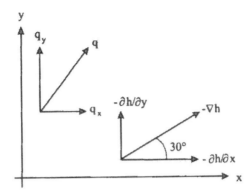

FIGURE 2.6. Hydraulic gradient and flow directions in an anisotropic hydraulic conductivity field.

Example: *In a two-dimensional flow field, the hydraulic conductivity matrix is*

$$\mathbf{K} = \begin{bmatrix} 10 & 6 \\ 6 & 20 \end{bmatrix} \text{ m/day}$$

A negative hydraulic gradient is found in the 30° direction with respect to the x-axis (see Figure 2.6), with the magnitude 0.01 m/m. Find the components of specific flux the flow direction.

The hydraulic gradient can be decomposed into

$$\frac{\partial h}{\partial x} = -0.01 \times \cos 30° = -0.00866$$

$$\frac{\partial h}{\partial y} = -0.01 \times \sin 30° = -0.005$$

Following Eq. (2.29), we obtain

$$
\begin{aligned}
q_x &= -(10\,\text{m/day}) \times (-0.00866) - (6\,\text{m/day}) \times (-0.005) \\
&= 0.117\,\text{m/day} \\
q_y &= -(6\,\text{m/day}) \times (-0.00866) - (20\,\text{m/day}) \times (-0.005) \\
&= 0.152\,\text{m/day}
\end{aligned}
$$

The flow direction is

$$\theta = \tan^{-1} \frac{0.152}{0.117} = 52.4°$$

and the magnitude is

$$|q| = \sqrt{0.117^2 + 0.152^2} = 0.192\,\text{m/day}$$

Equation (2.29) is a physical law and should remain valid under any rotation of the coordinate frame. To have such a property, **K** must be a *second order tensor*[5] and obey the following coordinate transformation law in the Cartesian system

$$K'_{ij} = \ell_{ik}\ell_{j\ell}K_{k\ell} \qquad \text{for} \quad i,j,k,\ell = 1,2,3 \qquad (2.31)$$

where $K_{k\ell}$ and K'_{ij} are respectively hydraulic conductivity tensors in the original coordinate system (x, y, z) and the new system (x', y', z'), ℓ_{ij} is the directional cosine

$$\ell_{ij} = \cos\theta_{ij} \qquad (2.32)$$

in which θ_{ij} is the angle between the ith-axis of the old system and the jth-axis of the new system. In the above, index notations have been utilized for a concise presentation. In particular, it should be noted that the *Einstein summation convention*, namely that repeated indices are summed, is implied in Eq. (2.31).

Based on thermodynamic considerations, it is generally agreed that the coefficient tensor for an energy dissipating system, such as thermal conductivity and hydraulic conductivity, should be symmetric. The hydraulic conductivity tensor hence has the symmetry property:

$$K_{ij} = K_{ji} \qquad \text{for} \quad i, j = 1, 2, 3 \qquad (2.33)$$

By the property of a symmetric tensor, it is well known that it is possible to rotate the coordinate frame to a set of mutually perpendicular *principal axes*, say, (x', y', z'), such that the transformed hydraulic conductivity tensor contains only three principal values as diagonal terms:

$$\mathbf{K}' = \begin{bmatrix} K'_{x'} & 0 & 0 \\ 0 & K'_{y'} & 0 \\ 0 & 0 & K'_{z'} \end{bmatrix} \tag{2.34}$$

In the above, we have shortened the number of indices for the diagonal terms to one, for convenience.

Given hydraulic conductivity matrix $\mathbf{K}$ in an arbitrary orientation, the problem of finding the principal axes and principal hydraulic conductivities is given by the solution of an *eigensystem*:

$$|\mathbf{K} - \lambda \mathbf{I}| = 0 \tag{2.35}$$

where $\mathbf{I}$ is the identity matrix, λ is a constant, and $|\cdots|$ denotes determinant. The three λ values that satisfy the above equation are *eigenvalues*. These values are the principal hydraulic conductivities. The three *eigenvectors* then define the principal axes.

Example: *In a given coordinate system, the hydraulic conductivity matrix is*

$$\mathbf{K} = \begin{bmatrix} 10 & 2 & 3 \\ 2 & 5 & 4 \\ 3 & 4 & 12 \end{bmatrix} \text{ m/day}$$

Find the principal axes and principal hydraulic conductivities.

The numerical methodology for solving eigensystems can be found in most numerical analysis books,[101] hence is not presented here. There also exists computer software, such as *Mathematica*™ or *Matlab*™, that can be directly used for this purpose. For example, the command

```
Eigensystem[N[{{10, 2, 3}, {2, 5, 4}, {3, 4, 12}}]]
```

in *Mathematica* returns the eigenvalues as well as the eigenvectors. The three principal directions are found to be $-0.51\,\mathbf{i} - 0.38\,\mathbf{j} - 0.77\,\mathbf{k}$, $0.85\,\mathbf{i} - 0.12\,\mathbf{j} - 0.51\,\mathbf{k}$, and $0.10\,\mathbf{i} - 0.92\,\mathbf{j} + 0.38\,\mathbf{k}$. The corresponding principal hydraulic conductivities are 16.0, 7.92, and 3.12 m/day, respectively. More detail of the software *Mathematica* will be given in a later section, Sec. 3.11.

Although the three-dimensional eigensystem is better solved by a numerical technique, the solution of a two-dimensional eigensystem can be explicitly obtained. In that case, Eq. (2.35) is expanded as

$$
\begin{vmatrix} K_{xx} - \lambda & K_{xy} \\ K_{yx} & K_{yy} - \lambda \end{vmatrix} = \lambda^2 - (K_{xx} + K_{yy})\lambda + (K_{xx}K_{yy} - K_{xy}^2)
$$

$$
= 0 \tag{2.36}
$$

The two eigenvalues are solved from the quadratic equation. Equating them to the principal hydraulic conductivities, we obtain

$$
K_{x'}' = \frac{K_{xx} + K_{yy}}{2} + \sqrt{\frac{(K_{xx} - K_{yy})^2}{4} + K_{xy}^2}
$$

$$
K_{y'}' = \frac{K_{xx} + K_{yy}}{2} - \sqrt{\frac{(K_{xx} - K_{yy})^2}{4} + K_{xy}^2} \tag{2.37}
$$

The angle of rotation needed to reach the principal axes is given as

$$
\theta = \frac{1}{2}\tan^{-1}\frac{2K_{xy}}{K_{xx} - K_{yy}} \tag{2.38}
$$

where θ is measured counterclockwise.

In the reversed setup, we are given the principal axes x' and y', and the principal values $K_{x'}'$ and $K_{y'}'$, and are asked to find the hydraulic conductivity tensor K_{ij} after a clockwise rotation of angle θ. We can use the following formulae

$$
K_{xx} = \frac{1}{2}\left(K_{x'}' + K_{y'}'\right) + \frac{1}{2}\left(K_{x'}' - K_{y'}'\right)\cos 2\theta
$$

$$
K_{yy} = \frac{1}{2}\left(K_{x'}' + K_{y'}'\right) - \frac{1}{2}\left(K_{x'}' - K_{y'}'\right)\cos 2\theta
$$

$$
K_{xy} = -\frac{1}{2}\left(K_{x'}' - K_{y'}'\right)\sin 2\theta \tag{2.39}
$$

to perform the conversion.

Once we are aligned with the principal axes, Darcy's law as shown in Eq. (2.29) simplifies to:

$$
\begin{aligned}
q_x &= -K_x \frac{\partial h}{\partial x} \\
q_y &= -K_y \frac{\partial h}{\partial y} \\
q_z &= -K_z \frac{\partial h}{\partial z}
\end{aligned}
\tag{2.40}
$$

In the above we have dropped the prime denoting principal axes for convenience.

It is important to emphasize that heterogeneity and anisotropy are two independent concepts. As a consequence, the following four combinations are possible:

- Homogeneous and isotropic: K is a constant.
- Heterogeneous and isotropic: $K = K(x, y, z)$.
- Homogeneous and anisotropic: K_{ij} are constants.
- Heterogeneous and anisotropic: $K_{ij} = K_{ij}(x, y, z)$.

2.4   Field Equations

We have so far introduced two sets of governing equations for groundwater flow—Darcy's law given as Eq. (2.29) and the continuity equation as Eq. (2.28). If the aquifer physical properties, S_s and K_{ij}, are given, Eqs. (2.29) and (2.28) are four equations with four unknowns, q_x, q_y, q_z, and h, forming a simultaneous solution system.

It is however more desirable to eliminate some variables to reduce the number of equations to be simultaneously solved. This can be achieved by substituting Eq. (2.29) into Eq. (2.28), to obtain a single three-dimensional field equation

$$
S_s \frac{\partial h}{\partial t} = \frac{\partial}{\partial x_i} \left(K_{ij} \frac{\partial h}{\partial x_j} \right)
\tag{2.41}
$$

which contains a single unknown h. In the above, tensor notation has been utilized to shorten the formula. In its full form, the right hand side must be expanded into 9 terms by summing over the indices $i, j = 1, 2, 3$. The presentation is more efficient if the coordinate system is aligned with the principal axes. Equation (2.41) then simplifies to

$$S_s \frac{\partial h}{\partial t} = \frac{\partial}{\partial x}\left(K_x \frac{\partial h}{\partial x}\right) + \frac{\partial}{\partial y}\left(K_y \frac{\partial h}{\partial y}\right) + \frac{\partial}{\partial z}\left(K_z \frac{\partial h}{\partial z}\right) \qquad (2.42)$$

For a homogeneous, anisotropic medium, it becomes

$$S_s \frac{\partial h}{\partial t} = K_x \frac{\partial^2 h}{\partial x^2} + K_y \frac{\partial^2 h}{\partial y^2} + K_z \frac{\partial^2 h}{\partial z^2} \qquad (2.43)$$

For a heterogeneous, isotropic medium, it takes the form

$$S_s \frac{\partial h}{\partial t} = \nabla \cdot (K \nabla h) \qquad (2.44)$$

where $\nabla\cdot$ is the *divergence operator*, whose definition is given as Eq. (2.48) at the end of the section. For a homogeneous, isotropic medium, we find further simplification

$$\frac{S_s}{K} \frac{\partial h}{\partial t} = \nabla^2 h \qquad (2.45)$$

where ∇^2 is the *Laplacian operator*, see Eq. (2.49). Equation (2.45) is known as a *diffusion equation* in mathematics. Finally, when a steady state exists, Eq. (2.45) reduces to the well known *Laplace equation*:

$$\nabla^2 h = 0 \qquad (2.46)$$

Depending on the physical parameters, one of the above equations, (2.41) to (2.46), should be used for the solution of piezometric head in an initial and boundary value problem.

Example: *A homogeneous, isotropic, confined aquifer of constant thickness is cut through by two parallel rivers. Figure 2.7 gives a*

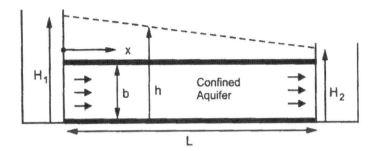

FIGURE 2.7. Flow in a confined aquifer between two rivers.

cross-sectional view. The hydraulic conductivity is $K = 10$ m/day, thickness is $b = 4$ m, specific storage is $S_s = 5 \times 10^{-4}$ m^{-1}, and the distance between two rivers is $L = 1$ km. Initially, water levels in the rivers are $H_1 = 10$ m and $H_2 = 5$ m with a steady state flow established in the aquifer. At $t = 0$, water level in the left river is suddenly lowered by 3 m. Solve for the resultant transient flow in the aquifer.

In view of the geometry, the problem can be treated as a one-dimensional problem. First consider the steady state flow condition prior to the river stage change. The appropriate governing equation to use is Eq. (2.46). Expressed in one-dimension, it is

$$\frac{d^2 h_1}{dx^2} = 0$$

where h_1 is used to denote the piezometric head. The above equation has the simple general solution

$$h_1 = Ax + B$$

Using the two boundary conditions $h_1 = H_1$ at $x = 0$ and $h_1 = H_2$ at $x = L$, we obtain the initial steady state solution

$$h_1 = \frac{H_2 - H_1}{L} x + H_1$$

The flow exchanged between the two rivers per meter of river length is

$$Q = -Kb\frac{H_2 - H_1}{L}$$

$$= -\frac{10\frac{m}{day} \times 4m \times (5m - 10m)}{1000m} = 0.2\ m^3/m \cdot day$$

When the boundary condition on the left side is suddenly changed, a transient response is induced in the aquifer. The one-dimensional diffusion equation is given as

$$\frac{S_s}{K}\frac{\partial h}{\partial t} - \frac{\partial^2 h}{\partial x^2} = 0$$

Here, for convenience, we shall solve for the head perturbed from the steady state,

$$h_2 = h - h_1$$

Substitute the above into the diffusion equation, and noticing that h_1 already satisfies the diffusion equation, we find that h_2 is governed by the identical diffusion equation

$$\frac{S_s}{K}\frac{\partial h_2}{\partial t} - \frac{\partial^2 h_2}{\partial x^2} = 0$$

The initial condition is $h_2 = 0$ at $t = 0$, and boundary conditions are $h_2 = \Delta H = -3$ m at $x = 0$ and $h_2 = 0$ at $x = L$.

The problem is analogous to that of heat conduction in a bar—the bar is initially at a constant (zero) temperature, and then suddenly cooled at one end. The solution is available in many mathematics textbooks;[33] hence we present the solution without a derivation

$$\frac{h_2}{\Delta H} = \frac{L - x}{L} - \frac{2}{\pi}\sum_{n=1}^{\infty}\frac{1}{n}\exp\left(-\frac{n^2\pi^2 Kt}{S_s L^2}\right)\sin\left(\frac{n\pi x}{L}\right)$$

The final solution is the sum of the two parts, h_1 and h_2, hence

$$h = \frac{H_2 - H_1 - \Delta H}{L}x + H_1 + \Delta H$$

$$-\frac{2\Delta H}{\pi}\sum_{n=1}^{\infty}\frac{1}{n}\exp\left(-\frac{n^2\pi^2 Kt}{S_s L^2}\right)\sin\left(\frac{n\pi x}{L}\right)$$

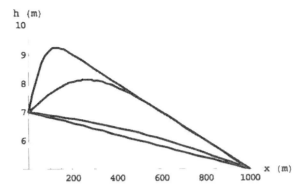

FIGURE 2.8. Transient head distribution in the aquifer. (From top curve down: 0.1, 1, 10, and 100 days since river stage change.)

In Figure 2.8 we plot the transient head distribution in the aquifer at time $t = 0.1$, 1, 10 and 100 days.

The aquifer discharge can be found by differentiating the above expression:

$$Q = -Kb \left[\frac{H_2 - H_1 - \Delta H}{L} - \frac{2\Delta H}{L} \sum_{n=1}^{\infty} \exp\left(-\frac{n^2\pi^2 Kt}{S_s L^2}\right) \cos\left(\frac{n\pi x}{L}\right) \right]$$

Particularly, at the left side of the aquifer, $x = 0$, the above equation becomes

$$Q = -Kb \left[\frac{H_2 - H_1 - \Delta H}{L} - \frac{2\Delta H}{L} \sum_{n=1}^{\infty} \exp\left(-\frac{n^2\pi^2 Kt}{S_s L^2}\right) \right]$$

Figure 2.9 shows this discharge as a function of time. We observe that the discharge is initially negative, meaning that the river is gaining water from the aquifer. But after $t = 5.7$ day, the flow reverses. At steady state, the left river loses 0.08 m^3/m·day to the aquifer.

For future reference, it is useful to summarize the differential operators that will be used throughout the book, namely the gradient,

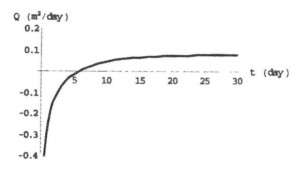

FIGURE 2.9. Discharge from the left river to the aquifer as a function of time.

the divergence, and the Laplacian operators, in various coordinate systems. Given ϕ as a scalar function, and $\mathbf{A}$ as a vector function of space, the gradient operator is

$$
\begin{aligned}
\nabla \phi &= \frac{\partial \phi}{\partial x}\mathbf{i} + \frac{\partial \phi}{\partial y}\mathbf{j} + \frac{\partial \phi}{\partial z}\mathbf{k} \qquad\qquad \text{in Cartesian coordinates} \\
&= \frac{\partial \phi}{\partial r}\mathbf{e}_r + \frac{1}{r}\frac{\partial \phi}{\partial \theta}\mathbf{e}_\theta + \frac{\partial \phi}{\partial z}\mathbf{e}_z \qquad \text{in cylindrical coordinates} \\
&= \frac{\partial \phi}{\partial R}\mathbf{e}_R + \frac{1}{R}\frac{\partial \phi}{\partial \theta}\mathbf{e}_\theta + \frac{1}{R\sin\theta}\frac{\partial \phi}{\partial \varphi}\mathbf{e}_\varphi
\end{aligned}
$$

$$\text{in spherical coordinates} \quad (2.47)$$

the divergence operator is

$$
\begin{aligned}
\nabla \cdot \mathbf{A} &= \frac{\partial A_x}{\partial x} + \frac{\partial A_y}{\partial y} + \frac{\partial A_z}{\partial z} \qquad\qquad \text{in Cartesian coordinates} \\
&= \frac{1}{r}\frac{\partial r A_r}{\partial r} + \frac{1}{r}\frac{\partial A_\theta}{\partial \theta} + \frac{\partial A_z}{\partial z} \qquad \text{in cylindrical coordinates} \\
&= \frac{1}{R^2}\frac{\partial R^2 A_R}{\partial R} + \frac{1}{R\sin\theta}\frac{\partial A_\theta \sin\theta}{\partial \theta} + \frac{1}{R\sin\theta}\frac{\partial A_\varphi}{\partial \varphi}
\end{aligned}
$$

$$\text{in spherical coordinates} \quad (2.48)$$

and the Laplacian operator is

$$
\begin{aligned}
\nabla^2 \phi &= \frac{\partial^2 \phi}{\partial x^2} + \frac{\partial^2 \phi}{\partial y^2} + \frac{\partial^2 \phi}{\partial z^2} \qquad \text{in Cartesian coordinates} \\[2mm]
&= \frac{1}{r}\frac{\partial}{\partial r}\left(r\frac{\partial \phi}{\partial r} \right) + \frac{1}{r^2}\frac{\partial^2 \phi}{\partial \theta^2} + \frac{\partial^2 \phi}{\partial z^2} \quad \text{in cylindrical coordinates} \\[2mm]
&= \frac{1}{R^2}\frac{\partial}{\partial R}\left(R^2 \frac{\partial \phi}{\partial R} \right) + \frac{1}{R^2 \sin\theta}\frac{\partial}{\partial \theta}\left(\sin\theta \frac{\partial \phi}{\partial \theta} \right) \\[2mm]
&\quad + \frac{1}{R^2 \sin^2 \theta}\frac{\partial^2 \phi}{\partial \varphi^2} \qquad \text{in spherical coordinates} \quad (2.49)
\end{aligned}
$$

2.5 Flow Net

After solving a groundwater flow problem in a two-dimensional geometry, it is of interest to plot lines of constant piezometric head and also lines showing flow directions. These lines generally help the understanding of the flow pattern. For isotropic aquifer under steady state condition without recharge, these two sets of lines form an orthogonal pattern of small rectangles known as the *flow net*.

In the past, the manual procedure of drawing a flow net by trial-and-error was an important technique for solving groundwater flow problems.[20] However, with the advent of electronic computer and numerical methods, this solution technique became obsolete. Nevertheless, the drawing of a flow net after the piezometric head solution has been obtained remains an important visualization technique of illustrating groundwater flow patterns. A brief derivation of the flow net relation is presented below.

First, we discuss the existence of a stream function. Assuming steady state and two-dimensional geometry, the continuity equation (2.28) becomes

$$
\frac{\partial q_x}{\partial x} + \frac{\partial q_y}{\partial y} = 0 \qquad (2.50)
$$

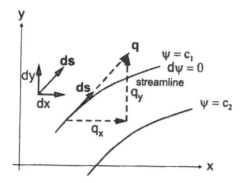

FIGURE 2.10. Relation between streamline and specific discharge vector.

A solution to the above equation is

$$q_x = -\frac{\partial \psi}{\partial y}$$

$$q_y = \frac{\partial \psi}{\partial x} \tag{2.51}$$

The above statement can be proven by the substitution of Eq. (2.51) into (2.50). It is easily shown that Eq. (2.50) is indeed satisfied. The function ψ is a *generating function* that generates the specific discharge components as shown in Eq. (2.51).

In a two-dimensional space, $\psi(x, y) = $ constant defined a family of lines (see Figure 2.10). Along these lines the increment of ψ is zero

$$d\psi = 0 \tag{2.52}$$

Assume that

$$\mathbf{ds} = dx\,\mathbf{i} + dy\,\mathbf{j} \tag{2.53}$$

is a vector tangential to a constant ψ-line. Its cross product with the specific discharge vector $\mathbf{q}$ gives

$$\begin{aligned}
\mathbf{q} \times \mathbf{ds} &= (q_x\,dy - q_y\,dx)\,\mathbf{k} \\
&= -\left(\frac{\partial \psi}{\partial y}\,dy + \frac{\partial \psi}{\partial x}\,dx\right)\mathbf{k} \\
&= -d\psi\,\mathbf{k} = 0
\end{aligned} \tag{2.54}$$

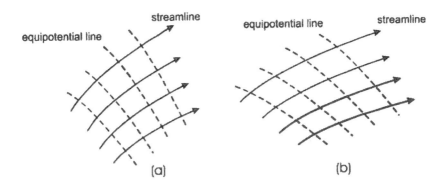

FIGURE 2.11. Streamlines and equipotential lines: (a) Isotropic medium (orthogonal) (b) Anisotropic medium (non-orthogonal).

In the above we have used Eqs. (2.51) to (2.53). Equation (2.54) indicates that **q** is in the same direction as **ds**. Hence the specific discharge vector is always tangential to the constant ψ-lines. For this reason, these lines are called *streamlines*. The function ψ is called the *stream function*.

Next we examine the family of lines given by $h(x, y) = $ constant, known as the *equipotential lines*. The gradient vector ∇h is a vector that is normal to equipotential lines. Similarly, $\nabla \psi$ is a vector normal to streamlines. The dot product of these two vectors gives

$$
\begin{aligned}
\nabla h \cdot \nabla \psi &= \frac{\partial h}{\partial x}\frac{\partial \psi}{\partial x} + \frac{\partial h}{\partial y}\frac{\partial \psi}{\partial y} \\
&= \frac{\partial h}{\partial x} q_y - \frac{\partial h}{\partial y} q_x \\
&= -\frac{\partial h}{\partial x} K \frac{\partial h}{\partial y} + \frac{\partial h}{\partial y} K \frac{\partial h}{\partial x} \\
&= 0
\end{aligned}
\tag{2.55}
$$

which means that they are perpendicular to each other. Consequently, equipotential lines and streamlines are orthogonal to each other forming the so-called "flow net", as demonstrated in Figure 2.11(a). We notice in Eq. (2.55) that hydraulic conductivity need not be homogeneous, but must be isotropic. When anisotropy is present, the flow net is no longer orthogonal, as demonstrated in Figure 2.11(b).

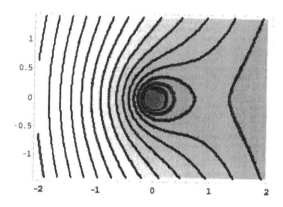

FIGURE 2.12. Equipotential lines of a pumping well in a uniform flow field.

Example: *Show equipotential lines and streamlines of a pumping well in a uniform groundwater flow field.*

The piezometric head of a uniform flow field is given by

$$h = -\frac{U}{K}x$$

From Darcy's law, it is clear that

$$q_x = U$$
$$q_y = 0$$

This represents a uniform flow in the x-direction of magnitude U. In Sec. 3.3, it will be shown that a steady state pumping well causes a depression in piezometric head of the amount

$$s = \frac{Q_w}{2\pi T}\ln\frac{R}{r}$$

where $r = \sqrt{x^2 + y^2}$ is the radial distance, Q_w, T, and R are respectively the well discharge, the transmissivity, and the radius of influence. Subtracting the depression, the resultant head is

$$h = -\frac{U}{K}x - \frac{Q_w}{2\pi T}\ln\frac{R}{r}$$

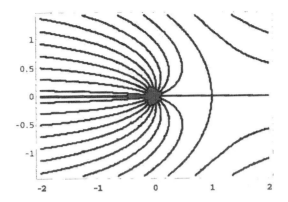

FIGURE 2.13. Streamlines of a pumping well in a uniform flow field.

For illustration purpose, arbitrary values are chosen for the parameters such that we write

$$h = -x + \ln r = -x + \frac{1}{2} \ln \left(x^2 + y^2 \right)$$

Setting h to different constants, contour lines can be drawn as shown in Figure 2.12. The stream function can be expressed as

$$\psi = \tan^{-1} \frac{y}{x} - y$$

The streamlines are plotted in Figure 2.13. Although these two figures are not superimposed, we can observe that these two family of lines are orthogonal to each other.

Example: *In an isotropic and uniform groundwater flow field where the equipotential lines are parallel to each other, it is possible to determine the groundwater flow direction based on the static water levels in three observation wells forming a triangle.*

Three wells located at (x_1, y_1), (x_2, y_2), and (x_3, y_3) have water levels observed at h_1, h_2 and h_3, respectively (see Figure 2.14). Using a graphical method, we connect the well locations into a triangle. Due to a constant gradient, it is possible to linearly interpolate the water level along the sides of the triangle, as marked in small dots in

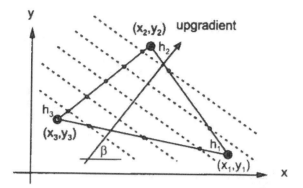

FIGURE 2.14. Determination of flow direction based on three observation wells.

Figure 2.14. When these dots of the same water level are connected, equipotential lines are formed, which are shown in dashed lines. The gradient direction is perpendicular to the equipotential lines. The flow direction is in the opposite, i.e. the down gradient, direction. Once such a graph is constructed, we can easily work out the angle β marking the gradient direction by inspecting the geometry.

It is also possible to directly calculate the flow direction. We notice that for a potential field with its gradient pointing to the β direction, the piezometric head takes the bilinear form

$$h = A\left(x\cos\beta + y\sin\beta\right) + B$$

where A gives the magnitude of gradient, and B determines a base line value. These two constants, together with the angle β, are to be determined from the water level data. Substituting the three well locations and piezometric head values, we obtain

$$h_1 = A\left(x_1\cos\beta + y_1\sin\beta\right) + B$$

and so forth. The three unknowns can be solved. Particularly, the angle β is determined as

$$\beta = \tan^{-1}\frac{(h_1 - h_2)(x_2 - x_3) - (h_2 - h_3)(x_1 - x_2)}{(h_2 - h_3)(y_1 - y_2) - (h_1 - h_2)(y_2 - y_3)}$$

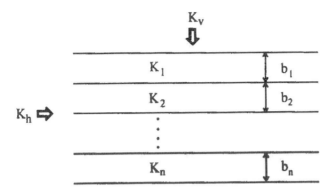

FIGURE 2.15. Equivalent horizontal and vertical hydraulic conductivity of a layered system.

For example, for wells located at (1000 m, 100 m), (600 m, 700 m), (100 m, 300 m), as shown in Figure 2.14, with water levels above a certain datum as 7.5 m, 9.5 m, and 4.5 m, respectively, we calculate $\beta = 53.7°$.

2.6   Equivalent Aquifer Properties

For a layered geological formation, there are occasions where it is not of interest to resolve the detail of the layering. A number of layers may be lumped into a single one for hydraulic purposes. It then becomes necessary to find the equivalent homogeneous properties of the lumped layer that gives the same hydraulic performance of the multilayered system. In other words, we seek the *equivalent hydraulic conductivity* and *equivalent specific storage* of the system.

Consider a layered system as shown in Figure 2.15, where K_1, K_2, ..., K_n are hydraulic conductivities, and b_1, b_2, ..., b_n are thickness, respectively of layers 1, 2, ..., n. Assume first a horizontal flow. Since the head varies only in the horizontal direction, each layer has the same head gradient. Given that head gradient $\Delta h/\Delta L$, the flow in each layer can be calculated according to Darcy's law, and then summed to give the total discharge (per unit width of aquifer perpendicular to the paper):

$$Q = \sum_{i=1}^{n} Q_i = -\sum_{i=1}^{n} b_i K_i \frac{\Delta h}{\Delta L} \qquad (2.56)$$

For a homogeneous layer of the same total thickness, $b = \sum_{i=1}^{n} b_i$, and a hydraulic conductivity K_h, the discharge is

$$Q = -b K_h \frac{\Delta h}{\Delta L} \qquad (2.57)$$

For these two systems to have the same discharge under the same hydraulic gradient, we find

$$K_h = \frac{1}{b} \sum_{i=1}^{n} b_i K_i \qquad (2.58)$$

K_h hence is the *equivalent horizontal hydraulic conductivity*. The relation in Eq. (2.58) is known as an *algebraic mean*.

When the discharge of a whole layer is concerned, we observe that the quantities K and b always appear together as a product. This suggests that we can combine them into a quantity T, known as the *transmissivity*, such that

$$T = b K_h = \sum_{i=1}^{n} b_i K_i = \sum_{i=1}^{n} T_i \qquad (2.59)$$

In the limiting case, consider that the hydraulic conductivity is a continuous function of the vertical direction, $K = K(z)$. Taking b_i in Eq. (2.58) as infinitesimal quantities, we find the formula for the equivalent horizontal hydraulic conductivity as

$$T = \int_0^b K(z) \, dz \qquad (2.60)$$

The concept of transmissivity will be examined further in Sec. 3.1.

For a flow in the vertical direction (see Figure 2.15), we seek the *equivalent vertical hydraulic conductivity* K_v. In this case, the head changes in the vertical direction, but the specific discharge q must be the same in each layer due to the continuity requirement, assuming a steady state condition. The total head drop Δh is the sum of head drop Δh_i of the individual layers,

$$\Delta h = \sum_{i=1}^{n} \Delta h_i = -\sum_{i=1}^{n} \frac{qb_i}{K_i} \tag{2.61}$$

In the last relation, we have applied Darcy's law. For an equivalent homogeneous layer of thickness b, hydraulic conductivity K_v, and total head drop Δh, Darcy's law gives

$$\Delta h = -\frac{qb}{K_v} \tag{2.62}$$

Equating Eqs. (2.61) to (2.62), we find the equivalent vertical hydraulic conductivity

$$K_v = \frac{b}{\sum_{i=1}^{n} \frac{b_i}{K_i}} \tag{2.63}$$

Such a relation is known as a *harmonic mean* in mathematics. For the case of a continuously varied hydraulic conductivity, $K = K(z)$, the limit of Eq. (2.63) gives

$$K_v = \frac{b}{\int_0^b \frac{1}{K(z)} \, dz} \tag{2.64}$$

It is a well known fact in algebra that the algebraic mean is always greater than or equal to the harmonic mean. This leads to the conclusion that $K_h \geq K_v$.

A layered system hence always exhibits an apparent anisotropy, with the horizontal hydraulic conductivity greater than the vertical one. This physical characteristic can be understood without a mathematical proof. When a hydraulic gradient is applied in the horizontal direction, the flow automatically seeks the path of least resistance and avoids low transmissivity layers. A few layers of large transmissivity can dominate the flow process by allowing a large quantity of flow to pass through. On the other hand, for flow in the vertical direction, the low hydraulic conductivity layers become the bottleneck of the process by constricting the flow, as the same quantity of flow must pass through every layer.

We next investigate the equivalent specific storage. Since storage is a transient phenomenon, we need to examine the diffusion equation. We assume a horizontal flow through a layered system similar to that shown in Figure 2.16. For simplicity, consider only two layers. If there exists a barrier between the layers, a transient head difference across the layers will develop. However, since the barrier does not exist, leakage between the two layers will take place in an attempt to equilibrate any head difference. We assume that the aquifers are relatively thin and are in perfect contact such that the head in the vertical direction is instantly equilibrated. In other words, the head varies only in the horizontal, x-direction, and not in the vertical direction. Hence we can write two one-dimensional diffusion equations, one for each layer, as follows

$$S_{s(1)}b_1 \frac{\partial h}{\partial t} - K_1 b_1 \frac{\partial^2 h}{\partial x^2} + q_\ell = 0 \qquad (2.65)$$

$$S_{s(2)}b_2 \frac{\partial h}{\partial t} - K_2 b_2 \frac{\partial^2 h}{\partial x^2} - q_\ell = 0 \qquad (2.66)$$

In the above, q_ℓ denotes the necessary leakage flux that ensures identical head in the two layers. We notice that the gain in one layer is the loss in the other. Since the head is identical, we can add these two equations to yield

$$\left(S_{s(1)}b_1 + S_{s(2)}b_2\right) \frac{\partial h}{\partial t} - \left(K_1 b_1 + K_2 b_2\right) \frac{\partial^2 h}{\partial x^2} = 0 \qquad (2.67)$$

Here we see the role of not only the equivalent specific storage, but also the equivalent hydraulic conductivity shown in Eq. (2.58). The above analysis can be easily generalized to an n-layer system. Hence the *equivalent horizontal specific storage* $S_{s(h)}$ is given as

$$S_{s(h)} = \frac{1}{b} \sum_{i=1}^{n} S_{s(i)}b_i \qquad (2.68)$$

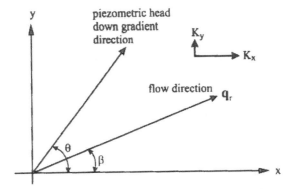

FIGURE 2.16. Flow direction and directional hydraulic conductivity in an anisotropic field.

Similar to transmissivity, we have introduced a quantity known as the *storativity*, S,

$$S = S_{s(h)}b \qquad (2.69)$$

which will be more closely examined in Sec. 3.1.

The above derivation involves the equivalent property in the horizontal direction. In the vertical direction, unfortunately, there does not exist a concept that will lead to a simple expression for equivalent vertical specific storage. The equivalent quantity is dependent on the hydraulic conductivity and boundary condition, and is a function of time. However, under the condition that the top or the bottom of the layered system is impermeable such that at large time a uniform drawdown across the layers is found, a simple relation exists[59]

$$S_{s(v)} = \frac{1}{b} \sum_{i=1}^{n} S_{s(i)}b_i \qquad (2.70)$$

which is the same as Eq. (2.68).

We have shown in the above equivalent aquifer properties in the presence of heterogeneity. Next, it is of interest to examine the equivalent hydraulic conductivity when there exists anisotropy. Figure 2.16 shows a horizontal flow field, where we have aligned the x and y axes with the principal hydraulic conductivity directions. A negative hydraulic gradient is observed in the direction marked by the angle θ. The flow direction is generally different. If we assume that $K_x > K_y$, the flow direction r will be swayed toward the x-axis, as schematically shown in Figure 2.16 and marked by the angle β. The specific flux $\mathbf{q_r}$ has two components q_x and q_y. The angle β is found as

$$\beta = \tan^{-1} \frac{q_y}{q_x} \tag{2.71}$$

The angle θ is given by

$$\theta = \tan^{-1} \frac{\partial h/\partial y}{\partial h/\partial x} \tag{2.72}$$

Since x and y are principal axes, Darcy's law gives

$$q_x = -K_x \frac{\partial h}{\partial x}$$
$$q_y = -K_y \frac{\partial h}{\partial y} \tag{2.73}$$

Substituting Eqs. (2.72) and (2.73) into Eq. (2.71), we obtain

$$\beta = \tan^{-1} \left(\frac{K_y}{K_x} \tan\theta \right) \tag{2.74}$$

Hence, given the head gradient and the ratio of anisotropic hydraulic conductivity, the flow direction can be found from the above formula.

Once the flow direction is known, we find it convenient to define a *directional hydraulic conductivity* K_r such that the following apparent isotropic flow relation holds

$$q_r = -K_r \frac{\partial h}{\partial r} \tag{2.75}$$

Following the chain rule, we can write $\partial h/\partial r$ as

$$\frac{\partial h}{\partial r} = \frac{\partial x}{\partial r} \frac{\partial h}{\partial x} + \frac{\partial y}{\partial r} \frac{\partial h}{\partial y}$$
$$= \cos\beta \frac{\partial h}{\partial x} + \sin\beta \frac{\partial h}{\partial y} \tag{2.76}$$

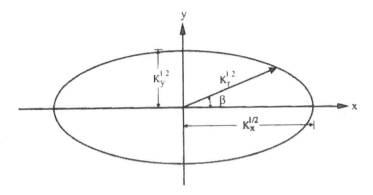

FIGURE 2.17. Directional hydraulic conductivity ellipse.

where we note that

$$\frac{\partial x}{\partial r} = \cos \beta$$

$$\frac{\partial y}{\partial r} = \sin \beta \qquad (2.77)$$

We also need these equations

$$q_x = q_r \cos \beta$$

$$q_y = q_r \sin \beta \qquad (2.78)$$

Finally, with the substitution of Eqs. (2.73), (2.76), and (2.78) into Eq. (2.75), and the cancellation of q_r, we find the following formula for K_r:

$$K_r = \frac{1}{\dfrac{\cos^2 \beta}{K_x} + \dfrac{\sin^2 \beta}{K_y}} \qquad (2.79)$$

This is the equivalent hydraulic conductivity in the flow direction such that Eq. (2.75) gives the correct specific discharge as if the medium were isotropic.

Equation (2.79) reveals an interesting property of the directional hydraulic conductivity. In a two-dimensional plane, if we plot in polar coordinates a trajectory using $K_r^{1/2}$ as the radial distance and β as

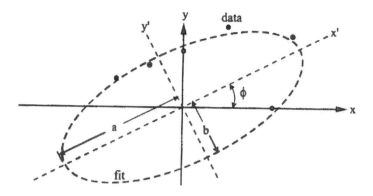

FIGURE 2.18. Best-fit ellipse of the directional hydraulic conductivity measurements.

the polar angle (see Figure 2.17), we obtain an ellipse. This can be proven by substituting $K_r = r^2$ in Eq. (2.79), and replacing $r \cos \beta$ by x, and $r \sin \beta$ by y. We thus obtain

$$\frac{x^2}{K_x} + \frac{y^2}{K_y} = 1 \tag{2.80}$$

As illustrated in Figure 2.17, this *directional hydraulic conductivity ellipse* has major and minor axes given by $K_x^{1/2}$ and $K_y^{1/2}$. Along a given angle β, the radial distance drawn from the origin to the ellipse gives the value of $K_r^{1/2}$ in that direction.

The ellipse shown in Figure 2.17 can be used as a visual tool to determine the principal directions based on field measurements of directional hydraulic conductivities. The directional measurements of apparent hydraulic conductivities can be plotted as points in a map as shown in Figure 2.18. Since the principal axes are not known *a priori*, the ellipse is inclined at an angle. A best-fit ellipse through these data points can identify the principal directions as well as the principal hydraulic conductivity values, as schematically shown in Figure 2.18.

Example: *What is the minimum number of directional hydraulic conductivity measurements required to determine the anisotropic field? Derive the analytical formulae for its determination.*

The question can be answered from a geometry point of view. An ellipse centered at the origin is uniquely define by three characteristics: its major axis a, minor axis b, and the angle of inclination ϕ, as seen in Figure 2.18. Hence three data points are required to produce a unique fit.

The three data points are expressed as $r_1 = K_1^{1/2}$, $r_2 = K_2^{1/2}$, and $r_3 = K_3^{1/2}$, with corresponding angles β_1, β_2, and β_3. To slightly simplify the algebraic work, we choose the x-axis to coincide with the first direction, hence setting $\beta_1 = 0$. The equation for the ellipse shown in Figure 2.18 is

$$\frac{r^2 \cos^2 (\theta - \phi)}{a^2} + \frac{r^2 \sin^2 (\theta - \phi)}{b^2} = 1$$

Substituting the three pairs of data into the above we obtain

$$b^2 \cos^2 \phi + a^2 \sin^2 \phi \;=\; \frac{a^2 b^2}{r_1^2}$$

$$b^2 \cos^2 (\beta_2 - \phi) + a^2 \sin^2 (\beta_2 - \phi) \;=\; \frac{a^2 b^2}{r_2^2}$$

$$b^2 \cos^2 (\beta_3 - \phi) + a^2 \sin^2 (\beta_3 - \phi) \;=\; \frac{a^2 b^2}{r_3^2}$$

We can solve the first two equations to obtain

$$a \;=\; \left[\frac{r_1^2 r_2^2 \sin(2\phi - \beta_2) \sin \beta_2}{r_1^2 \sin^2 \phi - r_2^2 \sin^2(\phi - \beta_2)} \right]^{1/2}$$

$$b \;=\; \left[\frac{r_1^2 r_2^2 \sin(2\phi - \beta_2) \sin \beta_2}{r_2^2 \cos^2(\phi - \beta_2) - r_1^2 \cos^2 \phi} \right]^{1/2}$$

Using these in the third equation, we obtain a transcendental equation in ϕ:

$$2r_1^2 r_2^2 \sin(2\phi - \beta_2) \sin \beta_2 - 2r_1^2 r_3^2 \sin(2\phi - \beta_3) \sin \beta_3$$
$$+ r_2^2 r_3^2 \left[\cos(2\phi - 2\beta_3) - \cos(2\phi - 2\beta_2) \right] = 0$$

This equation cannot be solved in closed form for ϕ; hence a numerical root finder will be used to find ϕ. After ϕ is found, a and b are easily evaluated by back substitution.

A numerical example is examined below. Assume that three directional measurements are made in directions $\beta_1 = 0°$, $\beta_2 = 60°$, $\beta_3 = 120°$, giving $K_1 = 300$ m/day, $K_2 = 180$ m/day, $K_3 = 102$ m/day, or $r_1 = 17.3$, $r_2 = 13.4$, and $r_3 = 10.1$. These numbers are substituted in the equation for ϕ, and a root $\phi = 20.1°$ is found. Substituting into the equations for a and b we find $a = 20.2$ and $b = 9.99$, or $K'_{x'} = 408$ m/day and $K'_{y'} = 99.8$ m/day.

For homogeneous, but anisotropic medium, Eq. (2.43) can be transformed into a governing equation for isotropic medium through a geometric transformation. First we choose one of the principle directions along which we want the physical dimension to be preserved. Choosing z as that direction, a new coordinate system

$$
\begin{aligned}
X &= \sqrt{K_z/K_x}\, x \\
Y &= \sqrt{K_z/K_y}\, y \\
Z &= z
\end{aligned}
\tag{2.81}
$$

is introduced. Equation (2.43) is then rewritten in the new coordinates as

$$
\frac{S_s}{K_z}\frac{\partial h}{\partial t} = \frac{\partial^2 h}{\partial X^2} + \frac{\partial^2 h}{\partial Y^2} + \frac{\partial^2 h}{\partial Z^2}
\tag{2.82}
$$

which is equivalent to Eq. (2.45). The above transformation shows that homogeneous, anisotropic problems can be solved as homogeneous, isotropic ones. A flow net constructed in the transformed domain is orthogonal. However, since the coordinates are transformed, it should be cautioned that the geometry is distorted. Normal flux boundary condition is also transformed. The normal derivative may no longer be normal to the transformed boundary. More detail of solutions of this type can be found in Bear.[12]

2.7   Flow Refraction at Interface

Similar to the refraction law in optics, when a flow line crosses the interface between two media, a sudden change in flow direction results.

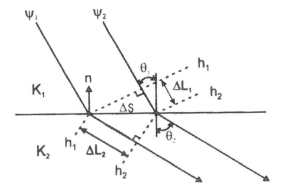

FIGURE 2.19. Flow across hydraulic conductivity interface.

Referring to Figure 2.19, we observe two parallel streamlines crossing the interface of two media of hydraulic conductivity K_1 and K_2,
respectively. The dashed lines indicate equipotential lines. For continuity, the normal flux q_n leaving zone 1 must equal to that entering
zone 2. Based on Darcy's law and the proper vector component, we
have

$$K_1 \frac{\Delta h}{\Delta L_1} \cos\theta_1 = K_2 \frac{\Delta h}{\Delta L_2} \cos\theta_2 \tag{2.83}$$

where $\Delta h = h_2 - h_1$. From the geometry in Figure 2.19, ΔL_1 and
ΔL_2 are related to Δs as

$$\Delta s = \frac{\Delta L_1}{\sin\theta_1} = \frac{\Delta L_2}{\sin\theta_2} \tag{2.84}$$

Combining Eqs. (2.83) and (2.84), we obtain a relation similar to
Snell's law in optics.

$$\frac{K_1}{K_2} = \frac{\tan\theta_1}{\tan\theta_2} \tag{2.85}$$

Based on the above relation, we observe that a flow line entering a
low hydraulic conductivity zone from a high hydraulic conductivity
one bends towards the line perpendicular to the interface. For example, for a permeability contrast of $K_1/K_2 = 10$, and an incident
angle $\theta_1 = 70°$ in the high hydraulic conductivity zone, θ_2 in the low

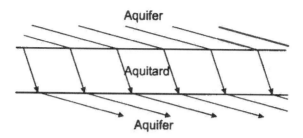

FIGURE 2.20. Refraction of streamline in aquifer-aquitard system.

conductivity zone is about 15°. Figure 2.20 illustrates flow lines in an aquifer-aquitard-aquifer system with a 10:1:10 hydraulic conductivity contrast. We observe that the streamlines in the aquitard bent toward the vertical direction. This is one of the reasons that aquitard flow is modeled as nearly vertical in multilayered aquifer theory. A further demonstration will be given in a later section, Sec. 4.1.

Chapter 3
AQUIFER THEORY

3.1 Confined Aquifer

A *confined aquifer* is a water containing and conducting formation bounded at the top and the bottom by impermeable formations. Aquifers are typically horizontal layers with large horizontal to vertical aspect ratios. Figure 3.1 gives a schematic view of a confined aquifer with a greatly exaggerated vertical scale. Due to the typical aquifer geometry, streamlines are assumed to be nearly horizontal. Since equipotential lines are perpendicular to the streamlines, they become nearly vertical. Changes of piezometric head in the vertical (z) direction hence are negligible. Mathematically, this *nearly horizontal flow assumption* allows us to reduce the three-dimensional spatial functional dependence of piezometric head to just two-dimensional:

$$h = h(x, y, t) \tag{3.1}$$

Another view of this simplification is based on the *hydrostatic assumption*. If streamlines are nearly horizontal, the vertical acceleration component is negligible. The pressure distribution is approximately hydrostatic in the vertical direction. Based on the definition Eq. (2.1), a gain or loss in piezometric head due to an elevation change, Δz, is compensated by the loss or gain of pressure, $\Delta p = \gamma \Delta z$. It is clear that h is independent of the vertical location if the pressure is hydrostatic. Equation (3.1) offers an opportunity to construct a two-dimensional theory for confined aquifer that is

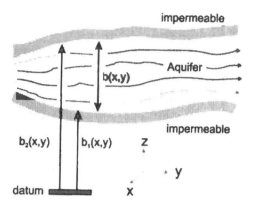

FIGURE 3.1. A confined aquifer.

much easier to handle both analytically and numerically. The theory also conforms with the traditional ways that aquifer parameters are measured in the field. An aquifer theory based on the assumption of Eq. (3.1) is known as the *hydraulic approach*.

To obtain a two-dimensional theory, it is necessary to suppress the z-dependence of not only the piezometric head, but also other parameters such as hydraulic conductivity and specific storage. This is achieved by integrating in the vertical direction to find their average. For example, the x-component specific flux is expressed in this form, taking into consideration the functional dependence of piezometric head,

$$q_x(x, y, z, t) = -K_x(x, y, z)\frac{\partial h(x, y, t)}{\partial x} \tag{3.2}$$

Here we assume that the vertical direction is one of the material principle axes, and that x and y designate the other two principal directions. The above equation can be integrated with respect to z from the bottom of the aquifer $b_1(x, y)$ to the top of the aquifer $b_2(x, y)$ (see Figure 3.1) to yield

$$Q_x(x, y, t) = -T_x(x, y)\frac{\partial h(x, y, t)}{\partial x} \tag{3.3}$$

where

$$Q_x(x, y, t) = \int_{b_1(x,y)}^{b_2(x,y)} q_x(x, y, z, t)\, dz \qquad (3.4)$$

is the x-component discharge per unit aquifer width (in the y-direction), which has the dimension of $[L^2/T]$, and

$$T_x(x, y) = \int_{b_1(x,y)}^{b_2(x,y)} K_x(x, y, z)\, dz \qquad (3.5)$$

defines the x-component *transmissivity* (cf. Eq. (2.60)), also of the dimension $[L^2/T]$. Likewise, integrating the y-component Darcy's law produces Q_y and T_y. Expressed in shorthand, the equivalent Darcy's law becomes

$$\begin{aligned} Q_x &= -T_x \frac{\partial h}{\partial x} \\ Q_y &= -T_y \frac{\partial h}{\partial y} \end{aligned} \qquad (3.6)$$

The continuity equation (2.28) can be similarly integrated to give

$$S \frac{\partial h}{\partial t} = -\frac{\partial Q_x}{\partial x} - \frac{\partial Q_y}{\partial y} + w \qquad (3.7)$$

In the above

$$S(x, y) = \int_{b_1(x,y)}^{b_2(x,y)} S_s(x, y, z)\, dz \qquad (3.8)$$

is the *storage coefficient*, or *storativity* (cf. Eq. (2.68)). Parallel to the definition of specific storage in Eq. (2.26), S is defined as the volume of fluid released from a unit horizontal area of aquifer due to a unit decline in piezometric head, which is dimensionless.

We also find from the integration,

$$-\int_{b_1(x,y)}^{b_2(x,y)} \frac{\partial q_z(x, y, z, t)}{\partial z}\, dz$$

$$= -q_z(x, y, z, t)\big|_{z=b_2(x,y)} + q_z(x, y, z, t)\big|_{z=b_1(x,y)}$$

$$= w(x, y, t) \qquad (3.9)$$

the quantity w, which is the net specific discharge received as "leakage" from layers immediately above and below the aquifer. The negative sign in Eq. (3.9) is a result of q_z being positive when pointing upward in the z-direction. For confined aquifer, w is zero. However, we shall retain this term anticipating the leaky aquifer theory to be developed in Chapter 4.

For the special case that the aquifer material is homogeneous, i.e., K_x, K_y, and S_s are constants, transmissivity and storativity can be found as

$$
\begin{aligned}
T_x(x, y) &= K_x\, b(x, y) \\
T_y(x, y) &= K_y\, b(x, y) \\
S(x, y) &= S_s\, b(x, y)
\end{aligned}
\tag{3.10}
$$

where

$$
b(x, y) = b_2(x, y) - b_1(x, y)
\tag{3.11}
$$

is the aquifer thickness. This shows that an aquifer can be heterogeneous even if its material properties are not. Also in this case, we observe that q_x and q_y are functions of x and y only, hence

$$
\begin{aligned}
Q_x(x, y, t) &= q_x(x, y, t)\, b(x, y) \\
Q_y(x, y, t) &= q_y(x, y, t)\, b(x, y)
\end{aligned}
\tag{3.12}
$$

Combining the vertically integrated continuity equation (3.7) and Darcy's law (3.6), we obtain the two-dimensional equation:

$$
S\frac{\partial h}{\partial t} = \frac{\partial}{\partial x}\left(T_x\frac{\partial h}{\partial x}\right) + \frac{\partial}{\partial y}\left(T_y\frac{\partial h}{\partial y}\right) + w
\tag{3.13}
$$

This is the *governing equation for confined aquifer flow.*

At this point, it is of interest to introduce the definition of a *drawdown*, a quantity preferred by field hydrogeologists. Drawdown is defined as the head difference between an initial equilibrated state $h(x, y, 0) = h_o(x, y)$ and the current piezometric head $h(x, y, t)$. This is seen in Figure 3.2 as the depression part,

$$
s(x, y, t) = h_o(x, y) - h(x, y, t)
\tag{3.14}
$$

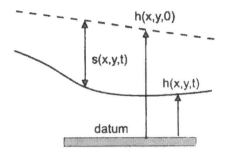

FIGURE 3.2. Definition of drawdown.

This quantity is preferred because it avoids the use of a datum upon which the piezometric head is based. Rather, the undisturbed water level in an observation well is used as a reference point. Any disturbance in the field is measured as the difference from the original water level.

Substituting Eq. (3.14) into Eq. (3.13) and noting that h_o satisfies the steady state equation

$$\frac{\partial}{\partial x}\left(T_x \frac{\partial h_o}{\partial x}\right) + \frac{\partial}{\partial y}\left(T_y \frac{\partial h_o}{\partial y}\right) = 0 \tag{3.15}$$

we obtain an equation similar to Eq. (3.13):

$$S\frac{\partial s}{\partial t} = \frac{\partial}{\partial x}\left(T_x \frac{\partial s}{\partial x}\right) + \frac{\partial}{\partial y}\left(T_y \frac{\partial s}{\partial y}\right) - w \tag{3.16}$$

This is the governing equation of confined aquifer flow based on the drawdown variable.

For different parameter assumptions, we can present the following special cases. For isotropic, but heterogeneous aquifer, we have

$$S\frac{\partial s}{\partial t} = \frac{\partial}{\partial x}\left(T \frac{\partial s}{\partial x}\right) + \frac{\partial}{\partial y}\left(T \frac{\partial s}{\partial y}\right) - w \tag{3.17}$$

Here we emphasize that heterogeneity need not be caused by hydraulic conductivity, as a variable aquifer thickness can have the same effect. For anisotropic, but homogeneous aquifer, Eq. (3.16) becomes

$$S\frac{\partial s}{\partial t} = T_x \frac{\partial^2 s}{\partial x^2} + T_y \frac{\partial^2 s}{\partial y^2} - w \tag{3.18}$$

When an aquifer is isotropic and homogeneous, we obtain the two-dimensional diffusion equation

$$\nabla^2 s - \frac{S}{T}\frac{\partial s}{\partial t} = \frac{w}{T} \tag{3.19}$$

where $\nabla^2 = \partial^2/\partial x^2 + \partial^2/\partial y^2$ is the two-dimensional Laplacian operator. Further assuming steady state, we find the Poisson equation

$$\nabla^2 s = \frac{w}{T} \tag{3.20}$$

Without recharge, Eq. (3.20) becomes the Laplace equation

$$\nabla^2 s = 0 \tag{3.21}$$

3.2 Unconfined Aquifer

An *unconfined aquifer* is the uppermost, water bearing, and conducting formation. Its bottom is impermeable and its top is exposed to the atmosphere. Water seeps down by the action of gravity until the lower part of the formation becomes saturated. There exists a *free surface* (or *phreatic surface, water table*) where the pressure is equal to atmospheric pressure (zero). For water supply purposes, the presence of an unsaturated zone above the phreatic surface is typically ignored, or lumped into an equivalent *capillary fringe*.

Parallel to the development of confined aquifer theory in Sec. 3.1, it is desirable to derive two-dimensional governing equations based on the similar hydraulic approach. The nearly horizontal flow assumption in this particular case is referred to as the *Dupuit assumption* honoring the historical contribution of Dupuit (see Sec. 1.2).

Referring to Figure 3.3, we shall assume that the aquifer bottom is horizontal and is selected as the datum. The streamlines are assumed to be nearly horizontal, despite the fact that the change in elevation of the phreatic surface is the sole mechanism that drives the flow. The piezometric head hence is independent of z and $h = h(x,y)$. We notice that the quantity h plays the dual role of being a piezometric head, and also being the elevation of the phreatic surface. This is evident from the definition of piezometric head Eq. (2.1) with pressure set to zero.

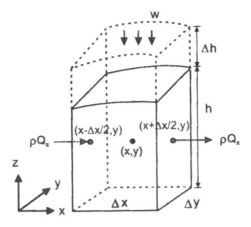

FIGURE 3.3. Continuity equation for unconfined aquifer.

The nearly horizontal flow assumption used here is justified because the gradient of water table is quite small in a large scale problem, normally involving slopes in the range of 1/100 to 1/1000. Exceptions however exist near some localized perturbation zones, such as near pumping wells or seepage surfaces. Rapid change of water table can be observed and this assumption can introduce larger error in these localized regions.

From Darcy's law we find

$$q_x(x, y, z, t) = -K_x(x, y, z)\frac{\partial h(x, y, t)}{\partial x} \qquad (3.22)$$

The above can be integrated with respect to z to obtain

$$Q_x(x, y, t) = -K_x(x, y)\, h(x, y, t)\frac{\partial h(x, y, t)}{\partial x} \qquad (3.23)$$

where the discharge per unit width is given by

$$Q_x(x, y, t) = \int_0^{h(x,y)} q_x(x, y, z, t)\, dz \qquad (3.24)$$

and $K_x(x, y)$ stands for the vertically averaged hydraulic conductivity

$$K_x(x, y) = \frac{1}{h(x, y, t)}\int_0^{h(x,y)} K_x(x, y, z)\, dz \qquad (3.25)$$

In shorthand, Eq. (3.23) and its y-component can be written as

$$Q_x = -K_x h \frac{\partial h}{\partial x} = -\frac{K_x}{2} \frac{\partial h^2}{\partial x}$$

$$Q_y = -K_y h \frac{\partial h}{\partial y} = -\frac{K_y}{2} \frac{\partial h^2}{\partial y} \qquad (3.26)$$

In the above, we have used the conversion

$$h \frac{\partial h}{\partial x} = \frac{1}{2} \frac{\partial h^2}{\partial x} \qquad (3.27)$$

following the chain rule.

Consider next the mass conservation principle. Based on Figure 3.3, we can take the similar Taylor series expansion as demonstrated in Eq. (2.18) and integrate with respect to z to yield

$$\text{net flux in the } x\text{-direction} = -\int_0^{h(x,y)} \frac{\partial \rho q_x}{\partial x} \Delta x \Delta y \, dz$$

$$= -\frac{\partial \rho Q_x}{\partial x} \Delta x \Delta y \qquad (3.28)$$

As before, the spatial variation of ρ will be ignored. A similar equation can be written in the y-direction.

In the z-direction, we consider a downward infiltration of rate w, due to recharge mechanisms such as precipitation. The infiltration has the dimension of velocity and the gain of mass in z-direction is $\rho w \Delta x \Delta y$. Summing up contributions in all three directions, the result is equated to the rate of change of mass in the prism, which gives

$$-\left(\frac{\partial \rho Q_x}{\partial x} + \frac{\partial \rho Q_y}{\partial y} \right) \Delta x \Delta y + \rho w \Delta x \Delta y = \frac{\rho S_y \Delta h \, \Delta x \Delta y}{\Delta t} \qquad (3.29)$$

On the right hand side, we note that $\Delta h \Delta x \Delta y$ represents the volume change due to a water table rising by the amount of Δh. Since not the entire volume is occupied by water, a volume fraction S_y, known as the *specific yield*, is multiplied.

Specific yield, similarly defined as the storativity for confined aquifer, is the volume of fluid released from a unit horizontal area of

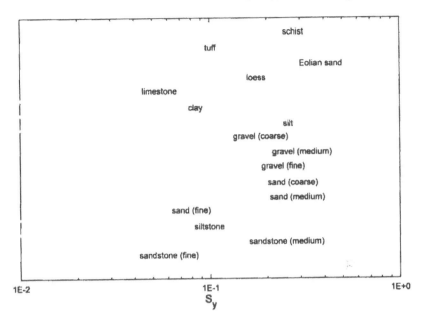

FIGURE 3.4. Range of specific yield for various geological materials. (*Source*: After Morris & Johnson[86] and Mercer, et al.[83])

unconfined aquifer due to a unit decline of head. From physical consideration, these two storage mechanisms are very different. Storativity of confined aquifer involves the compressibility of the fluid and the frame due to pressure release. Specific yield is largely the consequence of draining of pores due to the drop of water table by the action of gravity. When actual draining is present, aquifer and fluid compressibility effect is mostly negligible. S_y has the value close to porosity ϕ. It is however less than porosity because water can be held in the form of pendular saturation (unconnected rings of water between grains) and hygroscopic water (moisture held to solid surface by molecular attraction) that cannot be drained by gravity alone. Typical values of storativity are of the order 10^{-6}–10^{-4}, whereas specific yield can be as large as 20–30%. In Figure 3.4 we present the ranges of specific yield for a number of different aquifer materials.

Dividing Eq. (3.29) through by $\rho \Delta x \Delta y$ and taking limit, we obtain the continuity equation

$$-\left(\frac{\partial Q_x}{\partial x} + \frac{\partial Q_y}{\partial y}\right) + w = S_y \frac{\partial h}{\partial t} \qquad (3.30)$$

Substituting in Eq. (3.26), we find the counterpart of Eq. (3.13) as

$$S_y \frac{\partial h}{\partial t} = \frac{\partial}{\partial x}\left(K_x h \frac{\partial h}{\partial x}\right) + \frac{\partial}{\partial y}\left(K_y h \frac{\partial h}{\partial y}\right) + w \qquad (3.31)$$

Again, it is desirable to express the above into a drawdown variable. In this case, however, a *modified drawdown* s^*, as first suggested by Jacob,[73] needs to be used:

$$s^* = \frac{H^2 - h^2}{2H} = s - \frac{s^2}{2H} \qquad (3.32)$$

where H is the unperturbed water table thickness. Equation (3.31) can be transformed into

$$S_y^* \frac{\partial s^*}{\partial t} = \frac{\partial}{\partial x}\left(T_x \frac{\partial s^*}{\partial x}\right) + \frac{\partial}{\partial y}\left(T_y \frac{\partial s^*}{\partial y}\right) - w \qquad (3.33)$$

where

$$\begin{aligned} T_x &= K_x H \\ T_y &= K_y H \end{aligned} \qquad (3.34)$$

are transmissivities for unconfined aquifer, and

$$S_y^* = \frac{S_y}{\sqrt{1 - 2\left(s^*/H\right)}} \qquad (3.35)$$

is the modified specific yield. Equation (3.33) is the unconfined aquifer counterpart of Eq. (3.16).

Equation (3.33) is clearly nonlinear. Under the condition of small drawdown, namely $s^* \ll H$, which is a condition typically satisfied everywhere except near certain local perturbations, we note that

$$S_y^* \approx S_y \qquad (3.36)$$

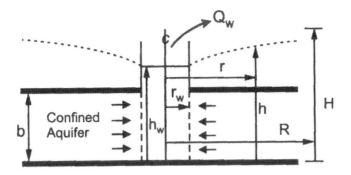

FIGURE 3.5. Pumping well in a confined aquifer.

Equation (3.33) is then linearized to become

$$S_y \frac{\partial s^*}{\partial t} = \frac{\partial}{\partial x}\left(T_x \frac{\partial s^*}{\partial x}\right) + \frac{\partial}{\partial y}\left(T_y \frac{\partial s^*}{\partial y}\right) - w \qquad (3.37)$$

Or, consistent with the approximation of Eq. (3.36), we can further write

$$s^* \approx s \qquad (3.38)$$

The above equation is now formally identical to Eq. (3.17), except for the parameter S_y taking place of S. Other simplifications such as isotropy and homogeneity follow. For example, for a homogeneous, isotropic aquifer, Eq. (3.37) takes the form of a simple diffusion equation

$$\nabla^2 s^* - \frac{S_y}{T}\frac{\partial s^*}{\partial t} = \frac{w}{T} \qquad (3.39)$$

3.3   Thiem Solution

Consider a fully penetrating pumping well located in a homogeneous, isotropic, confined aquifer of infinite horizontal extent (see Figure 3.5). The aquifer has a constant thickness b and hydraulic conductivity K, thus giving a transmissivity $T = Kb$. The well is pumped at a constant rate of Q_w until a steady state is reached. The piezometric head as a function of radial distance from the well is sought.

A confined aquifer is analogous to a pressurized conduit, in which water pressure can be raised to above hydrostatic level. An unconfined aquifer, on the other hand, is similar to an open channel, in which the pressure at the free surface (water table) must be atmospheric. Figure 3.5 schematically marks the piezometric head location as a dashed line.

Under the current assumptions of homogeneous, isotropic medium in steady state without recharge, the governing equation for confined aquifer, Eq. (3.13), is simplified to

$$\frac{\partial^2 h}{\partial x^2} + \frac{\partial^2 h}{\partial y^2} = 0 \tag{3.40}$$

In view of the axial symmetry of the problem, it is more convenient to write Eq. (3.40) in polar coordinates (see Eq. (2.49)):

$$\frac{1}{r}\frac{d}{dr}\left(r\frac{dh}{dr}\right) = 0 \tag{3.41}$$

The equation can be integrated twice to give

$$h = a \ln r + b \tag{3.42}$$

We need two boundary conditions to determine the two constants. A number of choices are available. Two such boundary conditions are given as

$$\begin{aligned} h &= h_1 & \text{at } r = r_1 \\ Q_w &= 2\pi T r \frac{dh}{dr} & \text{at all } r \end{aligned} \tag{3.43}$$

The first condition can be obtained by measuring water level in an observation well. The second condition is based on the consideration of continuity—the well discharge must be equal to the discharge across a cylindrical surface enclosing the well

$$Q = -q_r A = K\frac{dh}{dr} \cdot 2\pi r b \tag{3.44}$$

where q_r is the radial component of specific discharge, and A is the circumferential area of the cylinder centered at the well. The inflow and outflow to this volume must be the same, or else a gain or loss of water volume will result, which is in violation of the steady state assumption.

Using these two conditions in Eq. (3.43), we obtain the solution of piezometric head due to a constant rate pumping in a confined, homogeneous, isotropic aquifer:

$$h - h_1 = \frac{Q_w}{2\pi T} \ln \frac{r}{r_1} \qquad (3.45)$$

Equation (3.45) shows that given a pumping rate Q_w alone, the piezometric head is not determined. An observation well at $r = r_1$ with $h = h_1$ is needed to define the piezometric head profile. This result is counter intuitive as we expect that the pumping rate alone should determine the head profile. This paradox is caused by the "steady state" assumption. We shall shed more light to this situation as we continue on the examination of aquifer theory.

We note in the boundary condition Eq. (3.43) that the pumping well itself can be used as an observation well. In that case, Eq. (3.45) becomes

$$h - h_w = \frac{Q_w}{2\pi T} \ln \frac{r}{r_w} \qquad (3.46)$$

If observations are made at two locations, r_1 and r_2, where $h = h_1$ and h_2, respectively, Eq. (3.45) becomes

$$h_2 - h_1 = \frac{Q_w}{2\pi T} \ln \frac{r_2}{r_1} \qquad (3.47)$$

which can be used to determine the transmissivity

$$T = \frac{Q_w \ln (r_2/r_1)}{2\pi (h_2 - h_1)} \qquad (3.48)$$

Although it has been widely observed in the field that at large pumping time, piezometric head profiles can be fitted by a logarithmic profile, a close examination of the solution Eq. (3.45) reveals a difficulty. Logarithmic function is singular at $r \to 0$ and $r \to \infty$. The limit at $r \to 0$ does not pose a problem as it is excluded from the solution region by the finite well radius. At large distance from the well, however, Eq. (3.45) suggests that h approaches infinity. This is physically not possible. This pathological behavior is again caused by the steady state assumption made in the beginning. Consider an aquifer that is initially at a constant head everywhere and is without external replenishment. As water is continuously removed from the aquifer by the action of pumping, a steady state piezometric head cannot be maintained! The artificial enforcement of a steady-state condition hence creates a mathematical solution rising without limit in the far field.

A remedy, or an artificial fix, to this paradox is to define a *radius of influence* R such that the water table is undisturbed at and beyond R. Assigning the undisturbed head to the notation H, Eq. (3.45) can be written as

$$H - h = \frac{Q_w}{2\pi T} \ln \frac{R}{r} \qquad \text{for} \quad r_w \leq r \leq R \qquad (3.49)$$

If the value of R is known, the need for a head observation is eliminated. Much effort has gone into finding empirical formulae for R under various conditions. One of the first to make such attempts is Adolph Thiem (see Sec. 1.2); hence Eq. (3.49) is commonly referred to as the *Thiem equation*. It is also called the *equilibrium equation* in contrast to the *non-equilibrium equation* of Theis, to be introduced in Sec. 3.5.

A few of the empirical and semi-empirical equations are shown below:[12]

$$R = 3000 \, s_w K^{1/2} \qquad (3.50)$$

$$R = b \left(\frac{K}{2w} \right)^{1/2} \qquad (3.51)$$

$$R = 2.45 \left(\frac{Tt}{S} \right)^{1/2} \qquad (3.52)$$

where $s_w = H - h_w$ is the drawdown in the well, w is the recharge intensity, T and S are transmissivity and storativity as defined before, and t is time. We note from the above that a constant R does not exist after all. It is either tied to a head observation in the well,

Eq. (3.50), a replenishment rate, Eq. (3.51), or is dependent on time, Eq. (3.52).

Despite its shortcomings, the Thiem equation, (3.49) or (3.45), is still a good representation of the piezometric head profile. This is especially so if a head observation is made at a certain distance, and the head profile is not to be extrapolated much beyond that distance. The reason for such an agreement will become clear once the non-equilibrium pumping well solution is introduced in Sec. 3.5.

Before leaving this section, it is of interest to rewrite the above results in terms of the drawdown variable. Following Eq. (3.14), the drawdown for unconfined aquifer is defined as

$$s = H - h \tag{3.53}$$

Equations (3.47) and (3.49) are re-introduced as

$$s_1 - s_2 = \frac{Q_w}{2\pi T} \ln \frac{r_2}{r_1} \tag{3.54}$$

$$s = \frac{Q_w}{2\pi T} \ln \frac{R}{r} \tag{3.55}$$

In Eq. (3.54) we notice the reversal of r and r_1 in their position in the logarithm. We shall refer to these as the *Thiem solution*.

Example: *A confined aquifer is pumped at a rate of 100 m³/hr for an extensive period of time such that there is no appreciable change in water level. A drawdown of 2 m is measured in an observation well 100 m away from the pumping well. Recorded water level in the pumping well is not utilized because it is unreliable due to the operation of the pump and the well loss. A radius of influence is estimated to be between 1 km to 1.5 km. What is the transmissivity of the aquifer?*

Transmissivity can be calculated using Eq. (3.55). Assuming $R = 1$ km, it is calculated as

$$T = \frac{100\frac{m^3}{hr} \times \ln(1000m/100m)}{2 \times 3.14 \times 2m} = 440\frac{m^2}{day}$$

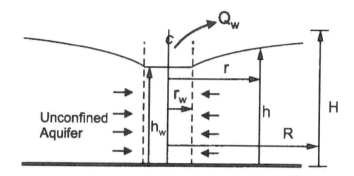

FIGURE 3.6. Pumping well in an unconfined aquifer.

For $R = 1.5$ km, we obtain $T = 517$ m^2/day. This shows that a 50% uncertainty in estimating radius of influence causes only a 17% difference in calculated transmissivity.

3.4   Dupuit Solution

In the preceding section, the aquifer is confined. The case of a fully penetrating pumping well located in a homogeneous, isotropic, unconfined aquifer of infinite horizontal extent without surface recharge (see Figure 3.6) is treated here. The aquifer has a hydraulic conductivity K, and is pumped at a constant rate of Q_w. Again, a "steady state" solution is sought. The governing equation is simplified from Eq. (3.31) and is written in axial symmetry as

$$\frac{1}{r}\frac{d}{dr}\left(r\frac{dh^2}{dr}\right) = 0 \qquad (3.56)$$

Comparing to Eq. (3.41), we note that in the above an h^2 replaces the h variable in the previous equation.

The boundary conditions corresponding to Eq. (3.43) are

$$h = h_1 \qquad\qquad\qquad \text{at } r = r_1$$

$$Q_w = 2\pi K r h \frac{dh}{dr} = \pi K r \frac{dh^2}{dr} \quad \text{at all } r \qquad (3.57)$$

In the discharge formula, we notice that the confined aquifer thickness b contained in transmissivity in Eq. (3.43) is replaced by the piezometric head h here.

Following the same procedure as in the last section, the solution of piezometric head is found as

$$h^2 - h_1^2 = \frac{Q_w}{\pi K} \ln \frac{r}{r_1} \tag{3.58}$$

or

$$H^2 - h^2 = \frac{Q_w}{\pi K} \ln \frac{R}{r} \tag{3.59}$$

The above equations are attributed to Dupuit (see Sec. 1.2). We note that these equations contain the same steady state assumption as the Thiem solution, hence suffers the same paradox as stated earlier.

To switch to the drawdown notation, we adopt the definition of modified drawdown s^* as shown in Eq. (3.32). We then produce from Eqs. (3.58) and (3.59) the pair of equation identical to Eqs. (3.54) and (3.55)

$$s_1^* - s_2^* = \frac{Q_w}{2\pi T} \ln \frac{r_2}{r_1} \tag{3.60}$$

$$s^* = \frac{Q_w}{2\pi T} \ln \frac{R}{r} \tag{3.61}$$

Here we recall that the transmissivity is approximated as $T = KH$.

3.5   Theis Solution

As illustrated in the preceding sections, the steady-state assumption for a pumped aquifer is an artificial one that leads to a paradox. This deficiency was realized by Theis in 1931:[122]

Theis did a geologic reconnaissance, and during August he obtained depth-to-water data on about 200 wells. On November 17, after irrigation withdrawals had ceased in the area, he conducted a 7-hour aquifer test and then calculated permeability using the Thiem equilibrium method. He was not pleased with the results. He concluded that the test (based on a 650 gallon-per-minute pumping well and four observation wells) gave results so divergent that they could not be used to establish the permeability of the aquifer in the area.

Subsequent development led to the well known Theis solution as described in Sec. 1.2.

In the following, we shall present two different ways of deriving the Theis solution as they provide different physical and mathematical insight. Both approaches are later needed in various occasions for the derivation of other aquifer solutions.

Consider a confined, homogeneous and isotropic aquifer of infinite horizontal extent. The governing equation without vertical recharge, according to Eq. (3.19), is

$$\nabla^2 s - \frac{S}{T}\frac{\partial s}{\partial t} = 0 \tag{3.62}$$

The Laplacian operator in the above can be expressed in cylindrical (polar) coordinates following Eq. (2.49). Assuming axial symmetry, Eq. (3.62) simplifies to

$$\frac{1}{r}\frac{\partial}{\partial r}\left(r\frac{\partial s}{\partial r}\right) - \frac{S}{T}\frac{\partial s}{\partial t} = 0 \tag{3.63}$$

Similar to Eq. (3.43), a fully penetrating pumping well with constant discharge Q_w is simulated by the boundary condition

$$2\pi r T \frac{\partial s}{\partial r}\bigg|_{r=r_w} = -Q_w \tag{3.64}$$

However, unlike Eq. (3.43), which is valid for all r, here the condition is valid only at the well radius, $r = r_w$, due to the unsteady condition. Also, at infinite distance, the aquifer is undisturbed. Hence

$$s \to 0 \quad \text{as} \quad r \to \infty, \quad \text{for all} \quad t < \infty \tag{3.65}$$

Solution of Eq. (3.63) is sought with boundary conditions (3.64) and (3.65), and the following initial condition

$$s = 0 \quad \text{at} \quad t = 0 \tag{3.66}$$

To solve Eq. (3.63), we first apply the *Laplace transform* (see Appendix A for a brief review of the Laplace transform). Following Eqs. (A.3) and (A.4), we obtain the equation

$$\frac{1}{r}\frac{d}{dr}\left(r\frac{d\tilde{s}}{dr}\right) - \frac{Sp}{T}\tilde{s} = 0 \tag{3.67}$$

where the tilde indicates the Laplace transform and p is the transform parameter. This procedure eliminates the time variable such that the drawdown variable $s = s(x, y, t)$ changes from a function of space and time to a function of space only, $\tilde{s} = \tilde{s}(x, y)$. When Eq. (3.67) is written in the standard form

$$r^2\frac{d^2\tilde{s}}{dr^2} + r\frac{d\tilde{s}}{dr} - \frac{Sp}{T}r^2\tilde{s} = 0 \tag{3.68}$$

it is recognized as the *modified Bessel equation*. [52] Its general solution is[52]

$$\tilde{s} = C_1\,I_0\left(\sqrt{\frac{pS}{T}}\,r\right) + C_2\,K_0\left(\sqrt{\frac{pS}{T}}\,r\right) \tag{3.69}$$

where I_0 is the *modified Bessel function of the first kind of order 0*, and K_0 is the *modified Bessel function of the second kind of order 0*. In Figures 3.7 and 3.8 we plot the two functions, together with their derivatives

$$\frac{d\,I_0(\eta)}{d\eta} = I_1(\eta)$$
$$\frac{d\,K_0(\eta)}{d\eta} = -K_1(\eta) \tag{3.70}$$

We note the asymptotic behaviors of these functions: as $\eta \to 0$,

$$\begin{aligned}
I_0(\eta) &\to 1 \\
I_1(\eta) &\to 0 \\
K_0(\eta) &\to -\ln\eta \\
K_1(\eta) &\to 1/\eta
\end{aligned} \tag{3.71}$$

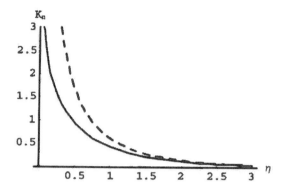

FIGURE 3.7. Modified Bessel function of the second kind (K_0 in solid line; K_1 in dashed line).

and as $\eta \to \infty$,

$$I_0(\eta), \ I_1(\eta) \ \rightarrow \ \frac{e^{\eta}}{\sqrt{2\pi\eta}} \to \infty$$

$$K_0(\eta), \ K_1(\eta) \ \rightarrow \ \sqrt{\frac{\pi}{2\eta}}\, e^{-\eta} \to 0 \qquad (3.72)$$

Based on these asymptotic behaviors, it is easily shown that C_1 in Eq. (3.69) is zero, following the boundary condition (3.65). We hence rewrite Eq. (3.69) as

$$\tilde{s} = C_2 \, K_0 \left(\sqrt{\frac{pS}{T}}\, r \right) \qquad (3.73)$$

Next, we apply the condition Eq. (3.64) at $r = r_w$, and obtain

$$\tilde{s} = \frac{Q_w}{4\pi T} \frac{2}{p} \frac{K_0 \left(\sqrt{\frac{pS}{T}}\, r \right)}{\sqrt{\frac{pS}{T}}\, r_w \, K_1 \left(\sqrt{\frac{pS}{T}}\, r_w \right)} \qquad (3.74)$$

The above is the solution of drawdown for a finite radius well. As the well radius is typically very small, the drawdown at a large distance, $r \gg r_w$ is unaffected by the value of r_w. We can therefore take

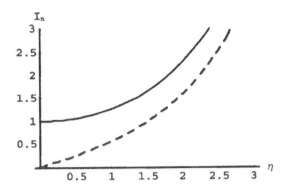

FIGURE 3.8. Modified Bessel function of the first kind (I_0 in solid line; I_1 in dashed line).

the approximation of $r_w \to 0$ without affecting the solution. From Eq. (3.71) we observe that $\eta K_1(\eta) \to 1$ as $\eta \to 0$. Using this condition in Eq. (3.74), we obtain the Theis solution in the Laplace transform space as

$$\tilde{s} = \frac{Q_w}{4\pi T} \frac{2}{p} K_0\left(\sqrt{\frac{pS}{T}} r\right) \tag{3.75}$$

To obtain solution in the time domain, the *inverse Laplace transform* needs to be performed. This final step is deferred. Here we present the second approach of obtaining the Theis solution as follows.

The second approach utilizes the concept of *Green's function*. We start from the governing equation (3.19), which is repeated below

$$\nabla^2 s - \frac{S}{T}\frac{\partial s}{\partial t} = \frac{w}{T} \tag{3.76}$$

We notice the presence of a recharge term w, which will be used to simulate a pumping well. We assume that the fluid is extracted from a circular area of radius r_w at a uniform rate (see Figure 3.9). To account for the total discharge Q_w, the recharge is given by

$$w = -\frac{Q_w}{\pi r_w^2} \tag{3.77}$$

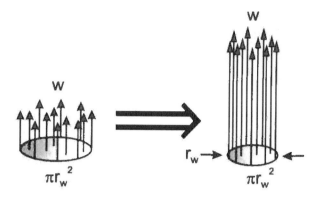

FIGURE 3.9. Taking limit of Dirac delta function.

Here a negative sign is present because Q_w is an extraction and w is a recharge. Similar to the previous approach, it is desirable to take the well radius to the limit, $r_w \to 0$, as it has little effect on drawdown at large distance. This is achieved by introducing the mathematical concept of a *Dirac delta function*.

Consider a function defined in a circular area with $A = \pi r_w^2$. The function takes the constant value of $1/\pi r_w^2$ inside the circle, and 0 outside. The integration of this function over the area is clearly unity. The circle can be shrunk by taking $r_w \to 0$. In this process, the function approaches infinity inside the circle. However, the integration of the function still produces unity. This process is illustrated in Figure 3.9. This function is known as the two-dimensional Dirac delta function $\delta(\mathbf{x} - \mathbf{x}_o)$ with the following properties:

$$
\begin{aligned}
\delta(\mathbf{x} - \mathbf{x}_o) &\to \infty && \text{at} && \mathbf{x} = \mathbf{x}_o \\
\delta(\mathbf{x} - \mathbf{x}_o) &= 0 && \text{for} && \mathbf{x} \neq \mathbf{x}_o \\
\int_A \delta(\mathbf{x} - \mathbf{x}_o) \, d\mathbf{x} &= 1 && \text{for} && \mathbf{x}_o \in A \\
\int_A \delta(\mathbf{x} - \mathbf{x}_o) \, d\mathbf{x} &= 0 && \text{for} && \mathbf{x}_o \notin A
\end{aligned}
\tag{3.78}
$$

In the above $\mathbf{x} = (x, y)$, $\mathbf{x}_o = (x_o, y_o)$, and $\mathbf{x}_o$ is the singular point where the delta function is located. The volume preserving property of this function can be used to simulate the process of reducing the well radius, yet maintaining the same pumping rate in the solution presented below.

It is clear now that if the above limiting process is applied to Eq. (3.77), we obtain

$$w = -Q_w \delta(\mathbf{x} - \mathbf{0}) \tag{3.79}$$

where we have assumed that the well is located at the origin $(0,0)$. Substituting the above into the right-hand-side of Eq. (3.76) gives

$$\nabla^2 s - \frac{S}{T} \frac{\partial s}{\partial t} = -\frac{Q_w}{T} \delta(\mathbf{x} - \mathbf{0}) \, \mathrm{H}(t - 0) \tag{3.80}$$

In the above we have further introduced a *Heaviside unit step function* $\mathrm{H}(t - 0)$ with the following properties

$$\begin{aligned} \mathrm{H}(t-0) &= 1 \quad \text{for} \quad t > 0 \\ \mathrm{H}(t-0) &= 0 \quad \text{for} \quad t \le 0 \end{aligned} \tag{3.81}$$

This function is necessary to ensure that the aquifer is undisturbed for all $t \le 0$. At $t = 0^+$, a constant rate pumping Q_w is suddenly started.

The Laplace transform will be used to solve Eq. (3.80). Using properties found in Appendix A, the result is

$$\nabla^2 \tilde{s} - \frac{pS}{T} \tilde{s} = -\frac{Q_w}{pT} \delta(\mathbf{x} - \mathbf{0}) \tag{3.82}$$

This is known as a *modified Helmholtz equation*.[53] Solution of this kind, namely a partial differential equation with a Dirac delta function on the right hand side, is known as *Green's function*.[52] When there is no boundary condition present, as in the current case, it is called a *free-space Green's function*.

In standard form, Eq. (3.82) can be expressed as

$$\nabla^2 \phi - k^2 \phi = -2\pi \delta(\mathbf{x} - \mathbf{0}) \tag{3.83}$$

Solution of the above can be found in many engineering mathematics books.[52] Not surprisingly, it is in the form of the modified Bessel function

$$\phi = \mathrm{K}_0(kr) \tag{3.84}$$

where $r = |\mathbf{x} - \mathbf{x}_o| = \sqrt{(x - x_o)^2 + (y - y_o)^2}$ is the radial distance measured from the singular point $\mathbf{x}_o$, which in the present case is set at the origin $(0, 0)$. Comparing Eqs. (3.82) and (3.83) and adjusting to a proper multiplication factor, we conclude that

$$\tilde{s} = \frac{Q_w}{4\pi T} \frac{2}{p} K_0 \left(\sqrt{\frac{pS}{T}} r \right) \tag{3.85}$$

This expression is the same as Eq. (3.75). These two approaches hence yield the same result.

Now it is necessary to complete the last step of the solution, the inverse transform of Eq. (3.85). According to Table A-1, we find the following formula:

$$\mathcal{L}^{-1} \left\{ \frac{2}{p} K_0 \left(2\sqrt{p} \right) \right\} = E_1 \left(\frac{1}{t} \right) \tag{3.86}$$

where $E_1 (\xi)$ is the *exponential integral* defined as[1]

$$E_1 (\xi) = \int_\xi^\infty \frac{e^{-x}}{x} \, dx = -E_i (-\xi) \tag{3.87}$$

In the above, E_i is another definition of exponential integral:

$$E_i (\xi) = \int_{-\infty}^\xi \frac{e^x}{x} \, dx \tag{3.88}$$

To continue with the Laplace inversion, we also use Eq. (A.6)

$$\mathcal{L}^{-1} \left\{ \tilde{f}(ap) \right\} = \frac{1}{a} f \left(\frac{t}{a} \right) \tag{3.89}$$

with a scaling factor $a = r^2 S/4T$, and finally obtain

$$s = \frac{Q_w}{4\pi T} E_1 \left(\frac{r^2 S}{4Tt} \right) \tag{3.90}$$

as the solution.

Following the development of aquifer theory, it is customary to replace E_1 by a notation W, and express Eq. (3.90) as

$$s = \frac{Q_w}{4\pi T} W(u) \tag{3.91}$$

where $W(u)$ is known as the *Theis well function*. Obviously we have

$$W(u) = \int_u^\infty \frac{e^{-u}}{u} du \tag{3.92}$$

with the notation

$$u = \frac{r^2 S}{4Tt} \tag{3.93}$$

Equation (3.91) is the well known *Theis solution*, or the *nonequilibrium solution*, in contrast to the Thiem equilibrium solution.

The specific discharge in the radial direction is found by differentiating the drawdown. We obtain

$$q_r = T\frac{\partial s}{\partial r} = -\frac{Q_w}{2\pi r} \exp\left(-\frac{r^2 S}{4Tt}\right) \tag{3.94}$$

It is observed that at a fixed location r, the value of q_r changes from 0 at $t = 0$ to $-Q_w/2\pi r$ as $t \to \infty$.

3.6   Evaluation of Well Function

The well function $W(u)$ is classified as a *special function* in mathematics. An issue concerning practical applications is its efficient evaluation. The traditional approach to solve this problem is by tabulation. Most groundwater books contain tables of well functions. However, this approach has several shortcomings:

- Tables are of limited resolution. Interpolation is needed.
- Many well functions yet to be introduced contain multiple parameters. Tabulation is either inefficient or impractical.
- In a computer simulation, tabular input is cumbersome. It is desirable to instantly evaluate the function from a formula.

Hence it is helpful to discuss mathematical algorithms for the evaluation of the function on demand.

The Theis well function $W(u)$ is associated with the exponential integral E_1, whose mathematical property is well known. Based on a mathematical handbook,[1] it can be expressed in an infinite series as

$$
\begin{aligned}
W(u) &= -0.5772157 - \ln u + u - \frac{u^2}{2 \times 2!} + \frac{u^3}{3 \times 3!} - \frac{u^4}{4 \times 4!} + \cdots \\
&= -0.5772157 - \ln u - \sum_{n=1}^{\infty} (-1)^n \frac{u^n}{n \times n!}
\end{aligned}
\tag{3.95}
$$

The series is convergent for all values of u. However, we observe that as u gets large, the individual terms get large too. The function actually goes to zero as $u \to \infty$ by the alternating nature of the series. It is obvious that the error caused by round-off will very soon overwhelm the solution. We hence limit the use of Eq. (3.95) to only cases of $u \leq 1$.

For $u > 1$, we can use the following polynomial and rational approximation[1]

$$
W(u) = \frac{e^{-u}}{u} \frac{u^2 + a_1 u + a_2}{u^2 + b_1 u + b_2}
\tag{3.96}
$$

in which

$$
\begin{aligned}
a_1 &= 2.334733 \quad a_2 = 0.250621 \\
b_1 &= 3.330657 \quad b_2 = 1.681534
\end{aligned}
\tag{3.97}
$$

The relative error of this equation is of the order 10^{-5} for all values of $u > 1$, hence is satisfactory for the present applications. A *Fortran* program *TheisW.for* that calculates the well function by automatically selecting the proper formula is listed in Appendix B.

Example: *A well is pumping 40 m^3/hr from a confined aquifer with $T = 1000$ m^2/day and $S = 0.0001$. What is the drawdown at an observation well located 1 km away, at $t = 10$ hr?*

First we calculate

$$
u = \frac{(1000 \, \text{m})^2 \times 0.0001}{4 \times 1000 \, \frac{\text{m}^2}{\text{day}} \times 10 \, \text{hr}} = 0.06
$$

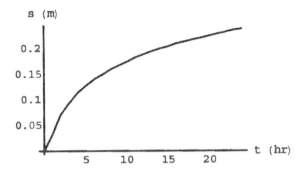

FIGURE 3.10. Drawdown versus time.

The drawdown is

$$s = \frac{40\,\mathrm{m^3/hr}}{4\pi \times 1000\,\frac{\mathrm{m^2}}{\mathrm{day}}}\,W(0.06) = 0.0764\,\mathrm{m} \times 2.295 = 0.175\,\mathrm{m}$$

In Figure 3.10 the drawdown is shown for the first 24 hr.

Example: *If we define the radius of influence R as the radius around the well such that a fixed percentage (say 99%) of water withdrawn from the well is produced within this radius, find the radius of influence according to the Theis solution.*

At a given time t, we can find the volume of water withdrawn from the aquifer within the radius R by integrating the cone of depression formed by the drawdown as

$$V(R,t) = S \int_0^R s \cdot 2\pi r \; dr$$

Here we note the presence of the storage coefficient S which represents the fraction of water that can be produced per unit horizontal aquifer area and per unit decline of head. Substituting the Theis solution Eq. (3.91) into the above, and carrying out the integration, we obtain

$$V(R,t) = Q_w t \left[1 - e^{-u_R} + u_R W(u_R)\right]$$

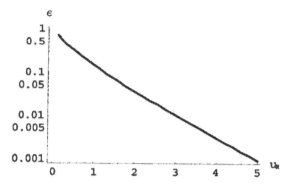

FIGURE 3.11. Pumped volume outside of the radius of influence as percentage of total pumped volume, ϵ, versus u_R.

where

$$u_R = \frac{R^2 S}{4Tt}$$

The water withdrawn from the whole aquifer at any given time is

$$V(\infty, t) = Q_w t$$

The percentage of volume that is included in R is $V(R,t)/V(\infty,t)$, and the percentage that is excluded is

$$\epsilon = 1 - \frac{V(R,t)}{V(\infty,t)} = e^{-u_R} - u_R W(u_R)$$

It is a function of the variable u_R only. In Figure 3.11 we plot ϵ versus u_R in semi-log scale. For $\epsilon = 1\%$, i.e. 99% of water is extracted within R, we read $u_R = 3.05$. By the definition of u_R, the radius of influence corresponding to this value is

$$R = 3.49 \left(\frac{Tt}{S}\right)^{1/2}$$

This shows that the radius of influence is not a constant. Rather, it increases with the square root of time. The above formula is consistent with Eq. (3.52), differing only by a constant factor.

3.7   Cooper-Jacob Solution

For the purpose of determining aquifer parameters, an approximate solution of Eq. (3.91) was proposed by Cooper and Jacob.[35] When u is small, which implies that the radial distance r is small and/or the time t is large, the well function in Eq. (3.95) can be approximated by the first two terms only

$$W(u) \approx -\gamma - \ln u = \ln \frac{0.562}{u} \qquad (3.98)$$

where $\gamma = 0.5772157\ldots$ is the Euler number. For $u < 0.03$, the error of the above formula is less than 1%. For $u = 0.1$, the error is about 5%. Substituting the above and the definition of u into Eq. (3.91) we obtain the *Cooper-Jacob solution*

$$s = \frac{Q_w}{4\pi T} \ln \frac{2.25Tt}{r^2 S} \qquad (3.99)$$

This expression can be used in place of the Theis solution Eq. (3.91) if the pumping time is large enough such that the condition $u < 0.03$ is satisfied.

If the above solution is used to represent drawdown at two observation wells located at radial distances r_1 and r_2, at the same time t_1, we have

$$
\begin{aligned}
s_1 &= \frac{Q_w}{4\pi T} \ln \frac{2.25Tt_1}{r_1^2 S} \\
s_2 &= \frac{Q_w}{4\pi T} \ln \frac{2.25Tt_1}{r_2^2 S}
\end{aligned}
\qquad (3.100)
$$

Subtracting the two equations and utilizing the operational properties of logarithmic function, we obtain

$$s_1 - s_2 = \frac{Q_w}{2\pi T} \ln \frac{r_2}{r_1} \qquad (3.101)$$

This expression is exactly the Thiem solution, Eq. (3.54). This shows that although the Thiem solution is based on the improper physical assumption of steady state, it is actually the approximate solution of a transient drawdown at large pumping time given by the

Cooper-Jacob solution. This provides an explanation of the paradox discussed in Sec. 3.3. Hence the Thiem solution gives the correct result if it is properly applied.

We notice that both the Thiem solution and the Cooper-Jacob solution have the deficiency that as $r \to \infty$, $s \to -\infty$ (i.e. $h \to \infty$). However, there is no contradiction in the Cooper-Jacob solution, as the solution is to be applied only if u is small. As $r \to \infty$, this condition is violated and the solution is not supposed to be valid. For the Thiem solution, this condition is not evident in its derivation. Hence one must be careful in its application.

Equation (3.101), as well as the Thiem solution Eq. (3.54), can be used to determine the aquifer transmissivity as

$$T = \frac{Q_w \ln(r_2/r_1)}{2\pi(s_1 - s_2)} \tag{3.102}$$

In the above, we assume that the drawdowns at two observation wells are measured at the same time. The Cooper Jacob solution Eq. (3.99) also offers the opportunity of making two drawdown measurements in the same observation well at two different times, t_1 and t_2, such that

$$s_1 - s_2 = \frac{Q_w}{4\pi T} \ln \frac{t_1}{t_2} \tag{3.103}$$

Hence transmissivity can be determined from a single well as

$$T = \frac{Q_w \ln(t_1/t_2)}{4\pi(s_1 - s_2)} \tag{3.104}$$

This capability is not present in the Thiem solution. However, it should be cautioned that despite the availability of the two formulae, Eqs. (3.102) and (3.104), transmissivity is rarely determined this way. It normally relies on the recording of a series of drawdown versus time data, and is determined by a least square fit procedure. This is discussed in the next section.

Example: *Solve the same example problem in the preceding section by the Cooper-Jacob solution.*

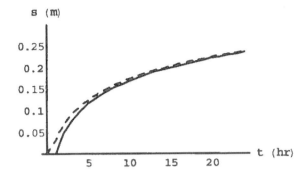

FIGURE 3.12. Comparison of drawdown between Cooper-Jacob (solid line) and Theis (dashed line) solutions.

In Figure 3.12 the calculated drawdown based on the Cooper-Jacob solution, Eq. (3.99), and the Theis solution, Eq. (3.91), are compared for the first 24 hr. We notice that the condition $u < 0.1$ is translated into $t > 6$ hr. Indeed the Cooper-Jacob solution compares well with the Theis solution, except for the first few hours.

3.8 Type Curve

Once we are given a complete set of parameters that include transmissivity T and storativity S, the Theis solution Eq. (3.91) can be used as a *forward solution* to forecast drawdown under various operation conditions. However, before the equation can be put into such use, we must first determine the aquifer parameters. Direct measurement using aquifer core samples is normally not feasible. Aquifer parameters are typically determined by performing field pumping tests. The procedure of using observed drawdown data to infer aquifer properties is known as an *inverse solution*.

To find the two parameters T and S, in principle only two observed drawdown data points are needed. This is however not acceptable because pumping test data contain noise. Parameters determined this way are prone to large amount of random error. To control the

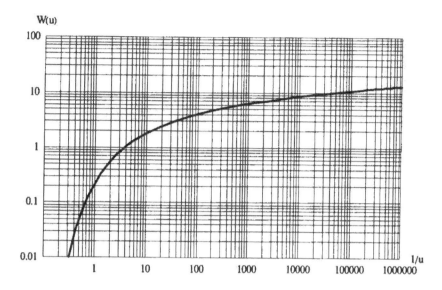

FIGURE 3.13. Theis type curve.

error, more data points are typically acquired. Optimal values of T and S are sought such that the overall discrepancy between the theoretical drawdown and the group of data is minimized.

Due to the relative complexity of the Theis solution, a mathematical procedure based on optimization is not easy. As a remedy, Theis in 1936 (as reported by Jacob[72] and Wenzel[121]) devised an ingenious graphical procedure. This method, known as the *type-curve matching method*, has become immensely popular. Subsequent investigators have also developed a number of its variations and extensions.

In the graphical procedure, the theoretical curve $W(u)$ is plotted versus $1/u$ on a log-log paper, see Figure 3.13. We note that $1/u$ instead of u is chosen because the time parameter t is directly proportional to $1/u$. This curve is known as a *type curve*. Next, the drawdown data is prepared into a *data curve* as s versus t/r^2 in a separate log-log paper of the same scale, see Figure 3.14. These two papers are superimposed and moved with their axes parallel to each other. A "match" is called when the curve is observed to best fit the data points. The distances translated between the two set of axes are read. Following certain formulae, T and S can be determined from

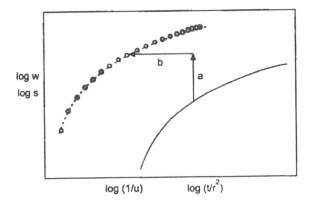

FIGURE 3.14. Theis type curve matching.

the translations. A theoretical background of this process is given below.

Applying logarithmic function on both sides of the Theis solution, Eq. (3.91), we obtain

$$\log_{10} s = \log_{10} W + \log_{10} \frac{Q_w}{4\pi T}$$
$$= \log_{10} W + a \qquad (3.105)$$

where we define

$$a = \log_{10} \frac{Q_w}{4\pi T} \qquad (3.106)$$

Also, applying logarithmic function to the definition u in Eq. (3.93) and rearranging, we find

$$\log_{10} \frac{t}{r^2} = \log_{10} \frac{1}{u} - \log_{10} \frac{4T}{S}$$
$$= \log_{10} \frac{1}{u} - b \qquad (3.107)$$

in which

$$b = \log_{10} \frac{4T}{S} \qquad (3.108)$$

We observe that a point on the type curve is given by the coordinates $(\log_{10} 1/u, \ \log_{10} W)$, whereas that on the data curve is given by

$(\log_{10} t/r^2, \log_{10} s)$. Equations (3.105) and (3.107) indicate that if the y-coordinate of a point on the type curve is translated by the vertical distance a, and its x-coordinate by the horizontal distance $-b$, it matches a point on the data curve. As a and b are constants, this match is in fact found for all points on both curves. The task now is to find the unknown distances a and b by manually moving the two sheets of paper until a best match is found visually. This procedure is schematically illustrated in Figure 3.14 where the solid line is the type curve, circular symbols are data points representing the data curve. The dashed curve shows the final position of type curve when a match is called. With these distances known, T and S can be solved from Eqs. (3.106) and (3.108) as

$$T = \frac{Q_w}{4\pi \cdot 10^a} \qquad (3.109)$$

$$S = \frac{4T}{10^b} \qquad (3.110)$$

Despite the popularity of this technique, we note however a few shortcomings:

- In the presence of random data error, the process of calling a "match" is quite subjective.

- Once the aquifer theory goes beyond the Theis theory, type curve may no longer be a single curve. A family of curves can exist that makes an objective matching even more difficult.

For these reasons, we shall use the mathematical process of nonlinear least square optimization to perform parameter determination. This subject is covered in Chapter 10.

3.9 Constant Drawdown Well (Jacob–Lohman Solution)

The Theis solution is based on a constant well discharge. However, under certain field conditions, a constant well discharge cannot be maintained. For example, the well may not have enough penetration depth to allow for the progressive decline of the water table in the well. In this case, it is likely that a constant drawdown is maintained in the well with a diminishing discharge. Another possibility occurs in a *self-flowing well*, sometimes referred to as *artesian well*. In a self-flowing well, the piezometric head is above the ground level; hence water does not need to be lifted. If the well is not capped, then the discharging head is a constant, equal to the ground level. This type of solution involving constant drawdown is provided by Jacob and Lohman.[75]

Given the same aquifer geometry as the Theis problem, we assume that the aquifer is initially undisturbed, i.e. $s = 0$ at $t = 0$. The boundary conditions are: for $t > 0$,

$$
\begin{aligned}
s &= s_w & \text{at} \quad r &= r_w \\
s &\to 0 & \text{as} \quad r &\to \infty
\end{aligned}
\tag{3.111}
$$

where r_w is the well radius and s_w is the constant drawdown in the well.

The governing equation subject to the Laplace transform is the same as Eq. (3.67). The boundary conditions become

$$
\begin{aligned}
\tilde{s} &= \frac{s_w}{p} & \text{at} \quad r &= r_w \\
\tilde{s} &\to 0 & \text{as} \quad r &\to \infty
\end{aligned}
\tag{3.112}
$$

Similar to the Theis solution, the solution is given by Eq. (3.73). Using the first condition in Eq. (3.112), we can solve for the coefficient C_2 and the drawdown is

$$
\tilde{s} = \frac{s_w}{p} \frac{K_0\left(\sqrt{pS/T}\, r\right)}{K_0\left(\sqrt{pS/T}\, r_w\right)}
\tag{3.113}
$$

Next, we apply the Laplace inverse transform to Eq. (3.113) and symbolically express the result as

$$
s = s_w F(u_w, \rho)
\tag{3.114}
$$

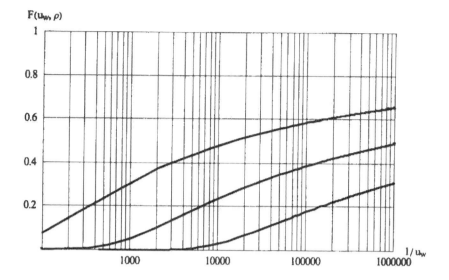

FIGURE 3.15. Well function $F(u_w, \rho)$ versus $1/u_w$. (Curves from left to right: for $\rho = 10$, 30 and 100.)

where $F(u_w, \rho)$ is a constant-drawdown well function and

$$u_w = \frac{r_w^2 S}{4Tt} \tag{3.115}$$

$$\rho = \frac{r}{r_w} \tag{3.116}$$

The value of F is obviously between 0 and 1 because the drawdown in the aquifer cannot exceed that in the well, s_w, and it approaches 0 at large distance. The analytical expression of $F(u_w, \rho)$ as found by Bochever[17] is rather complex and is not of interest to be presented here. Instead, it will be evaluated by the numerical inversion of the Laplace transform, which is demonstrated in the next section. In Figure 3.15 we present the function $F(u_w, \rho)$ for three values of ρ, 10, 30, and 100.

In addition to the drawdown, it is of interest to find the well discharge, which is given as

$$\tilde{Q}_w \;=\; -2\pi r T \left.\frac{\partial \tilde{s}}{\partial r}\right|_{r=r_w}$$

$$=\; 2\pi T s_w \,\frac{\sqrt{pS/T}\, r_w \, \mathrm{K}_1\left(\sqrt{pS/T}\, r_w\right)}{p\,\mathrm{K}_0\left(\sqrt{pS/T}\, r_w\right)} \tag{3.117}$$

Same as the Theis solution, we can take the limit $r_w \to 0$ in the above. By the asymptotic behavior of $\eta \mathrm{K}_1(\eta) \to 1$ as $\eta \to 0$ (see Eq. (3.71)), Eq. (3.117) simplifies to

$$\tilde{Q}_w = \frac{2\pi T s_w}{p\,\mathrm{K}_0\left(\sqrt{pS/T}\, r_w\right)} \tag{3.118}$$

Jacob and Lohman[75] found the inverse of Eq. (3.118) which can be written in the following form:

$$Q_w(t) = \frac{4\pi T s_w}{G(u_w)} \tag{3.119}$$

where

$$\frac{1}{G(u_w)} = \frac{1}{2\pi u_w} \int_0^\infty \xi \exp\left(-\frac{\xi^2}{4u_w}\right) \left\{\frac{\pi}{2} + \tan^{-1}\left[\frac{\mathrm{Y}_0(\xi)}{\mathrm{J}_0(\xi)}\right]\right\} d\xi \tag{3.120}$$

In the above, J_0 and Y_0 are respectively the Bessel function of the first and the second kind of order zero.

The function $G(u_w)$, expressed in a different form, has been tabulated by Jacob and Lohman,[75] and in the same form as the above, by Vuković and Soro.[117] In Figure 3.16 the function $G(u_w)$ is plotted for the range $10 < 1/u_w < 10^6$. Also plotted for comparison is the Theis type curve $W(u)$, in dashed line. We notice that for $u_w < 0.001$ ($1/u_w > 1000$) we can approximate

$$G(u_w) \approx W(u_w) \tag{3.121}$$

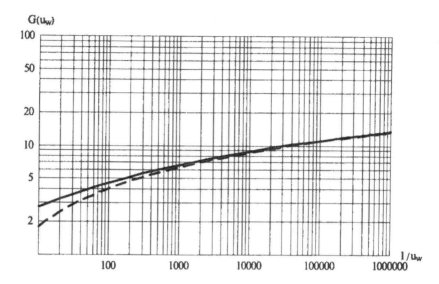

FIGURE 3.16. The Jacob-Lohman well function $G(u_w)$ (solid line) compared to Theis well function $W(u_w)$ (dashed line).

with an error less than 5%. For practical applications, as indicated by Vuković and Soro,[117] the condition of $u_w < 0.001$ is typically satisfied after a few seconds of pumping. Hence Eq. (3.121) offers a practical way of evaluating $G(u_w)$.

Although Eqs. (3.113) and (3.114) give the constant well drawdown solution, we may take a different view by considering them as a variable discharge solution. This can be achieved by eliminating the well drawdown s_w between Eqs. (3.113) and (3.118) to give

$$\tilde{s} = \frac{p\tilde{Q}_w}{4\pi T} \frac{2}{p} \mathrm{K}_0 \left(\sqrt{\frac{pS}{T}}\, r \right) \tag{3.122}$$

This Laplace transform solution is valid for all variable discharge cases, not just for the current case. We may test this statement by checking the Theis solution, in which Q_w is a constant. Its Laplace transform is given by $\tilde{Q}_w = Q_w/p$ (see Table A-1). Equation (3.122) hence reduces to Eq. (3.75). In the constant drawdown case, Q_w is a function of time and $\tilde{Q}_w$ has been solved as Eq. (3.118). Equation

(3.122) is also valid for any variable discharge $Q_w = Q_w(t)$, as long as its corresponding Laplace transform $\widetilde{Q}_w$ is known.

To obtain the solution in time, we need to perform Laplace inversion of Eq. (3.122). This can be achieved by utilizing the *convolutional theorem*, given as Eq. (A.9) in Appendix A:

$$\mathcal{L}^{-1}\left\{\tilde{f}_1(p)\,\tilde{f}_2(p)\right\} = \int_0^t f_1(t-\tau)\,f_2(t)\,d\tau \qquad (3.123)$$

We can consider the right hand side of Eq. (3.122) as the product of two Laplace transform quantities:

$$\tilde{s} = \frac{\widetilde{Q}_w}{4\pi T}\cdot 2\mathrm{K}_0\left(\sqrt{\frac{pS}{T}}\,r\right) \qquad (3.124)$$

The second part can be inverted with the assistance of the formula in Table A-1:

$$\mathcal{L}^{-1}\left\{2\mathrm{K}_0\left(2\sqrt{p}\right)\right\} = \frac{1}{t}\exp(-\frac{1}{t}) \qquad (3.125)$$

and also Eq. (A.6). Finally, with the application of the convolutional theorem Eq. (3.122), the drawdown solution in time is

$$s(r,t) = \frac{1}{4\pi T}\int_0^t \frac{Q_w(\tau)}{t-\tau}\exp\left[-\frac{r^2 S}{4T(t-\tau)}\right]\,d\tau \qquad (3.126)$$

The above form is known as a *convolutional integral.* Equation (3.126) is also valid for all $Q_w(t)$. For the constant drawdown case, $Q_w(t)$ is given by Eq. (3.119).

More detail of using Eqs. (3.122) and (3.126) for predicting drawdown under variable pumping rate condition will be discussed in Sec. 5.1.

3.10   Numerical Inversion of Laplace Transform

We have observed that the Laplace transform reduces the time differential operator to an algebraic operation and the solution is simplified. The trade-off is that as the last step, the solution needs to be inverted to restore the time variable. Analytical Laplace inversion is normally not an easy job. With a few exceptions, such as that for the Theis solution, most of the aquifer solutions are difficult to invert. Once inverted, the resulting expression is often cumbersome and inefficient to evaluate, as evident in the Jacob-Lohman case presented in the preceding section.

To overcome this difficulty, we propose to numerically invert the Laplace transform, thus circumventing the analytical inversion and its subsequent evaluation. There exists a large number of numerical Laplace inversion algorithms and a list of them has been compiled.[96, 97] The accuracy and efficiency of some of the algorithms have been evaluated.[29, 40] Among them, the *Stehfest method* [110] is one that is widely adopted in engineering applications. This algorithm is presented below.

We denote $F(p)$ as the Laplace transform of the function $f(t)$. The Stehfest algorithm states that the function $f(t)$ can be calculated at any $t > 0$ using a number of discrete $F(p)$ values using the following approximate formula:

$$f(t) \approx \frac{\ln 2}{t} \sum_{i=1}^{n} c_i \, F\left(\frac{i \ln 2}{t}\right) \qquad (3.127)$$

where the coefficients c_i are given by

$$c_i = (-1)^{i+\frac{n}{2}} \sum_{k=\left[\frac{i+1}{2}\right]}^{\min(i,\frac{n}{2})} \frac{k^{\frac{n}{2}}(2k)!}{\left(\frac{n}{2}-k\right)! \, k! \, (k-1)! \, (i-k)! \, (2k-i)!} \qquad (3.128)$$

In the above, n is the number of terms in the series, which must be even, and the square brackets in the lower limit of the summation sign gives the greatest integer not larger than its argument (e.g., $\left[\frac{3}{2}\right] = 1$).

Example: *Calculate the Theis well function $W(u)$ using numerical Laplace inversion by Stehfest algorithm.*

According to Eq. (3.85), the well function can be obtained from the inverse transform

$$W\left(\frac{r^2 S}{4Tt}\right) = \mathcal{L}^{-1}\left\{\frac{2}{p} K_0\left(\sqrt{\frac{pS}{T}} \, r\right)\right\}$$

using variables (t, p) as the Laplace transform pair. The above formula, however, requires the knowledge of the parameters r, S, and T. Since the well function can be expressed as a function of one parameter u only, it is more convenient to conduct the inversion in this dimensionless parameter. The inverse transform formula Eq. (3.86) can be repeated here with new notations:

$$\mathcal{L}^{-1}\left\{\frac{2}{p^*} K_0\left(2\sqrt{p^*}\right)\right\} = E_1\left(\frac{1}{t^*}\right) = W\left(\frac{1}{t^*}\right) = W(u)$$

Here we use (t^*, p^*) as the transform pair to distinguish them from the physical variable t and its conjugate p. We note that if we interpret t^* as $1/u$, we can get $W(u)$ directly from the above inverse formula. This means that in the inverse transform subroutine that computes $W(u)$ at a given u, the value of $1/u$ should be used for the variable t^*.

We calculate below, for example, $W(0.01)$, which means $u = 0.01$ or $t^* = 1/u = 100$. The function $F(p^*)$ to use in the Stehfest formulae Eqs. (3.127) is

$$F(p^*) = \frac{2}{p^*} K_0\left(2\sqrt{p^*}\right)$$

Equation (3.127) shows that p^*, the argument of F, should be replaced by $i \, (\ln 2) \, /t^*$. Using 4 terms in the approximation, we obtain

$$W(0.01) \approx \frac{\ln 2}{100} \sum_{i=1}^{4} c_i \times \frac{2}{i \frac{\ln 2}{100}} K_0\left(2\sqrt{i \frac{\ln 2}{100}}\right)$$

$$= \frac{\ln 2}{100}(-2 \times 556.542 + 26 \times 230.502 - 48 \times 135.487$$

$$+24 \times 92.147)$$

$$= 4.076$$

TABLE 3.1. Comparison of the Theis well function evaluated using 4-term Stehfest approximation with the exact solution.

u	10^{-15}	10^{-14}	10^{-13}	10^{-12}	10^{-11}
$W(u)$ Stehfest	34.00	31.70	29.39	27.09	24.79
$W(u)$ Exact	33.96	31.66	29.36	27.05	24.75
u	10^{-10}	10^{-9}	10^{-8}	10^{-7}	10^{-6}
$W(u)$ Stehfest	22.49	20.18	17.88	15.58	13.28
$W(u)$ Exact	22.45	20.15	17.84	15.54	13.24
u	10^{-5}	10^{-4}	10^{-3}	10^{-2}	10^{-1}
$W(u)$ Stehfest	10.97	8.672	6.370	4.076	1.863
$W(u)$ Exact	10.94	8.633	6.332	4.038	1.823

This can be compared with the exact solution $W(0.01) = 4.038$. The error is less than 1%. In Table 3.1, $W(u)$ is computed for a range of u values using the 4-term Stehfest approximation. We also show the exact solution for comparison. The table terminates at $u = 0.1$, as for $u > 0.1$, the accuracy is no longer satisfactory with only 4 terms in the series.

The above procedure is programmed into a *Fortran* program *TheisWL.for* and is provided in Appendix B. The Laplace inverse part is organized into subroutines such that they are "reusable" by other programs. The evaluation of Bessel functions I_0, I_1, K_0 and K_1 is based on polynomial and rational approximation[1] and is presented in function subroutines in Appendix B. In the *TheisWL.for* program, 10 terms are used in the Stehfest series for $u < 1.0$. For the range $1.0 \leq u < 10$, 18 terms are used. The program is not recommended for use for $u > 10$. As observed in Figure 3.10, when u gets large ($1/u$ gets small), the value of $W(u)$ becomes small very rapidly. Most published type curves terminate after $u > 3$ (or $1/u < 0.3$), where $W(u) \approx 0.01$. Some tables list u value up to 10 where $W(u)$ is of the order 10^{-6}. The current program covers these ranges well and can calculate $W(u)$ with a relative error no greater than 0.1%.

Example: *Calculate the Jacob-Lohman well functions $G(u_w)$ and $F(u_w, \rho)$ by Stehfest algorithm.*

Comparing Eqs. (3.118) with (3.119), we find by definition that

$$\frac{1}{G(u_w)} = \mathcal{L}^{-1}\left\{\frac{1}{2p\,K_0\left(\sqrt{pS/T}\,r_w\right)}\right\}$$

Defining a constant

$$a = \frac{r_w^2 S}{4T}$$

we can rewrite the inverse formula as

$$\frac{1}{G\left(\frac{a}{t}\right)} = \mathcal{L}^{-1}\left\{\frac{a}{2ap\,K_0\left(2\sqrt{ap}\right)}\right\}$$

Based on Eqs. (A.5) and (A.6), we find that there exists such a relation

$$\frac{1}{G\left(\frac{1}{t}\right)} = \mathcal{L}^{-1}\left\{\frac{1}{2p\,K_0\left(2\sqrt{p}\right)}\right\}$$

The above equation is free from physical parameters, hence can be used for the evaluation of well function based on the dimensionless parameter u_w. Defining $t^* = 1/u_w$, and p^* as its transformation pair, $G(u_w)$ can be obtained as

$$G(u_w) = G\left(\frac{1}{t^*}\right) = \left(\mathcal{L}^{-1}\left\{\frac{1}{2p^*\,K_0\left(2\sqrt{p^*}\right)}\right\}\right)^{-1}$$

A *Fortran* program JaLoG.for for evaluating $G(u_w)$ by numerical Laplace inversion is presented in Appendix B.

Similarly, the function $F(u_w, \rho)$ can be evaluated by the following inverse transform formula:

$$F(u_w, \rho) = F\left(\frac{1}{t^*}, \rho\right) = \mathcal{L}^{-1}\left\{\frac{K_0\left(2\rho\sqrt{p^*}\right)}{p^*\,K_0\left(2\sqrt{p^*}\right)}\right\}$$

which is programmed as JaLoF.for.

3.11   Computer Algebra

Most of the results presented in this book are based on analytical solutions. It is convenient to be able to use a computer program based on the symbolic manipulation of exact mathematical expressions. Such programs, generally known as *computer algebra*, have become popular. In this section, we shall demonstrate the use of these analytical tools for the evaluation of well functions. These tools will be used to construct macros, which, like *Fortran* subroutines, are higher level functions, for groundwater flow simulation.

Although there exist a number of the computer algebra programs, such as *Mathematica*™, *Maple*™, *MACSYMA*™, *Reduce*™, etc., in this book we utilize only one of them, *Mathematica*.[31, 125] As the language structure of these programs are similar, readers preferring other software should be able to convert or reconstruct these macros.

Example: *Construct a Mathematica macro to perform the Laplace inversion by the Stehfest algorithm.*

The constructed macro is referred to as *nlapinv.m* in Appendix C. We observe that the code is very short, as compared to the corresponding *Fortran* code. With the macro constructed, it can be used in a large extent as a black box. To use the macro, it first needs to be loaded into a work session. This preparatory procedure is explained in Appendix C. This function now becomes a recognizable function of the software. It can be used in many ways, such as evaluating the function, taking its derivative, plotting, etc. For example, to find the Laplace inverse of the function

$$\tilde{f}(p) = \frac{1}{(1+p)^2}$$

at a given time, say $t = 2$, the following command is entered:

```
In[1]:= NLapInv[1/(1+p)^2,p,2,20]
Out[1]= 0.270671
```

The first argument of *NLapInv* is the Laplace transform expression to be inverted, the second argument identifies the symbol p as the

Laplace transform parameter, the third argument declares $t = 2$, and the fourth selects 20 terms in the Stehfest series. The program then returns the evaluated result as 0.270671. For comparison, we present the exact inversion as

$$\mathcal{L}^{-1}\left\{\frac{1}{(1+p)^2}\right\} = t \exp(-t)$$

Substituting $t = 2$, we find that the two solutions agree with each other to all 6 digits as displayed in the above.

The above case is only a demonstration, as the exact inversion is known and there is no need for an approximate inversion. However, there are cases that the exact inversion is not available. There are also cases that the exact inversion is available, but too complex to be efficiently evaluated. An accurate and efficient approximate inversion algorithm is highly valuable in these cases.

In the next test we present

```
In[2]:= NLapInv[2 BesselK[0,2 Sqrt[p]/p,p,1/0.01,20]
Out[2]= 4.03794
```

We notice that the first argument is the Laplace transform corresponding to the Theis well function as demonstrated in the first example of the preceding section. Here the symbol p rather than p^* is used because it is nothing but a symbol. For $u = 0.01$, we use $t = 1/u = 1/0.01$, as explained in the preceding section. For the symbol t we again dropped the asterisk. The above command then returns the value of $W(0.01)$.

Example: *Given a confined aquifer of transmissivity $T = 1000$ m^2/day and storativity $S = 0.00005$. A pumping well of effective radius 0.2 m is pumped at a rate such that a constant drawdown of 5 m is maintained in the well. What is the pumping rate as a function of time? What is the drawdown in an observation well 100 m from the pumping well?*

Following the above example for numerical Laplace inverse transform, a *Mathematica* macro package can be constructed that defines a number of well functions and their drawdown solutions. Such

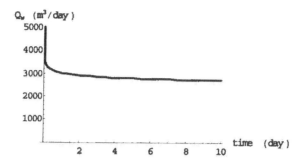

FIGURE 3.17. Discharge versus time in a constant drawdown well.

a package is provided in Appendix C as *wfield.m*. For instance, the Laplace inversion that leads to the Jacob-Lohman well function $W_2(u_w)$ has been programmed into the function WJacobLohman[uw]. Once the package *wfield.m* is loaded, to find the value of W_2 at an arbitrarily u_w, say 0.00005, we simply enter:

```
In[3]:= WJacobLohman[0.00005]
Out[3]= 9.53017
```

Based on the result in Eq. (3.119), we can use the function to find the well discharge at $t = 2$ day as

```
In[4]:= Qw = 4 Pi*1000*5/WJacobLohman[0.2^2*0.00005/
  (4*1000*2)]
Out[4]= 2907.14
```

In the above we have substituted the value $s_w = 5$ m and evaluated u_w in the argument of WJacobLohman[uw] according to the values given in the example. We note that consistent units must be used. The solution hence has the dimension of m^3/day.

Furthermore, we can plot the well discharge versus time for the first 10 days of pumping as

```
In[5]:= Plot[4 Pi*1000*5/WJacobLohman[0.2^2*0.00005/
  (4*1000*t)], {t,0,10}]
```

We observe that the argument of WJacobLohman[uw] does not need to be a numerical value. We can leave a symbol t in its definition,

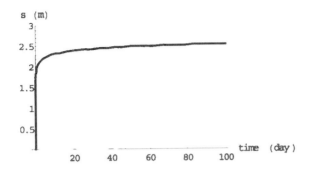

FIGURE 3.18. Drawdown in the observation well.

thus turning it into a function of time. We can then plot the function versus a range of time. After adding some graphic directives that define the axis labels etc., the result is presented as Figure 3.17. We notice how discharge decreases with time for the constant drawdown solution.

For the evaluation of drawdown, a function SJacobLohman[sw, T,S,rw,x,y,t] is created in *wfield.m*. We notice that all the input parameters that are needed to define the drawdown are presented as the arguments of the function. We also used Cartesian coordinates (x, y) instead of radial distance r for the convenience of two-dimensional simulations to be shown in Chapter 5.

Substituting in proper values, we can present drawdown versus time in the observation well using the following command:

```
In[6]:= Plot[SJacobLohman[5,1000,0.00005,0.2,100,0,t],{t,
    0,100}]
```

The result is shown in Figure 3.18.

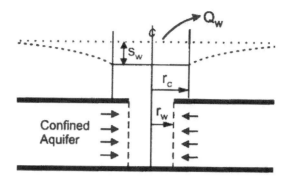

FIGURE 3.19. A large diameter well.

3.12 Large Diameter Well (Papadopulos-Cooper Solution)

For wells of large diameter, such as dug wells, the storage capacity of the well bore can become an important factor in predicting well drawdown. In the initial part of pumping, a significant portion of the discharge might be produced from the well bore, not from the aquifer. Papadopulos and Cooper[95] investigated this problem. Figure 3.19 defines the nomenclature for a large diameter well. The drawdown s_w is limited to the upper part of the well which has a larger radius r_c. The lower part of the well has an effective radius of r_w.

For an initially undisturbed aquifer which is suddenly pumped at a constant rate Q_w, this problem has the same governing equation, Eq. (3.63), as the Theis problem. In fact, the solution has the same, Eq. (3.73). However, the well discharge condition is different:

$$2\pi r_w T \left.\frac{\partial s(r,t)}{\partial r}\right|_{r=r_w} - \pi r_c^2 \frac{ds(r_w,t)}{dt} = -Q_w \qquad (3.129)$$

In the above, the second term on the left-hand-side accounts for the volume of water extracted from the well bore by the lowering of water surface. Using the Laplace transform, the above becomes

$$2\pi r_w T \left.\frac{\partial \tilde{s}(r,p)}{\partial r}\right|_{r=r_w} - \pi r_c^2 p\tilde{s}(r_w,p) = -\frac{Q_w}{p} \qquad (3.130)$$

Substituting Eq. (3.73) into the above, the coefficient C_2 can be solved and the drawdown is

$$\tilde{s} = \frac{Q_w}{4\pi T} \frac{4K_0\left(\sqrt{pS/T}\,r\right)}{p\left[2\sqrt{pS/T}\,r_w K_1\left(\sqrt{pS/T}\,r_w\right) + (pr_c^2/T)\,K_0\left(\sqrt{pS/T}\,r_w\right)\right]} \tag{3.131}$$

We can write the drawdown formula in the time domain as

$$s = \frac{Q_w}{4\pi T} W\left(u, \alpha, \rho\right) \tag{3.132}$$

where

$$u = \frac{r^2 S}{4Tt} \tag{3.133}$$

$$\alpha = \frac{r_w^2 S}{r_c^2} \tag{3.134}$$

$$\rho = \frac{r}{r_w} \tag{3.135}$$

The well function then corresponds to the following inversion

$$W\left(u, \alpha, \rho\right) = W\left(\frac{1}{t^*}, \alpha, \rho\right)$$

$$= \mathcal{L}^{-1}\left\{\frac{K_0\left(2\sqrt{p^*}\right)}{p^*\left[\left(\sqrt{p^*}/\rho\right)K_1\left(2\sqrt{p^*}/\rho\right) + (p^*/\alpha\rho^2)\,K_0\left(2\sqrt{p^*}/\rho\right)\right]}\right\} \tag{3.136}$$

If the well is of constant radius, $r_w = r_c$, then $\alpha = S$. The well function can be written as $W(u, S, \rho)$. This function is the same as the above and has been tabulated by Reed.[103] Here we provide a Fortran program and a Mathematica macro function as listed in Appendices B and C for its evaluation.

For drawdown in the well, $s_w(t) = s(r_w, t)$, Papadopulos and Cooper[95] presented the following solution

$$s_w = \frac{Q_w}{4\pi T} F\left(u_w, \alpha\right) \tag{3.137}$$

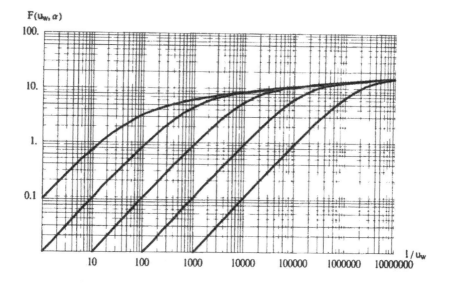

FIGURE 3.20. Papadopulos-Cooper type curve, from left to right, $\alpha = 10^{-1}$, 10^{-2}, 10^{-3}, 10^{-4} and 10^{-5}.

where

$$u_w = \frac{r_w^2 S}{4Tt} \tag{3.138}$$

and

$$F(u_w, \alpha) = \frac{32\alpha^2}{\pi^2} \int_0^\infty \left[1 - \exp\left(-\xi^2 / 4u_w\right)\right] \cdot$$
$$\left\{\xi^3 \left[\xi J_0(\xi) - 2\alpha J_1(\xi)\right]^2 + \left[\xi Y_0(\xi) - 2\alpha Y_1(\xi)\right]^2\right\}^{-1} d\xi \tag{3.139}$$

which has been tabulated by Papadopulos and Cooper.[95]

The function $F(u_w, \alpha)$ is obviously a special case of $W(u, \alpha, \rho)$ for $r = r_w$ and $\rho = 1$, thus

$$F(u_w, \alpha) = W(u_w, \alpha, 1) \tag{3.140}$$

If we choose to find it directly, it is found from the Laplace inversion:

$$F(u_w, \alpha) = F\left(\frac{1}{t^*}, \alpha\right)$$

$$= \mathcal{L}^{-1}\left\{\frac{K_0\left(2\sqrt{p^*}\right)}{p^*\left[\sqrt{p^*}\,K_1\left(2\sqrt{p^*}\right) + (p^*/\alpha)\,K_0\left(2\sqrt{p^*}\right)\right]}\right\} \qquad (3.141)$$

Example: *Plot $F(u_w, \alpha)$ for a range of α values.*

Using either the *Fortran* program *PaCoF.for* or the *Mathematica* function WPapaCooperF[uw,alpha] in *wfield.m* package, $F(u_w, \alpha)$ can be evaluated and plotted versus $1/u_w$. In Figure 3.20, we show these curves for $\alpha = 10^{-1}$, 10^{-2}, 10^{-3}, 10^{-4} and 10^{-5}.

Example: *A large diameter well in a confined aquifer has a radius $r_c = 1$ m, and an effective radius at the screened section $r_w = 60$ cm. The well is pumped at a constant discharge of 10 m³/hr. The aquifer is characterized by these parameters: $T = 300$ m²/day and $S = 0.0005$. What is the drawdown around the well 30 min after pumping?*

For drawdown in the well only, we use Eq. (3.137) with $\alpha = 1.8 \times 10^{-4}$ and $u_w = 7.2 \times 10^{-6}$. Using the utility mentioned in the preceding example, we can find $F\left(7.2 \times 10^{-6}, 1.8 \times 10^{-4}\right) = 9.42$. The drawdown in the well at 30 min is then

$$s_w = \frac{10 \text{ m}^3/\text{hr}}{4\pi \times 300 \text{ m}^2/\text{day}} \times 9.42 = 60 \text{ cm}$$

The full drawdown curve must be obtained by numerically inverting Eq. (3.131). Using either the *Fortran* program *PaCoW.for* or the *Mathematica* function SPapaCooper[Q,T,S,rw,rc,x,y,t], the drawdown in the 50 m range is plotted in Figure 3.21 in solid line. For comparison, the drawdown based on the Theis solution, Eq. (3.91), is presented in dashed line. The smaller drawdown of the large diameter well solution is a direct consequence of discharge supplied by the well storage.

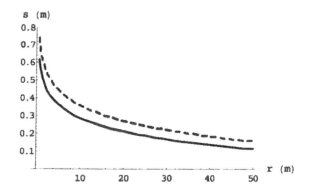

FIGURE 3.21. Drawdown versus radial distance for a large diameter well (solid line) as compared to Theis solution (dashed line).

3.13 Partially Penetrating Well (Hantush Solution)

A partially penetrating well is one whose screened section does not cover the whole thickness of an aquifer. This is often the case in practice. Figure 3.22 shows a well screened between the depths d and ℓ of a confined aquifer of thickness b. The solution of unsteady flow due to a step pumping of discharge Q_w is provided by Hantush.[60]

The aquifer is anisotropic with horizontal and vertical hydraulic conductivities as K_h and K_v, respectively. The transmissivity is defined based on the horizontal hydraulic conductivity, $T = K_h b$. The drawdown is given by

$$
s = \frac{Q_w}{4\pi T} \left[W(u) + \frac{2b}{\pi(\ell - d)} \sum_{n=1}^{\infty} \frac{1}{n} \left(\sin \frac{n\pi\ell}{b} - \sin \frac{n\pi d}{b} \right) \right.
$$
$$
\left. \cos \frac{n\pi(b - z)}{b} \, W \left(u, \sqrt{\frac{K_v}{K_h}} \frac{n\pi r}{b} \right) \right] \tag{3.142}
$$

in which z is the vertical distance measured from the bottom of the aquifer. In the above, $W(u)$ is the Theis well function as defined before (Eq. (3.91)), and $W(u, \beta)$ is the *Hantush-Jacob leaky aquifer well function* given by (see Sec. 4.6 for more detail):

$$
W(u, \beta) = \int_u^\infty \frac{1}{u} \exp \left(-u - \frac{\beta^2}{4u} \right) du \tag{3.143}
$$

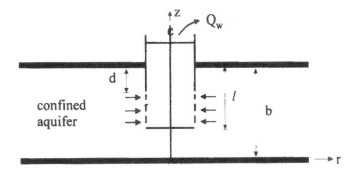

FIGURE 3.22. A partially penetrating well.

In the leaky aquifer solution, the parameter β is related to a leakage factor. Here, however, it simply represents

$$\beta = \sqrt{\frac{K_v}{K_h}} \frac{n\pi r}{b} \tag{3.144}$$

We shall discuss more about this leaky aquifer well function in Sec. 4.6. Here it is sufficient to show that the well function can be obtained from the Laplace inversion

$$W(u, \beta) = W\left(\frac{1}{t^*}, \beta\right) = \mathcal{L}^{-1}\left\{\frac{2}{p^*} K_0\left(\sqrt{4p^* + \beta^2}\right)\right\} \tag{3.145}$$

The *Fortran* program *HantJcbW.for* and the *Mathematica* function WHantushJacob[u,beta] are given in the Appendices for its evaluation. At a relatively large time, $t > bS/2K_v$, Hantush suggested that Eq. (3.142) can be approximated by

$$\begin{aligned} s = & \frac{Q_w}{4\pi T}\left[W(u) + \frac{4b}{\pi(\ell - d)} \sum_{n=1}^{\infty} \frac{1}{n}\left(\sin\frac{n\pi\ell}{b} - \sin\frac{n\pi d}{b}\right)\right. \cdot \\ & \left. \cos\frac{n\pi(b - z)}{b} K_0\left(\sqrt{\frac{K_v}{K_h}} \frac{n\pi r}{b}\right)\right] \end{aligned} \tag{3.146}$$

Example: *A well partially penetrates a confined aquifer of thickness 10 m. The screened portion is located between 1 m to 4 m from*

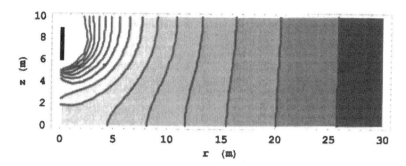

FIGURE 3.23. Drawdown caused by a partially penetrating well in a confined aquifer.

the aquifer top. Other parameters include well discharge of $Q_w =$ 30 m^3/hr, horizontal and vertical hydraulic conductivities $K_h = 80$ m/day and $K_v = 40$ m/day, and storativity $S = 0.0002$. Plot the drawdown around the well.

The drawdown after 10 days of pumping is evaluated based on Eq. (3.142), using either the *Fortran* program or the *Mathematica* function. The result is presented in contour lines in Figure 3.23. We observe that at a certain distance away from the well the contour lines become vertical, suggesting that the head is nearly independent of z. As commented by Hantush,[60] for

$$r > 1.5b\sqrt{\frac{K_h}{K_v}}$$

the partial penetration effect is insignificant, and the nearly horizontal flow condition is established. In the present case, this corresponds to $r > 21$ m. This condition is indeed verified.

3.14   Unconfined Aquifer Well (Neuman Solution)

For unconfined (water table) aquifer, the governing equation (3.33) is nonlinear, making its solution difficult. However, as suggested by Jacob,[73] under the condition of small drawdown, $s \ll H$, where H is

the undisturbed water table thickness, Eq. (3.33) can be linearized
to give

$$\nabla^2 s^* - \frac{S_y}{T} \frac{\partial s^*}{\partial t} = 0 \qquad (3.147)$$

Here we recall that

$$s^* = s - \frac{s^2}{2H} \qquad (3.148)$$

is the modified drawdown, $T = KH$ is the transmissivity, and S_y is
the specific yield. Equation (3.147) is of the same form as Eq. (3.62);
hence its solution is given by the Theis solution as follows,

$$s^* = \frac{Q_w}{4\pi T} W(u_y) \qquad (3.149)$$

where

$$u_y = \frac{r^2 S_y}{4T t} \qquad (3.150)$$

The drawdown formula for an unconfined aquifer is therefore
formally the same as that for a confined aquifer, except that the
modified drawdown s^* and the specific yield S_y replace the drawdown s and the storativity S. For very small drawdown, it is also
possible to replace s^* by s.

In field pumping tests, however, the drawdown data show that the
apparent aquifer response is different for small time as for large time.
Storage coefficient determined from small time data is significantly
smaller than that from large time. This phenomenon has been studied by Boulton[18] as a *delayed yield* mechanism. Boulton noted that
the draining of the pores by gravity pull is not instantaneous. He
assumed that the production of water from the specific yield S_y due
to a unit decline of head is spread over time. At very small times, the
contribution from gravity drainage is negligible. Hence the storage
effect is characterized by the storage coefficient S due to water and
aquifer compressibility, rather than the specific yield S_y. A theory is
devised to provide a smooth transition between S and S_y.

Neuman[89-91] on the other hand pointed out a deficiency in Boulton's theory. The delayed yield theory requires an empirical parameter known as the *delay index*, which is not explicitly associated with a physical parameter. Neuman demonstrated that the delayed yield is a natural consequence of correctly taking into account the dynamic free-surface condition which does not exist for a confined aquifer solution. The resultant type curves have similar characteristics as Boulton's, but are entirely based on physical parameters. We shall present below Neuman's version of drawdown in unconfined aquifers.

Neuman's solution for a pumping well in an unconfined aquifer with delayed yield is given as

$$s = \frac{Q_w}{4\pi T} W\left(u, u_y, \Gamma\right) \tag{3.151}$$

where

$$u = \frac{r^2 S}{4Tt} \tag{3.152}$$

$$u_y = \frac{r^2 S_y}{4Tt} \tag{3.153}$$

and Γ is a dimensionless parameter

$$\Gamma = \frac{r^2 K_v}{b^2 K_h} \tag{3.154}$$

with K_v and K_h the vertical and horizontal hydraulic conductivity, respectively, and b the initial saturated thickness of the aquifer. The well function is given by

$$W\left(u, u_y, \Gamma\right) = \int_0^\infty 4y\, J_0\left(y\, \Gamma^{1/2}\right) \sum_{n=0}^\infty a_n(y)\, dy \tag{3.155}$$

where J_0 is the Bessel function of the first kind of order zero, and

$$a_0 = \frac{\left\{ 1 - \exp\left[-\frac{\Gamma \left(y^2 - \gamma_0^2 \right)}{4u} \right] \right\} \tanh \gamma_0}{\left[y^2 + (1+\sigma)\gamma_0^2 - \frac{\left(y^2 - \gamma_0^2 \right)^2}{\sigma} \right] \gamma_0}$$

$$a_n = \frac{\left\{ 1 - \exp\left[-\frac{\Gamma \left(y^2 + \gamma_n^2 \right)}{4u} \right] \right\} \tan \gamma_n}{\left[y^2 - (1+\sigma)\gamma_n^2 - \frac{\left(y^2 + \gamma_n^2 \right)^2}{\sigma} \right] \gamma_n}; \quad n \geq 1 \quad (3.156)$$

In the above,

$$\sigma = \frac{u}{u_y} = \frac{S}{S_y} \qquad (3.157)$$

which is typically a small parameter, and γ_0 and γ_n are roots of the characteristic equations

$$\sigma\gamma_0 \sinh \gamma_0 - \left(y^2 - \gamma_0^2 \right) \cosh \gamma_0 = 0$$
$$\sigma\gamma_n \sin \gamma_n + \left(y^2 + \gamma_n^2 \right) \cos \gamma_n = 0; \quad n \geq 1 \quad (3.158)$$

This completes the definition of the solution.

The well function in Eq. (3.151) contains three dimensionless parameters, which makes any tabulation attempt impractical. Analogous to Boulton's type-curve,[18, 102] we can assume that $S_y \gg S$, hence $\sigma \to 0$. This allows Eq. (3.151) to reduce to two asymptotic families of type curves, $W(u, \Gamma)$ and $W(u_y, \Gamma)$, respectively known as type A and type B curves:[91]

$$W(u, \Gamma) = \int_0^\infty 64\, y\, J_0\left(y\,\Gamma^{1/2} \right)$$
$$\sum_{n=1}^\infty \frac{1 - \exp\left\{ -\frac{\Gamma}{16u} \left[4y^2 + (2n-1)^2\pi^2 \right] \right\}}{(2n-1)^2\pi^2 \left[4y^2 + (2n-1)^2\pi^2 \right]}\, dy \quad (3.159)$$

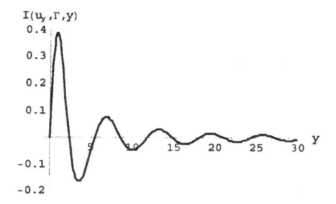

FIGURE 3.24. Integrand of Eq. 3.160 for $u = 10$ and $\Gamma = 1.0$.

$$W(u_y, \Gamma) = \int_0^\infty 4\,y\,J_0\left(y\,\Gamma^{1/2}\right) \left\{ \frac{\left[1 - \exp\left(-\frac{\Gamma y \tanh y}{4u_y}\right)\right]\tanh y}{2y^3} \right.$$

$$\left. + \sum_{n=1}^\infty \frac{16}{(2n-1)^2 \pi^2 \left[4y^2 + (2n-1)^2 \pi^2\right]} \right\}\,dy \quad (3.160)$$

These type A and type B curves can be used to determine aquifer parameters, respectively based on small time and large time pumping data.[50,91] Once the aquifer parameters are determined, either Eq. (3.151), or the asymptotic formulae

$$s = \frac{Q_w}{4\pi T}\,W(u, \Gamma) \qquad \text{for small time}$$

$$s = \frac{Q_w}{4\pi T}\,W(u_y, \Gamma) \qquad \text{for large time} \qquad (3.161)$$

can be used to predict the drawdown.

We note that the drawdown formulae Eqs. (3.151) and (3.161) are valid for small values of drawdown, $s \ll H$. For drawdown that is somewhat larger, the modified drawdown s^* in Eq. (3.148) should be used in place of s.

Example: *Develop a numerical algorithm that evaluates the Neuman well functions.*

In the previous well functions, the Laplace transform led to much simplified mathematical results. In the current case, however, the Laplace transform expressions are not in closed form and there is no advantage in using them. Therefore, we shall directly integrate Eqs. (3.155), (3.159), and (3.160) by Gaussian quadrature. In Figure 3.24 we plot the typical behavior of the integrand. We observe that it is alternating due to the presence of the Bessel function J_0. To obtain accurate result, zeroes of J_0 are sought by a Newton-Raphson scheme. Integration is performed by subintervals delimited by the even zeroes of $J_0(y\Gamma^{1/2})$, denoted as r_0, r_2, r_4, r_6, ... For convenience, we have defined $r_0 = 0$. Within each subinterval, Gaussian quadrature[101] is applied. For example, $W(u, u_y, \Gamma)$ is evaluated as:

$$W(u, u_y, \Gamma) \approx \sum_{m=1}^{N} \int_{r_{2(m-1)}}^{r_{2m}} I(u, u_y, \Gamma, y) \, dy$$

$$\approx \sum_{m=1}^{N} \sum_{j=1}^{N_G} w_j I(u, u_y, \Gamma, y_{mj})$$

in which I denotes the integrand of the well function, y_{mj} is the j-th Gaussian quadrature node located in subinterval m, w_j is the quadrature weights, N is the number of subintervals beyond which the integration is truncated, and N_G is the number of quadrature points.

There are several levels of summations in the above formula (the evaluation of integrand also involve summations), which can be time consuming. To speed up the convergence of the summation, the epsilon algorithm[79] is employed to extrapolate the series. Using a finite number of terms, the residual error can be estimated and corrected. This algorithm requires the storage of the mth partial sum of the series, which is denoted as $\varepsilon_0^{(m)}$. A recursive relation is defined as

$$\varepsilon_{i+1}^{(m)} = \varepsilon_{i-1}^{(m+1)} + \frac{1}{\varepsilon_i^{(m+1)} - \varepsilon_i^{(m)}}$$

which is initialized by $\varepsilon_{-1}^{(m)} = 0$. The term $\varepsilon_N^{(0)}$ then represents the extrapolated result based on the N terms in the series. This procedure is programmed as the function subroutine **epsilonn** in Appendix B.

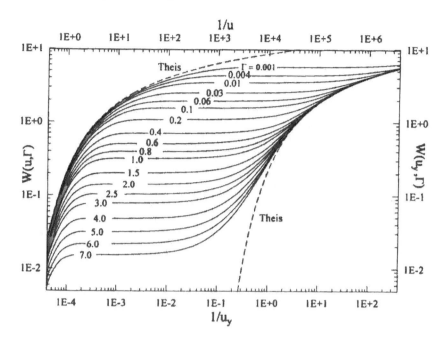

FIGURE 3.25. Neuman's unconfined aquifer well functions $W(u, \Gamma)$ and $W(u_y, \Gamma)$.

Two *Fortran* programs are given in Appendix B. *NeumanW1.for* performs the integration for the full type curves, Eq. (3.155). *NeumanW2.for* evaluates the two asymptotic type curves given by Eqs. (3.159) and (3.160). In Figure 3.25 the type A and type B curves, respectively based on Eqs. (3.159) and (3.160), are plotted. The two families of curves are separated by the labels denoting Γ values. In the same figure, we also plot the two Theis type curves using respectively S and S_y as storage coefficient, for comparison.

3.15 Summary

In this chapter, we have reviewed a number of drawdown solutions due to a pumping well located in an aquifer of infinite horizontal extent. The various conditions investigated include equilibrium versus non-equilibrium states, confined versus unconfined aquifers, constant discharge versus constant drawdown, full versus partial penetration, infinitesimal versus large diameter wells, with or without well storage, etc. Table 3.2 summarizes all the well functions developed in this chapter and their associated conditions. For each well function, *Fortran* and/or *Mathematica* programs are provided for its evaluation, as listed in Appendices B and C.

TABLE 3.2. Summary of non-leaky aquifer pumping well solutions.

Solution	Thiem	Dupuit
State	equilibrium	equilibrium
Aquifer	confined	unconfined
Leakage	no	no
Q_w	constant	constant
s_w	variable	variable
Penetration	full	full
Well Storage	no	no
Well Function	—	—

Solution	Theis	Cooper Jacob
State	non-equilibrium	non-equilibrium
Aquifer	confined	confined
Leakage	no	no
Q_w	constant	constant
s_w	variable	variable
Penetration	full	full
Well Storage	no	no
Well Function	$W(u)$	no

TABLE 3.2. (Continued)

Solution	Jacob Lohman	Papadopulos Cooper
State	non-equilibrium	non-equilibrium
Aquifer	confined	confined
Leakage	no	no
Q_w	variable	constant
s_w	constant	variable
Penetration	full	full
Well Storage	yes	no
Well Function	$F(u_w, \rho), G(u_w)$	$W(u, \alpha, \rho), F(u_w, \alpha)$
Solution	Hantush	Neuman
State	non-equilibrium	non-equilibrium
Aquifer	confined	unconfined
Leakage	no	no
Q_w	constant	constant
s_w	variable	variable
Penetration	partial	full
Well Storage	no	no
Well Function	—	$W(u, \Gamma), W(u_y, \Gamma)$

Chapter 4
LEAKY AQUIFER THEORY

4.1 Leaky Aquifer Systems

For the purpose of groundwater utilization, geological formations can be classified according to their hydraulic properties into *aquifers*, *aquitards*, and *aquicludes*. [12] Formations that contain and transmit a significant amount of water are called aquifers. Formations that are semi-permeable are known as aquitards. Those that practically do not transmit water are aquicludes. For convenience, several layers can be lumped into a single one based on their similar hydraulic properties. We also note that these definitions are only qualitative. Economical judgement of groundwater utilization, such as the rate of yield and cost of extraction for a certain purpose, can play a part in the definition.

Based on the hydraulic point of view, formations consist of aquifers that are separated by either aquitards or aquicludes. An aquifer that is bounded at the top and the bottom by impermeable layers and is hydraulically isolated from the rest is called a confined aquifer. An aquifer whose bottom is impermeable, but the top is in contact with the atmosphere, is an unconfined aquifer. In the preceding chapter, these single-layer aquifers have been discussed. On the other hand, if the bounding formations are semi-permeable, exchange of water with the aquitards and the immediate next aquifers can take place. This exchange is generally know as "leakage." Such aquifer-aquitard systems are called *leaky aquifer* systems, following the pioneering work of Jacob and Hantush (see Sec. 1.2).

In this chapter, we investigate the leaky aquifer theory, which involves the aquifer as well as the aquitard flow. Following the aquifer theory in Chapter 3, we shall inherit the nearly horizontal flow assumption for the aquifer flow. The flow in the aquitard is assumed to be triggered by the head difference between the adjacent aquifers. Due to the large horizontal to vertical aspect ratio of geological formations, the vertical head gradient in the aquitard is generally much greater than the horizontal one. Hence the equipotential lines in the aquitard are nearly horizontal, and the streamlines are nearly vertical. As an approximation, the leaky aquifer theory assumes that the flow in the aquitard takes place only in the vertical direction.

Another justification of assuming horizontal flow in aquifers and vertical flow in aquitards can be presented from the refraction law. As demonstrated in Sec. 2.7, the streamline entering from a high hydraulic conductivity medium (aquifer) at a large incident angle into a low hydraulic conductivity medium (aquitard) bends toward the vertical line. With a large hydraulic conductivity contrast, the flow in the low conductivity medium is nearly vertical (see Figure 2.20). Experience shows that the hydraulic conductivity contrast between the aquitard and aquifer should be at least 1 to 10 for these assumptions to be valid.[87]

In Figure 4.1 we present a simple leaky aquifer system. The aquifer shown is overlain by an aquitard and underlain by an impermeable layer. The top of the aquitard is maintained at a constant head, which is equivalent to a contact with a large hydraulic conductivity aquifer with unlimited water supply. We assume a two-dimensional geometry. A line drain is located on the left side of the aquifer ($x = 0$) and extends to the full height of the aquifer ($0 \leq y \leq 1$). We observe from the streamlines that water enters from the upper aquifer (not shown), passes through the aquitard, and enters the drain in the lower aquifer. In the three cases shown, the aquitard/aquifer hydraulic conductivity contrast is 1:1, 1:5, and 1:20, respectively. We observe that the horizontal and vertical streamline patterns respectively develop in the aquifer and aquitard as the hydraulic conductivity contrast increases. At the contrast 1:5, the leaky aquifer flow assumptions are barely acceptable. At the contrast 1:20, they become quite good.

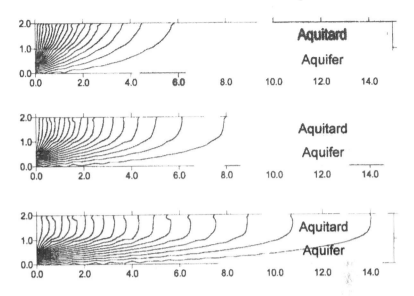

FIGURE 4.1. Streamline pattern for the aquifer-aquitard system. The hydraulic conductivity contrast is: top figure, 1:1; middle figure, 1:5; and bottom figure, 1:20.

4.2   Aquitard Flow

As discussed in the preceding section, the flow in the aquitard is assumed to be in the vertical direction. At a given location (x, y), the governing equation is dependent on the z-coordinate only. The three-dimensional field equation (2.41) reduces to

$$S'_s \frac{\partial h'}{\partial t} = K' \frac{\partial^2 h'}{\partial z^2} \tag{4.1}$$

Here we use a prime superscript to denote quantities associated with aquitards in order to distinguish them from quantities associated with aquifers. We have assumed in Eq. (4.1) that the aquitard is homogeneous in the vertical direction (but not necessarily in the horizontal direction).

Aquifer problems can be investigated as a perturbation from an equilibrated state. This applies to aquitard flow too. The corresponding quantity is the aquitard drawdown s' defined as the head deficit

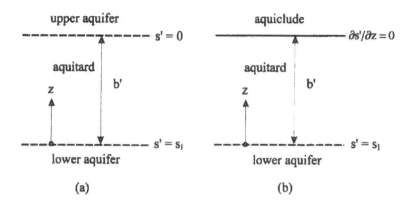

FIGURE 4.2. Two boundary value problems for aquitard flow.

from an equilibrated state h'_o (cf. Eq. (3.14)):

$$s'(z,t) = h'_o(z) - h'(z,t) \tag{4.2}$$

By equilibrium we mean steady state, hence

$$\frac{d^2 h'_o}{dz^2} = 0 \tag{4.3}$$

Equation (4.3) shows that h'_o must be a linear function

$$h'_o = az + b \tag{4.4}$$

changing from the head in the upper aquifer to that in the lower aquifer.

With the substitution of Eqs. (4.2) and (4.3), Eq. (4.1) becomes

$$S' \frac{\partial s'}{\partial t} = b' K' \frac{\partial^2 s'}{\partial z^2} \tag{4.5}$$

In the above, we have multiplied the whole equation by the thickness of the aquitard, b', and used the aquitard storativity

$$S' = b' S'_s \tag{4.6}$$

in place of the specific storativity, S'_s. Given proper initial and boundary conditions, Eq. (4.5) can be solved for the drawdown s'.

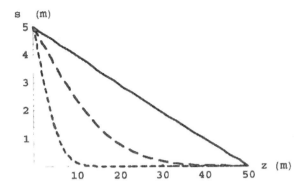

FIGURE 4.3. Drawdown distribution in aquitard (short dashed line: $t = 0.01$ day; long dashed line: $t = 0.1$ day; solid line: $t = 1$ day).

First, we consider the case of an upper aquifer and a lower aquifer separated by an aquitard (see Figure 4.2a). The aquitard is initially undisturbed, $s' = 0$. At $t = 0^+$, the lower aquifer is subjected to a step drawdown in time with magnitude s_l. The upper aquifer is assumed to be highly permeable with an infinite amount of water supply. Its head remains undisturbed at all times. These initial and boundary conditions can be summarized as

$$
\begin{aligned}
s'(z,0) &= 0 \\
s'(0,t) &= s_l \\
s'(b',t) &= 0
\end{aligned}
\tag{4.7}
$$

The mathematical problem of Eq. (4.5) subject to conditions Eq. (4.7) can be solved by the standard technique of separation of variables.[53] Without going into detail, this solution can be extracted from a mathematics book[19]

$$
\frac{s'(z,t)}{s_l} = 1 - \frac{z}{b'} - \frac{2}{\pi} \sum_{m=1}^{\infty} \frac{1}{m} \exp\left(-\frac{m^2 \pi^2 K' t}{S' b'}\right) \sin \frac{m \pi z}{b'}
\tag{4.8}
$$

The specific discharge in the aquitard is obtained by differentiating the above:

$$
q'_z(z,t) = -\frac{s_l K'}{b'} \left[1 + 2 \sum_{m=1}^{\infty} \exp\left(-\frac{m^2 \pi^2 K' t}{S' b'}\right) \cos \frac{m \pi z}{b'} \right]
\tag{4.9}
$$

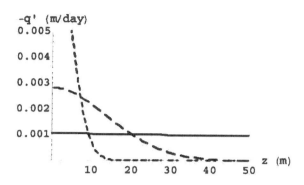

FIGURE 4.4. Specific flux distribution in aquitard (short dashed line: $t = 0.01$ day; long dashed line: $t = 0.1$ day; solid line: $t = 1$ day).

Example: *Given aquitard properties of $K' = 0.01$ m/day, $S' = 0.0005$, aquitard thickness $b' = 50$ m, and a drawdown in the lower aquifer $s_l = 5$ m, show the drawdown and specific discharge in the aquitard. Also evaluate the leakage volume.*

The drawdown and the specific discharge are evaluated using Eqs. (4.8) and (4.9) and are plotted in Figures 4.3 and 4.4 versus the aquifer depth z for three different times, $t = 0.01, 0.1$ and 1 day. We observe that after 1 day, the steady state is practically established. At the steady state, the drawdown is linearly distributed and the specific discharge is a constant, $q'_z(z, \infty) = -s_l K'/b' = -0.001$ m/day.

Equation (4.9) can be integrated with respect to time to give the depth of water (volume of water per unit horizontal area) leaving the aquitard and entering the lower aquifer, as:

$$d_l = \frac{s_l K'}{b'} \left\{ t + \frac{2 S' b'}{\pi^2 K'} \sum_{m=1}^{\infty} \frac{1}{m^2} \left[1 - \exp\left(-\frac{m^2 \pi^2 K' t}{S' b'} \right) \right] \right\}$$

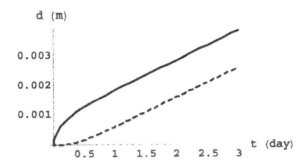

FIGURE 4.5. Depths of water leaked from the upper aquifer into aquitard, d_u (in dashed line), and that supplied from aquitard to lower aquifer, d_l (in solid line), as functions of time.

And the depth of water leaving the upper aquifer entering the aquitard is:

$$d_u = \frac{s_l K'}{b'} \left\{ t + \frac{2 S' b'}{\pi^2 K'} \sum_{m=1}^{\infty} \frac{(-1)^m}{m^2} \left[1 - \exp\left(-\frac{m^2 \pi^2 K' t}{S' b'} \right) \right] \right\}$$

In Figure 4.6 we plot these two depths as functions of time. Because the water supply in the upper aquifer is assumed to be unlimited (for example, continuously replenished by infiltration of precipitation), the leakage increases linearly with time at large times. We observe that there exists a permanent difference between the two curves and that more water leaves the aquitard than enters it. The difference corresponds to the depth of water extracted from the aquitard storage. As $t \to \infty$, its asymptotic value can be estimated as follows. We recall the definition of storativity in Sec. 3.1: storativity is the volume of water released per unit horizontal area due to a unit decline in head. We hence take the storativity and multiply it by the head drop s_l. This is dividing by 2 to obtain $s_l S'/2 = 0.00125$ m, which is correctly shown in Figure 4.6. This quantity is divided by two because the head drop is not uniform in the aquitard. It varies linearly from 0 to s_l.

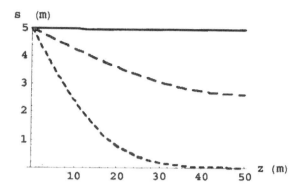

FIGURE 4.6. Drawdown distribution in aquitard (short dashed line: $t = 0.1$ day; long dashed line: $t = 1$ day; solid line: $t = 5$ day).

We notice that the volume of water derived from the aquitard storage is rather small. Its contribution can generally be neglected except for at small times. However, pumping tests are typically conducted for a relatively short period of time. The presence of aquitard storage can play a role in interpreting pumping tests.

The second case to examine is shown in Figure 4.2b. The top of the aquitard is sealed off by an impermeable layer, giving the boundary condition

$$\frac{\partial s'}{\partial z} = 0 \qquad \text{at } z = b' \tag{4.10}$$

The rest of the boundary and the initial conditions remain the same as the preceding problem. The solution can be found in a mathematics book:[19]

$$\frac{s'(z,t)}{s_l} = 1 - \frac{4}{\pi} \sum_{m=1}^{\infty} \frac{1}{2m-1} \exp\left[-\frac{(2m-1)^2 \pi^2 K' t}{4S'b'} \right] \cdot$$

$$\sin \frac{(2m-1)\pi z}{2b'} \tag{4.11}$$

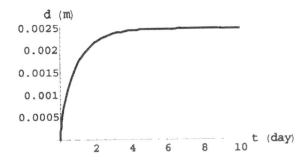

FIGURE 4.7. Depth of water leaked from the aquitard to the lower aquifer as a function of time.

Example: *Evaluate the drawdown for the above case using data provided in the preceding example. Also, derive the expression of leakage from the aquitard.*

In Figure 4.6 the drawdown is plotted versus aquitard depth for $t = 0.1$, 1 and 5 day. We observe that the drawdown in the aquitard is essentially uniform after 5 days.

The volume of leakage from aquitard to the lower aquifer can be derived by differentiating Eq. (4.11) with respect to z to obtain flux, and then integrating with respect to time to find the accumulated depth of water, which is given by

$$d_l = \frac{8s_l S'}{\pi^2} \sum_{m=1}^{\infty} \frac{1}{(2m-1)^2} \left\{ 1 - \exp\left[-\frac{(2m-1)^2 \pi^2 K' t}{4S' b'} \right] \right\}$$

The result is presented in Figure 4.7. We observe that after 5 days the aquitard practically has released all the water that can be produced from storage under the current drawdown: $d_l = s_l S' = 0.0025$ m.

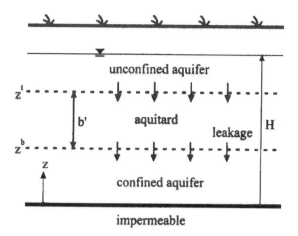

FIGURE 4.8. A leaky aquifer system.

4.3 Governing Equations

Consider a confined aquifer overlain by an aquitard and an unconfined aquifer (Figure 4.8). The unconfined aquifer has a large hydraulic conductivity and unlimited water supply such that it is maintained at a constant head H at all times. The governing equation for the lower aquifer is given by Eq. (3.16), except that the recharge term w is replaced by a leakage flux term q_l with a sign change

$$S\frac{\partial s}{\partial t} = \frac{\partial}{\partial x}\left(T_x\frac{\partial s}{\partial x}\right) + \frac{\partial}{\partial y}\left(T_y\frac{\partial s}{\partial y}\right) + q_l \qquad (4.12)$$

Here q_l is positive when pointing upward. Based on the consideration of mass conservation, the aquifer leakage is equated to the flux at the bottom of the aquitard, $z = z^b$,

$$q_l = K'\left.\frac{\partial s'}{\partial z}\right|_{z=z^b} \qquad (4.13)$$

The aquitard drawdown s' is governed by the one dimensional equation (4.5) introduced in the preceding section,

$$S'\frac{\partial s'}{\partial t} = b'K'\frac{\partial^2 s'}{\partial z^2} \qquad (4.14)$$

The boundary conditions for Eq. (4.14) are:

$$\begin{aligned}
s' &= 0; & \text{at } z = z^t \\
s' &= s(x, y, t); & \text{at } z = z^b
\end{aligned} \qquad (4.15)$$

in which s is the drawdown defined in Eq. (4.12). The first condition is due to the constant head in the unconfined aquifer. The second condition is required by the dynamic equilibrium at the aquifer-aquitard interface. It is clear that the aquifer and aquitard solutions are coupled. However, a few special cases are discussed as follows.

4.4 Steady-State Leaky Aquifer Flow

Under steady state condition, the aquitard flow equation (4.14) becomes

$$\frac{\partial^2 s'}{\partial z^2} = 0 \qquad (4.16)$$

With the boundary conditions Eq. (4.15), the aquitard drawdown is a linear function of depth

$$s' = \frac{s}{b'} \left(z^t - z \right) \qquad (4.17)$$

where $b' = z^t - z^b$ is the aquitard thickness. Based on Eq. (4.13), the leakage into the aquifer is

$$-q_l = \frac{K's}{b'} \qquad (4.18)$$

Substituting the above into Eq. (4.12), we have

$$\frac{\partial}{\partial x} \left(T_x \frac{\partial s}{\partial x} \right) + \frac{\partial}{\partial y} \left(T_y \frac{\partial s}{\partial y} \right) - \frac{K'}{b'} s = 0 \qquad (4.19)$$

If the aquifer is homogeneous and isotropic, the above equation can be written into a *modified Helmholtz equation*

$$\frac{\partial^2 s}{\partial x^2} + \frac{\partial^2 s}{\partial y^2} - \frac{s}{\lambda^2} = 0 \qquad (4.20)$$

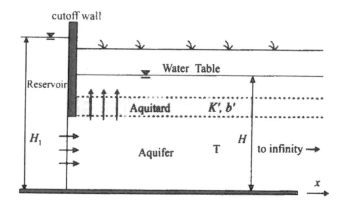

FIGURE 4.9. Leakage from a reservoir.

where

$$\lambda = \sqrt{\frac{b'T}{K'}} \tag{4.21}$$

is known as the *leakage factor*. The leakage factor is a combined property of aquifer and aquitard. It has the dimension of length and characterizes the exponential decay length of a local disturbance.

We can identify in Eq. (4.21) an aquitard property K'/b' which is known as the *leakage coefficient* or *leakance*. Its inverse, b'/K', is called the *hydraulic resistance*. The leakage coefficient has the meaning of aquitard flow across a unit horizontal area into the adjacent aquifer under a unit head difference between the top and bottom of aquitard.

Example: *A reservoir is cut off from leaking water into a phreatic aquifer by an impermeable wall (Figure 4.9). The reservoir however is in contact with a leaky aquifer at the bottom. Assume that the water level in the reservoir (H_1) and of the water table (H) are kept at constant and that the flow has become steady. What is the piezometric head distribution in the leaky aquifer? What is the rate of loss of water from the reservoir?*

Equation (4.20) can be reduced to its one-dimensional form:

$$\frac{d^2s}{dx^2} - \frac{s}{\lambda^2} = 0$$

The general solution of the above is

$$s = c_1\, e^{x/\lambda} + c_2\, e^{-x/\lambda}$$

The boundary conditions are

$$
\begin{aligned}
s &= -(H_1 - H) & \text{at } x = 0 \\
s &= 0 & \text{as } x \to \infty
\end{aligned}
$$

The solution is

$$s = -(H_1 - H)e^{-x/\lambda}$$

Or, expressed in terms of piezometric head, it is

$$h = H + (H_1 - H)e^{-x/\lambda}$$

We observe that the disturbance of aquifer head introduced from the left boundary diminishes exponentially as x increases. The leakage factor λ characterizes the length scale of this exponential decay.

The flux in the aquifer is found by differentiating the drawdown,

$$Q_x = \frac{T(H_1 - H)}{\lambda}e^{-\lambda x}$$

The rate of flow entering the aquifer from the reservoir per unit length of reservoir is found by substituting $x = 0$ in the above:

$$Q_l = \frac{T(H_1 - H)}{\lambda}$$

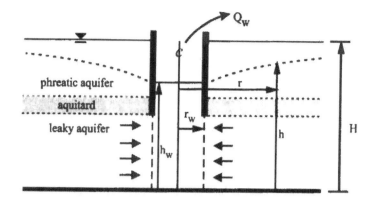

FIGURE 4.10. Pumping well in a leaky aquifer.

4.5 Steady-State Pumping Well (Jacob Solution)

For a steady-state pumping well located in a leaky aquifer (see Figure 4.10), the governing equation in radial coordinates is

$$\frac{1}{r}\frac{d}{dr}\left(r\frac{ds}{dr}\right) - \frac{s}{\lambda^2} = 0 \qquad (4.22)$$

Equation (4.22) can be transformed into the standard form of a well known *modified Bessel equation.* Its general solution is given by the Bessel functions:

$$s = c_1\,I_0\left(\frac{r}{\lambda}\right) + c_2\,K_0\left(\frac{r}{\lambda}\right) \qquad (4.23)$$

where I_0 and K_0 are respectively the modified Bessel function of the first kind and the second kind (see Sec. 3.5) of order zero.

Assume that the pumping well is cased in the upper aquifer and aquitard, but screened in the lower aquifer for the full depth (see Figure 4.10). Since water is extracted from the lower aquifer only, the following boundary conditions are imposed

$$-2\pi T r\frac{\partial s}{\partial r} = Q_w \qquad \text{at } r = r_w$$
$$s = 0 \qquad \text{as } r \to \infty \qquad (4.24)$$

Substituting Eq. (4.23) into the above, we obtain the pumping well solution as

$$s = \frac{Q_w}{2\pi T} \frac{K_0(r/\lambda)}{(r_w/\lambda)K_1(r_w/\lambda)} \tag{4.25}$$

The above solution assumes a finite well radius. For infinitesimal well radius, $r_w \to 0$, we apply the condition $K_1(\eta) \to 1/\eta$ as $\eta \to 0$ as shown in Eq. (3.71). Equation (4.25) reduces to

$$s = \frac{Q_w}{2\pi T} K_0\left(\frac{r}{\lambda}\right) \tag{4.26}$$

Equation (4.26) is the Jacob solution[74] of steady-state pumping well in a leaky aquifer.

To have consistent notation, we can write Eq. (4.26) as

$$s = \frac{Q_w}{4\pi T} W\left(\beta\right) \tag{4.27}$$

where

$$W\left(\beta\right) = W\left(\frac{r}{\lambda}\right) = 2\,K_0\left(\frac{r}{\lambda}\right) \tag{4.28}$$

will be called the *Jacob leaky aquifer well function*. Fortran subroutines evaluating Bessel functions are provided in Appendix B. The Bessel function $K_0(\eta)$ has been plotted in Figure 3.7. We observe that beyond $\eta = 2$, or $r = 2\lambda$, the drawdown rapidly diminishes to zero.

It is of interest to compare the steady state leaky aquifer solution with the steady state non-leaky aquifer solution (Thiem solution) presented in Sec. 3.3. Although both solutions exhibit logarithmic behavior as $r \to 0$ (see Eq. (3.71)), they behave differently as $r \to \infty$. As discussed earlier, the Thiem solution suffers from the paradox of an unbounded drawdown at large distance because a steady-state condition cannot be reasonably assumed without external replenishment. The Jacob solution, on the other hand, is correctly behaved, with $r \to \infty$, $s \to 0$. In this case, we realize that the water extracted from the lower aquifer is exactly compensated by the leakage from the upper aquifer. Hence a steady state can be maintained without violating the physical condition.

Example: *To protect a pumping well from withdrawing polluted water in the upper aquifer, it is desirable to delineate a "wellhead protection area" within which most of the leakage is derived. Find the radius around the well such that 95% of the leakage takes place within this area.*

Define a cylinder centered at the well with a radius r and the height of the aquifer. The flow across the cylindrical surface is

$$Q(r) = -2\pi r T \frac{\partial s}{\partial r} = Q_w \frac{r}{\lambda} K_1 \left(\frac{r}{\lambda} \right)$$

The difference between the well discharge Q_w and the above amount gives the leakage contributed by the phreatic aquifer above the cylinder. Hence, for the ratio

$$\frac{Q(r)}{Q_w} = \frac{r}{\lambda} K_1 \left(\frac{r}{\lambda} \right) = 0.05$$

we find $r/\lambda = 3.9$. This means that 95% of the leakage is derived from the phreatic aquifer within a radius of roughly four times the leakage factor λ.

4.6   Unsteady Well (Hantush-Jacob Solution)

The solution of unsteady drawdown due to a constant rate pumping well in a leaky aquifer is attributed to Hantush and Jacob.[57] In this solution, the aquitard storativity S' is assumed to be negligible. According to Eq. (4.14), the aquitard flow equation becomes steady state; hence the leakage is given by Eq. (4.18). Assuming homogeneity and isotropy, the governing equation under axisymmetry is

$$\frac{S}{T} \frac{\partial s}{\partial t} = \frac{1}{r} \frac{d}{dr} \left(r \frac{ds}{dr} \right) - \frac{s}{\lambda^2} \tag{4.29}$$

Applying the Laplace transform to the above, we obtain

$$\frac{1}{r}\frac{d}{dr}\left(r\frac{d\tilde{s}}{dr}\right) - \left(\frac{pS}{T} + \frac{1}{\lambda^2}\right)\tilde{s} = 0 \qquad (4.30)$$

This equation can be compared to that of the Theis solution, Eq. (3.67). They differ only by the coefficient multiplying the term $\tilde{s}$.

With the same condition of a step extraction rate Q_w at $r_w \to 0$, the drawdown solution can be found from Eq. (3.75) by replacing the argument of Bessel function with the appropriate coefficient:

$$\tilde{s} = \frac{Q_w}{4\pi T}\frac{2}{p}K_0\left(\sqrt{\frac{pS}{T} + \frac{1}{\lambda^2}}\,r\right) \qquad (4.31)$$

The solution in time is given by the Laplace inverse transform

$$s = \frac{Q_w}{4\pi T}\mathcal{L}^{-1}\left\{\frac{2}{p}K_0\left(\sqrt{\frac{pS}{T} + \frac{1}{\lambda^2}}\,r\right)\right\} \qquad (4.32)$$

The above inverse can be carried out analytically, leading to the well known Hantush-Jacob leaky aquifer solution

$$s = \frac{Q_w}{4\pi T}W(u, \beta) \qquad (4.33)$$

where $W(u, \beta)$ is the *Hantush-Jacob leaky aquifer well function,*

$$W(u, \beta) = \int_u^\infty \frac{1}{u}\exp\left(-u - \frac{\beta^2}{4u}\right)du \qquad (4.34)$$

with

$$u = \frac{r^2 S}{4Tt} \qquad (4.35)$$

$$\beta = \frac{r}{\lambda} = \sqrt{\frac{K'}{b'T}}\,r \qquad (4.36)$$

Although the well function $W(u, \beta)$ has been tabulated,[58] we seek an efficient way for its instant evaluation. The Laplace transform expression and the Stehfest inverse algorithm are used for this purpose.

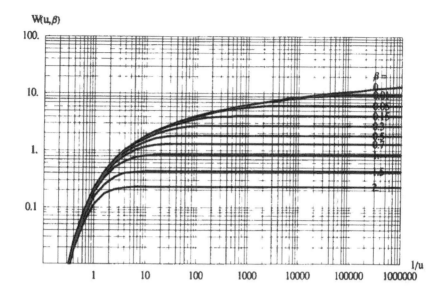

FIGURE 4.11. Hantush-Jacob leaky aquifer type curve.

Similar to the earlier treatment, rather than using the dimensional transform pair, (p, t), we shall use the dimensionless pair (p^*, t^*), to invert the well function, such that

$$W\left(u, \beta\right) = W\left(\frac{1}{t^*}, \beta\right) = \mathcal{L}^{-1}\left\{\frac{2}{p^*} K_0\left(\sqrt{4p^* + \beta^2}\right)\right\} \qquad (4.37)$$

The *Fortran* program *HantJcbW.for* and the *Mathematica* function WHantushJacob[u,beta] in the *WFfield.m* package are given in Appendices B and C for its evaluation.

Example: *Use the provided* Fortran *or* Mathematica *utilities to evaluate and plot the Hantush-Jacob leaky aquifer type curves.*

By choosing a set of $\beta = r/\lambda$ values, a family of type curves are presented in Figure 4.11. The uppermost curve, represented by $\beta = 0$, is the Theis type curve. As the Theis curve rises without a limit, the Hantush-Jacob type curves level off to constant values. In the limit, they approach the Jacob well function as defined in

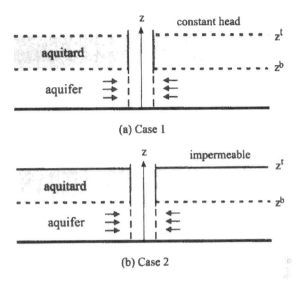

FIGURE 4.12. Two one-aquifer-one-aquitard systems with different boundary conditions.

Eq. (4.28):

$$\lim_{u \to 0} W(u, \beta) = W(\beta)$$

4.7 With Aquitard Storage (Hantush-Neuman Solution)

In the Hantush-Jacob solution derived in the preceding section, the aquitard storage has been ignored, rendering the aquitard response instantaneous. This is however not true, particularly at small pumping times. This restriction is lifted in this section.

Hantush[59] investigated this problem in two different types of boundary conditions, as illustrated in Figure 4.12. Case 1 assumes that the top of aquitard is in contact with an aquifer of constant head. In the second case, the top of aquitard is bounded by an impermeable layer.

The governing equations for leaky aquifer have been established as Eqs. (4.12)–(4.14). For a homogenous, isotropic aquifer under radial flow, Eqs. (4.12) and (4.13) can be combined to give

$$\frac{S}{T}\frac{\partial s}{\partial t} = \frac{1}{r}\frac{d}{dr}\left(r\frac{ds}{dr}\right) + \frac{K'}{T}\frac{\partial s'}{\partial z}\bigg|_{z=z^b} \tag{4.38}$$

Equation (4.38) cannot be solved alone. The aquitard drawdown equation (4.14) must be supplemented,

$$S'\frac{\partial s'}{\partial t} = b'K'\frac{\partial^2 s'}{\partial z^2} \tag{4.39}$$

The aquitard equation is subject to the initial

$$s'(r, z, 0) = 0 \tag{4.40}$$

and boundary conditions

$$\begin{aligned}
s'(r, z^b, t) &= s(r, t), & \text{case 1 and 2} \\
s'(r, z^t, t) &= 0, & \text{case 1} \\
\frac{\partial s'(r, z^t, t)}{\partial z} &= 0, & \text{case 2}
\end{aligned} \tag{4.41}$$

We notice that in the first equation of Eq. (4.41), the condition at $z = z^b$ is imposed by the requirement that the piezometric head must be continuous across the layer interface. Hence it is equated to the aquifer drawdown at that location.

Problems similar to those defined by Eqs. (4.39)–(4.41) have been solved in Sec. 4.2. These problems have the same governing equation (4.39), and initial and boundary conditions Eqs. (4.40) and (4.41), except for the boundary condition at $z = z^b$. In the earlier problems, a constant drawdown s_l was prescribed, whereas in the present problems a time varying drawdown $s(r, t)$ must be satisfied. Rather than solving this as a separate problem, this is a perfect situation to apply a mathematical technique known as the *Duhamel principle of superposition*,[33,53] which is briefly described as follows.

Assume that $U(z, t)$ is the solution of a linear partial differential equation with null initial and boundary conditions, but for one boundary condition, which is given as a unit step function in time $H(t - 0)$. If we solve a problem with the identical governing equation and initial and boundary conditions, except that the unit step function is replaced by a variable boundary condition $b(t)$ with its initial value $b(0) = 0$, then the solution is given by the *convolutional integral*:

$$V(z,t) = \int_0^t U(z,\tau)\frac{\partial b(t - \tau)}{\partial t} \, d\tau \qquad (4.42)$$

The function $U(z,t)$ based on the unit step boundary condition is a *step influence function*. The linearity of the system allows the solution to be obtained by superposition, which, in infinitesimal form, is an integration.

Using this principle, we can formally write down the solutions of the aquitard problems. For case 1, we have

$$s'(r,z,t) = \int_0^t \left[1 - \frac{z - z^b}{b'} - \frac{2}{\pi} \sum_{m=1}^{\infty} \frac{1}{m} \exp\left(-\frac{m^2\pi^2 K'\tau}{S'b'}\right) \right.$$
$$\left. \times \sin\frac{m\pi(z - z^b)}{b'} \right] \frac{\partial s(r, t - \tau)}{\partial t} \, d\tau \qquad (4.43)$$

and for case 2

$$s'(r,z,t) = \int_0^t \left\{ 1 - \frac{4}{\pi} \sum_{m=1}^{\infty} \frac{1}{2m - 1} \exp\left[-\frac{(2m - 1)^2\pi^2 K'\tau}{4S'b'}\right] \right.$$
$$\left. \times \sin\frac{(2m - 1)\pi(z - z^b)}{2b'} \right\} \frac{\partial s(r, t - \tau)}{\partial t} \, d\tau \quad (4.44)$$

In the above, the influence functions are respectively based on Eqs. (4.8) and (4.11). We notice the replacement of z by $z - z^b$ due to the definition of datum shown in Figures (4.2) and (4.12).

For the requirement in Eq. (4.38), we carry out the following operations: for case 1,

$$\frac{K'}{T}\frac{\partial s'}{\partial z}\bigg|_{z=z^b} = -\frac{K'}{b'T} \int_0^t \left[1 + 2\sum_{m=1}^{\infty} \exp\left(-\frac{m^2\pi^2 K'\tau}{S'b'}\right) \right]$$
$$\times \frac{\partial s(r, t - \tau)}{\partial t} \, d\tau \qquad (4.45)$$

and for case 2

$$\frac{K'}{T}\frac{\partial s'}{\partial z}\bigg|_{z=z^b} = -\frac{2K'}{b'T}\int_0^t \sum_{m=1}^\infty \exp\left[-\frac{(2m-1)^2\pi^2 K'\tau}{4S'b'}\right]$$
$$\times \frac{\partial s(r,t-\tau)}{\partial t}\,d\tau \qquad (4.46)$$

The above integration cannot be executed because $s(r,t)$ is not yet known. We can however substitute Eqs. (4.45) or (4.46) for the last term of Eq. (4.38). In doing so, we eliminate the aquitard drawdown variable. Equation (4.38) now is a single equation with a single unknown, and can be utilized for solution of s. However, the complexity of the equation makes a direct analytical solution improbable. We again resort to the Laplace transform to simplify the mathematics.

The Laplace transforms of Eqs. (4.38) and (4.39) are

$$\frac{1}{r}\frac{d}{dr}\left(r\frac{d\tilde{s}}{dr}\right) - \frac{pS}{T}\tilde{s} + \frac{K'}{T}\frac{\partial \tilde{s}'}{\partial z}\bigg|_{z=z^b} = 0 \qquad (4.47)$$

$$b'K'\frac{d^2\tilde{s}'}{dz^2} - pS'\tilde{s}' = 0 \qquad (4.48)$$

We first solve the aquitard problem using Eq. (4.48). Its general solution is

$$\tilde{s}' = C_1 \sinh\sqrt{\frac{pS'}{b'K'}}\,z + C_2 \cosh\sqrt{\frac{pS'}{b'K'}}\,z \qquad (4.49)$$

Using boundary conditions for case 1 and case 2 respectively lead to these solutions

$$\tilde{s}' = \tilde{s}\sinh\left[\sqrt{\frac{pS'}{b'K'}}\,(z^t - z)\right]\bigg/\sinh\sqrt{\frac{pS'b'}{K'}} \qquad (4.50)$$

$$\tilde{s}' = \tilde{s}\cosh\left[\sqrt{\frac{pS'}{b'K'}}\,(z^t - z)\right]\bigg/\cosh\sqrt{\frac{pS'b'}{K'}} \qquad (4.51)$$

In the above, we note that $b' = z^t - z^b$. We can further perform these operations respectively for case 1 and 2,

$$\frac{K'}{T}\frac{\partial \tilde{s}'}{\partial z}\bigg|_{z=z^b} = -\tilde{s}\frac{K'}{T}\sqrt{\frac{pS'}{b'K'}}\coth\sqrt{\frac{pS'b'}{K'}} \qquad (4.52)$$

$$\frac{K'}{T}\frac{\partial \tilde{s}'}{\partial z}\bigg|_{z=z^b} = -\tilde{s}\frac{K'}{T}\sqrt{\frac{pS'}{b'K'}}\tanh\sqrt{\frac{pS'b'}{K'}} \qquad (4.53)$$

These expressions are more manageable than the convolutional integrals, Eqs. (4.45) and (4.46). To be consistent with the notations in Chapter 6, we define the *memory functions* $\tilde{f}(p)$ and $\tilde{g}(p)$ as (see Sec. 6.4)

$$\tilde{f}(p) = \sqrt{\frac{S'b'}{pK'}}\coth\sqrt{\frac{pS'b'}{K'}} \qquad (4.54)$$

$$\tilde{g}(p) = \sqrt{\frac{S'b'}{pK'}}\tanh\sqrt{\frac{pS'b'}{K'}} \qquad (4.55)$$

Substituting Eqs. (4.52) to (4.55) into Eq. (4.47) to eliminate aquitard drawdown, we obtain for case 1

$$\frac{1}{r}\frac{d}{dr}\left(r\frac{d\tilde{s}}{dr}\right) - \left[\frac{pS}{T} + \frac{K'}{Tb'}p\tilde{f}(p)\right]\tilde{s} = 0 \qquad (4.56)$$

and for case 2

$$\frac{1}{r}\frac{d}{dr}\left(r\frac{d\tilde{s}}{dr}\right) - \left[\frac{pS}{T} + \frac{K'}{Tb'}p\tilde{g}(p)\right]\tilde{s} = 0 \qquad (4.57)$$

Now we are ready to solve the aquifer drawdown.

Comparing Eqs. (4.56) and (4.57) with those of Theis and the Hantush-Jacob solutions, Eqs. (3.67) and (4.30), we realize that they are the same modified Bessel equation with difference only in the coefficients. Hence the solution is

$$\tilde{s} = \frac{Q_w}{4\pi T}\frac{2}{p}K_0(kr) \qquad (4.58)$$

where

$$k = \sqrt{\frac{pS}{T} + \frac{K'}{Tb'}p\tilde{f}(p)} \qquad (4.59)$$

for case 1, and

$$k = \sqrt{\frac{pS}{T} + \frac{K'}{Tb'}p\tilde{g}(p)} \qquad (4.60)$$

for case 2.

We observe that the drawdown solutions for Theis, Hantush-Jacob, and the current cases are all of the same form, Eq. (4.58). The difference lies in the definition of the characteristic root k. We can observe the following limits. For case 1, when the aquitard storage becomes negligible, $S' \to 0$, as in the Hantush-Jacob case, Eq. (4.54) shows $\tilde{f}(p) \to 1/p$. Equation (4.59) becomes

$$k = \sqrt{\frac{pS}{T} + \frac{K'}{b'T}} = \sqrt{\frac{pS}{T} + \frac{1}{\lambda^2}} \qquad (4.61)$$

Equation (4.58) then reduces to the Hantush-Jacob solution Eq. (4.31). If the aquitard hydraulic conductivity further goes to zero, $K' \to 0$, k reduces to

$$k = \sqrt{\frac{pS}{T}} \qquad (4.62)$$

for both case 1 and 2. The Theis solution Eq. (3.75) emerges.

The analytical Laplace inverse transforms of Eqs. (4.58)–(4.60) are difficult to obtain. Hantush[59] could only find asymptotic solutions at large and small times. Neuman[87] on the other hand was able to derive the solution for case 1 as

$$s = \frac{Q_w}{4\pi T} \int_0^\infty \frac{2}{y}\left[1 - \exp\left(-\frac{K't}{b'S'}y^2\right)\right] \times$$
$$J_0\left(r\sqrt{\frac{K'S}{b'S'T}y^2 - \frac{K'}{b'T}y\cot y}\right) \, dy \quad (4.63)$$

where the Bessel function J_0 is set to zero if its argument becomes imaginary.

Although the time-domain solution is available for case 1, we shall rely on the numerical inversion of Eq. (4.58) for the evaluation of the well functions. In this regard, we define the well functions

$$s = \frac{Q_w}{4\pi T} W_1(u, \beta, \eta), \qquad \text{case 1} \tag{4.64}$$

$$s = \frac{Q_w}{4\pi T} W_2(u, \beta, \eta), \qquad \text{case 2} \tag{4.65}$$

We shall name W_1 and W_2 as the *Hantush-Neuman leaky aquifer well functions for case 1* and *case 2*, respectively. The dimensionless parameters are

$$u = \frac{r^2 S}{4Tt} \tag{4.66}$$

$$\beta = \frac{r}{\lambda} \tag{4.67}$$

$$\eta = \frac{S'}{S} \tag{4.68}$$

The well functions are obtained from these inverse formulae

$$W_1(u, \beta, \eta) = W_1(\frac{1}{t^*}, \beta, \eta)$$

$$= \mathcal{L}^{-1} \left\{ \frac{2}{p^*} K_0 \left(\sqrt{4p^* + 2\beta\sqrt{\eta p^*} \coth \frac{2\sqrt{\eta p^*}}{\beta}} \right) \right\} \tag{4.69}$$

$$W_2(u, \beta, \eta) = W_2(\frac{1}{t^*}, \beta, \eta)$$

$$= \mathcal{L}^{-1} \left\{ \frac{2}{p^*} K_0 \left(\sqrt{4p^* + 2\beta\sqrt{\eta p^*} \tanh \frac{2\sqrt{\eta p^*}}{\beta}} \right) \right\} \tag{4.70}$$

Fortran program *HantNeuW.for* and *Mathematica* functions WHantushNeuman1[u,beta,eta], WHantushNeuman2[u,beta,eta] are provided in Appendices B and C.

Example: *Examine the effect of aquitard storage and compare it with the Hantush-Jacob leaky aquifer solution (without aquitard storage).*

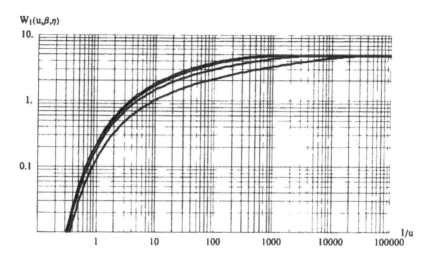

FIGURE 4.13. Hantush-Neuman well function $W_1(u, \beta, \eta)$ for $\beta = 0.1$ and 4 different η values. (From top curve down: $\eta = 0$ (Hantush-Jacob well function), 1, 10, and 100.)

Case 1 has the same aquifer-aquitard boundary conditions as the Hantush-Jacob leaky aquifer problem; hence it will be compared. Due to the number of parameters involved, we choose to fix β at 0.1. Using the *Fortran* or *Mathematica* programs provided, we evaluate the Hantush-Neuman well function Eq. (4.69) for four values of η ($= S'/S$), 0, 1, 10, and 100. This covers the range of aquitard storage from negligible (Hantush-Jacob solution) to very large compared to aquifer storage. The result is plotted in Figure 4.13. The top curve corresponding to $\eta = 0$ is the Hantush-Jacob type curve for $\beta = 0.1$. The second curve from the top is for $\eta = 1$, which almost coincides with the Hantush-Jacob curve. The difference however becomes significant for $\eta = 10$ and 100.

As time $(1/u)$ becomes large, all curves level off to a constant value. This limit is given by the Jacob steady-state leaky aquifer well function, Eq. (4.28). It, however, takes much longer time for the large aquitard storage curves to join the limit.

Although with the assistance of numerical tools we have now the capability to evaluate the full solutions, it is still of interest to discuss the asymptotic behavior of these aquifer systems following Hantush's work.[59,60] These asymptotic solutions serve to isolate the effects of various parameters in different time regimes. It might also be useful on the occasion of parameter determination. The simultaneous determination of a large number of parameters can be difficult and unreliable. The asymptotic solutions allow the use of data taken from different time regimes, which contain stronger signals of certain parametric behavior in that time range, for the more accurate determination of a smaller group of parameters.

To find the asymptotic time behavior, we note the following rules in the Laplace transform. As the Laplace transform parameter generally works as the inverse of time, the large time limit can be obtained by taking the limit of $p \to 0$, and the small time limit by taking $p \to \infty$. Based on Eqs. (4.54) and (4.55), we have these limits: for $t \to \infty$,

$$\lim_{p \to 0} p\tilde{f}(p) = 1 + \frac{pS'b'}{3K'} \tag{4.71}$$

$$\lim_{p \to 0} p\tilde{g}(p) = \frac{pS'b'}{K'} \tag{4.72}$$

and for $t \to 0$,

$$\lim_{p \to \infty} p\tilde{f}(p) = \lim_{p \to \infty} p\tilde{g}(p) = \sqrt{\frac{pS'b'}{K'}} \tag{4.73}$$

The characteristic roots k in Eqs. (4.59) and (4.60) become, as $t \to \infty$,

$$k = \sqrt{\left(1 + \frac{S'}{3S}\right)\frac{pS}{T} + \frac{K'}{Tb'}} \qquad \text{case 1} \tag{4.74}$$

$$k = \sqrt{\left(1 + \frac{S'}{S}\right)\frac{pS}{T}} \qquad \text{case 2} \tag{4.75}$$

and as $t \to 0$,

$$k = \sqrt{\frac{pS}{T} + \frac{1}{T}\sqrt{\frac{pS'K'}{b'}}} \qquad \text{case 1 and 2} \qquad (4.76)$$

Comparing Eqs. (4.74) and (4.75) with the earlier expressions, e.g., Eqs. (4.61) and (4.62), we conclude that at large times, the asymptotic drawdown formulae can be organized into:

$$s = \frac{Q_w}{4\pi T} W(u_1, \beta), \qquad \text{case 1} \qquad (4.77)$$

$$s = \frac{Q_w}{4\pi T} W(u_2), \qquad \text{case 2} \qquad (4.78)$$

where

$$u_1 = \left(1 + \frac{\eta}{3}\right) u \qquad (4.79)$$

$$u_2 = (1 + \eta) u \qquad (4.80)$$

In the above, $W(u_1, \beta)$ is the Hantush-Jacob well function, and $W(u_2)$ is the Theis well function. We notice that in case 1 the system at large time behaves as a leaky aquifer without aquitard storage, but with a larger, apparent aquifer storage of $S + (S'/3)$. In case 2, due to the impermeable top, the aquitard is rapidly depleted, and the system responds like a nonleaky Theis aquifer, but with an apparent aquifer storage of $S + S'$.

At small times, the characteristic root in Eq. (4.76) leads to a new well function:

$$s = \frac{Q_w}{4\pi T} H(u, \beta'), \qquad \text{case 1 and 2} \qquad (4.81)$$

where $H(u, \beta')$ is the *Hantush small time leaky aquifer well function* and

$$\beta' = \frac{\sqrt{\eta}\,\beta}{4} \qquad (4.82)$$

The well function can be obtained as

$$H(u, \beta') = H(\frac{1}{t^*}, \beta') = \mathcal{L}^{-1}\left\{\frac{2}{p^*} K_0\left(\sqrt{4p^* + 8\beta'\sqrt{p^*}}\right)\right\} \qquad (4.83)$$

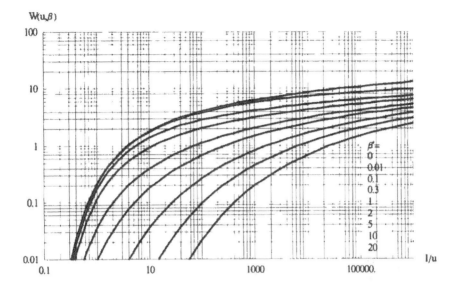

FIGURE 4.14. Hantush small time leaky aquifer type curve.

Fortran program *HantushH.for* and *Mathematica* function WHantush[u,beta] are found in Appendix B and C.

Hantush[59] also provided the solution in time:

$$H(u, \beta') = \int_{u}^{\infty} \frac{e^{-y}}{y} \operatorname{erfc}\left[\frac{\beta'\sqrt{u}}{\sqrt{y(y-u)}}\right] dy \qquad (4.84)$$

Figure 4.14 presents the Hantush type curve $H(u, \beta')$ for a number of β' values.

4.8 Summary

Similar to Chapter 3, we summarize the leaky aquifer well functions discussed in this chapter in a table (Table 4.1).

As a final remark, we note that although the geometry of the aquifer-aquitard system has been shown as an aquitard overlain an aquifer, all the solutions derived in this chapter remain valid if their positions are reversed, i.e. the aquifer is on top of the aquitard.

TABLE 4.1. Summary of leaky aquifer solutions.

Solution	Jacob	Hantush-Jacob
State	equilibrium	non-equilibrium
Aquifer	confined	confined
Leakage	yes	yes
Aquitard Storativity	no	no
Aquitard Boundary	no drawdown	no drawdown
Q_w	constant	constant
s_w	variable	variable
Penetration	full	full
Well Storage	no	no
Well Function	$W(\beta)$	$W(u,\beta)$

Solution	Hantush-Neuman case 1	Hantush-Neuman case 2
State	non-equilibrium	non-equilibrium
Aquifer	confined	confined
Leakage	yes	yes
Aquitard Storativity	yes	yes
Aquitard Boundary	no drawdown	impermeable
Q_w	constant	constant
s_w	variable	variable
Penetration	full	full
Well Storage	no	no
Well Function	$W_1(u,\beta,\eta)$	$W_2(u,\beta,\eta)$

TABLE 4.1. (Continued)

Solution	Hantush small time
State	non-equilibrium
Aquifer	confined
Leakage	yes
Aquitard Storativity	yes
Aquitard Boundary	—
Q_w	constant
s_w	variable
Penetration	full
Well Storage	no
Well Function	$H(u, \beta')$

Chapter 5

APPLICATION OF PUMPING WELL SOLUTIONS

5.1 Pumping at Variable Rate

The pumping well solutions constructed in the preceding chapter are based on linear or linearized governing equations. The implication of a linear system is that the solutions can be superimposed, both in time and in space. The superposition in time can be used to simulate wells that are pumped at variable, rather than constant, rates. Spatial superposition allows the simulation of well fields with multiple pumping wells. Also, a proper assemblage of fictitious image wells can be used to create certain physical boundary conditions. Some of these and other applications are illustrated in this chapter.

All the pumping well solutions obtained in Chapter 3 are based on a step-rise, constant pumping rate condition, except for the step drawdown solution presented in Sec. 3.9. A unit step rise pumping rate is expressed as:

$$Q_w = \mathrm{H}(t - t_o) \tag{5.1}$$

where H is the Heaviside unit step function, as illustrated in Eq. (3.81), and t_o is the pumping initiation time. The drawdown due to this unit step pumping can be expressed as

$$F(t) = \frac{1}{4\pi T} W(\cdots) \tag{5.2}$$

where F denotes a *step influence function*, and $W(\dots)$ is one of the suitable well functions derived in Chapters 3 and 4 (see Table 3.2 and 4.1).

If we take the time derivative of Eq. (5.1), we obtain a delta function as the pumping rate

$$Q_w = \delta(t - t_o) \tag{5.3}$$

The property of such function has been discussed in Sec 3.5. Physically, Eq. (5.3) suggests that a unit volume of water is removed from the aquifer instantly. The aquifer response corresponding to this pumping is obtained by differentiating the step influence function,

$$f(t) = \frac{dF(t)}{dt} \tag{5.4}$$

In the above, $f(t)$ is designated as the *impulse influence function*.

Using, for example, the Theis well function given as Eq. (3.92), we obtain the impulse influence function as

$$f(t) = \frac{1}{4\pi T t} \exp\left(-\frac{r^2 S}{4T t}\right) \tag{5.5}$$

That for the Hantush-Jacob solution, Eq. (4.34), is given by

$$f(t) = \frac{1}{4\pi T t} \exp\left(-\frac{r^2 S}{4T t} - \frac{K't}{b'S}\right) \tag{5.6}$$

Given the physical parameters S and T, and a fixed distance r, the typical behavior of the Theis impulse influence function versus time is plotted in Figure 5.1. We observe that the curve is led by a flat portion because it takes time for the propagation front of the drawdown to reach a given distance. The drawdown then rises up to a peak, and subsequently declines. As the pumping is an instantaneous removal of a fixed volume, not a continuous pumping, the drawdown eventually spreads out to the whole aquifer; hence the drawdown drops to zero.

Once the response due to a unit impulse is known, the drawdown due to an variable pumping schedule

$$Q_w = Q_w(t) \tag{5.7}$$

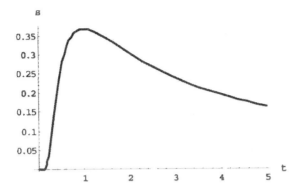

FIGURE 5.1. Behavior of the impulse influence function of Theis solution.

can be obtained from the mathematical operation known the Duhamel principle of superposition

$$s(t) = \int_0^t Q_w(\tau)\, f(t - \tau)\, d\tau \qquad (5.8)$$

We notice that the above equation is slightly different from Eq. (4.42) in that the discharge, rather than its derivative, is in used in the integrand. This is because the influence function in Eq. (4.42) is based on the step function, whereas f is based on the impulse function. With an integration by parts, and a proper handling of the integration limits, these two formulae are interchangeable.

For the case of the Theis well function, using Eq. (5.5) in Eq. (5.8) yields

$$s(r, t) = \frac{1}{4\pi T} \int_0^t \frac{Q_w(\tau)}{t - \tau} \exp\left[-\frac{r^2 S}{4T(t - \tau)} \right] d\tau \qquad (5.9)$$

The above formula, when integrated numerically, can be used to calculate the drawdown response of an arbitrary variable pumping rate in a confined aquifer.

Example: *Given a confined aquifer with transmissivity 5000 m²/day, storativity 0.0001, and a pumping rate*

$$Q_w = Q_o \left(1 - e^{-kt} \right)$$

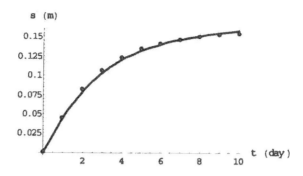

FIGURE 5.2. Drawdown due to a variable pumping rate. (Dots: solution by numerical integration of convolutional integral. Solid line: solution by Laplace transform technique.)

where $Q_o = 1000$ m^3/day, and $k = 0.5$ day^{-1}, what is the drawdown at the distance 200 m from the pumping well?

For a given time t, Eq. (5.9) can be numerically integrated. Gaussian quadrature is employed for this purpose. A *Fortran* program *VarPump.for* is provided in Appendix B for evaluating the drawdown. Figure 5.2 plots drawdown versus time as dot symbols. One hundred Gaussian points are needed to produce an accurate result. The solid line in Figure 5.2 is the solution obtained using a different technique to be described below.

For problems with different pumping schedules, the program *VarPump.for* can be easily modified. The user only needs to replace the pumping rate formula used in the function subroutine *qw* by his own.

To apply Eq. (5.8), the impulse influence function, i.e. the derivative of well function, is needed. Due to the complexity of some time domain well functions, the derivative may not always be easy to obtain. Therefore it is more convenient to use the well function itself in the convolutional integral. For this purpose, an alternative formula to Eq. (5.8) is devised below.

Performing integration by parts on Eq. (5.8), we obtain:

$$s(t) = Q_w(0)F(t) + \int_0^t \frac{\partial Q_w(\tau)}{\partial \tau} F(t - \tau)\, d\tau \qquad (5.10)$$

where $F(t)$ is now associated with the well function, see Eq. (5.2). We notice that the time derivative of pumping rate $Q_w(t)$ is used in the integrand. If the pumping rate contains jumps, such as an initial jump caused by $Q_w(0) \neq 0$, its contribution must be properly taken into account, shown as the first term on the right hand side of Eq. (5.10). Any other step rises at $t = t_1 > 0$ must be taken care of by adding such terms

$$\left[Q_w(t_1^+) - Q_w(t_1^-) \right] F(t - t_1)$$

to the right hand side of Eq. (5.10).

As demonstrated in Chapters 3 and 4, most of the well functions are too complicated for direct evaluation such that the Laplace transform and its numerical inverse are used for their calculation. In fact, the Laplace transform not only can simplify the well functions, it also removes the convolutional integral. This is based on the convolutional theorem as shown in Eq. (A.10) of Appendix A.

Applying the Laplace transform to Eq. (5.8), we can convert the integral into a product:

$$\tilde{s} = \widetilde{Q}_w(p)\, \tilde{f}(p) \qquad (5.11)$$

Realizing from Eqs. (5.4) and (A.3) that

$$\tilde{f}(p) = p\, \widetilde{F}(p) \qquad (5.12)$$

we can express Eq. (5.11) as

$$\tilde{s} = \widetilde{Q}_w(p)\, p\, \widetilde{F}(p) \qquad (5.13)$$

This is the formula for drawdown under variable rate pumping in the Laplace transform domain.

To find the solution in time, we perform numerical Laplace inversion to Eq. (5.13). Since numerical inversion is already needed in the evaluation of most well functions, there is no additional effort involved in the variable pumping rate solution. We should note, however, this technique requires that $Q(t)$ be given in an analytical expression such that its analytical Laplace transform expression can be found.

As an illustration, we take the Theis aquifer case. The Laplace transform drawdown solution is

$$\tilde{s} = \frac{\tilde{Q}_w}{2\pi T} K_0 \left(\sqrt{\frac{pS}{T}}\, r \right) \tag{5.14}$$

for any given variable pumping rate $Q_w(t)$ with its corresponding Laplace transform $\tilde{Q}_w$. In particular, for a step rise constant rate pumping, $Q_w H(t-0)$, where Q_w is a constant, its Laplace transform is Q_w/p (see Table A-1). Substituting this for $\tilde{Q}_w$ in Eq. (5.14), we obtain exactly the Theis solution given by Eq. (3.75).

Example: *Given the preceding example of variable pumping rate, find the drawdown versus time using the Laplace transform technique.*

The Laplace transform of the pumping rate is found by utilizing the Laplace transform table (Table A-1),

$$\tilde{Q}_w = Q_o \left(\frac{1}{p} - \frac{1}{p+k} \right)$$

The drawdown becomes

$$s(t) = \frac{Q_o}{2\pi T} \mathcal{L}^{-1} \left\{ \left(\frac{1}{p} - \frac{1}{p+k} \right) K_0 \left(\sqrt{\frac{pS}{T}}\, r \right) \right\}$$

The numerical Laplace inversion can be carried out using either a Fortran program in Appendix B, or the *NLapInv.m* Mathematica macro in Appendix C. The result of the inversion is shown in Figure 5.2 as the solid line. It compares well with the solution obtained by the numerical integration of the convolutional integral (in dots).

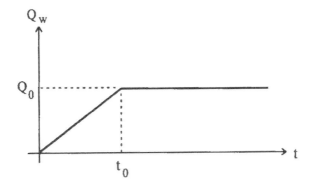

FIGURE 5.3. A variable pumping schedule.

Example: *Given a pumping schedule shown as the ramp function in Figure 5.3, find the drawdown formula by the Laplace transform technique.*

The pumping rate can be expressed in the time domain as

$$Q_w(t) = \frac{Q_o}{t_o}\left[t - (t - t_o)\,\mathrm{H}(t - t_o)\right]$$

We notice that the use of a Heaviside unit step function in the second term inside the square brackets renders the term zero when $t < t_o$. Utilizing relations in Appendix A, we obtain the Laplace transform of the above as

$$\tilde{Q}_w(p) = \frac{Q_o}{t_o\,p^2}\left(1 - e^{-pt_o}\right)$$

If the well is located in a leaky aquifer governed by the Hantush-Jacob drawdown equation (4.31), the drawdown corresponding to the given pumping schedule is

$$s = \frac{Q_o}{2\pi T\,t_o}\,\mathcal{L}^{-1}\left\{\frac{1}{p^2}\left(1 - e^{-pt_o}\right)\mathrm{K}_0\left(\sqrt{\frac{pS}{T} + \frac{1}{\lambda^2}}\,r\right)\right\}$$

Either *Fortran* or *Mathematica* can be used to evaluate the time domain solution.

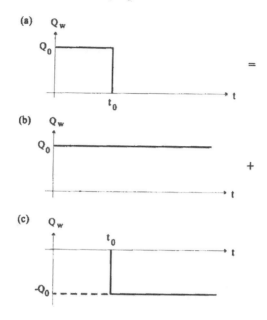

FIGURE 5.4. Simulating slug test by superposition: (a) = (b) + (c).

5.2 Superposition in Time

In the preceding section, we assume that the pumping rate is continuous in time. Whenever jumps are involved, their contribution needs to be separately treated and added to the convolutional integral. In practical operations, the pumping schedule often changes in step-increases or decreases, while remaining constant in between. In those cases, summation rather than integration is the more efficient algorithm for evaluating the drawdown due to variable rate pumping.

Consider first the *slug test* often performed by field hydrogeologists to determined aquifer parameters. The well is pumped at a constant rate, and then suddenly shuts off. The pumping schedule is illustrated in Figure 5.4(a). This schedule can be simulated as the summation of two constant rate pumping, one starting at $t = 0$, with rate Q_o, and the second starting at $t = t_o$, with rate $-Q_o$, as illustrated in

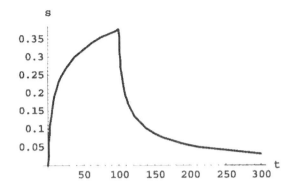

FIGURE 5.5. Drawdown from a slug test.

Figure 5.4(b) and (c). The drawdown can be evaluated as

$$s(r,t) = \frac{Q_o}{4\pi T} W\left(\frac{r^2 S}{4Tt}, \dots\right) - \frac{Q_o}{4\pi T} W\left(\frac{r^2 S}{4T(t-t_o)}, \dots\right) \quad (5.15)$$

where $W(u, \dots)$ represents the various constant pumping well functions. We also note that the well functions are interpreted as
$W(u, \dots) = 0$ for $u < 0$. In other words, $W(u, \dots)$ must return
the value 0 for any t less than the pumping initiation time t_o. Figure
5.5 gives a typical drawdown response to a slug test.

We next investigate a pumping schedule made of step changes as
shown in Figure 5.6. Again, the drawdown can be calculated by the
superposition of constant rate pumping solutions:

$$
\begin{aligned}
s(r,t) &= \frac{Q_1}{4\pi T} W\left(\frac{r^2 S}{4Tt}\right) + \frac{Q_2 - Q_1}{4\pi T} W\left(\frac{r^2 S}{4T(t-t_1)}\right) \\
&\quad + \frac{Q_3 - Q_2}{4\pi T} W\left(\frac{r^2 S}{4T(t-t_2)}\right) + \dots \\
&= \sum_{i=1}^{n} \frac{Q_i - Q_{i-1}}{4\pi T} W\left(\frac{r^2 S}{4T(t-t_{i-1})}\right) \quad \text{for } t > t_{n-1} \;\; (5.16)
\end{aligned}
$$

In the above, we assign $Q_0 = 0$ and $t_0 = 0$. We also interpret
that each well function contains a Heaviside unit step function such
that $W(u) = 0$ for $u < 0$.

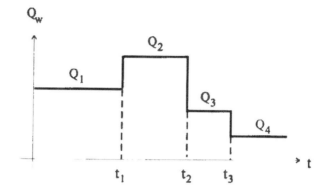

FIGURE 5.6. A step pumping schedule.

Example: *A confined aquifer has a transmissivity 2000 m²/day and a storativity 0.00005. The aquifer is pumped with a schedule shown in Figure 5.6, in which Q_1, Q_2, Q_3 are respectively 700, 1000, and 500 m³/day, and t_1, t_2, t_3 are respectively 10, 18, and 24 day. What is the drawdown in an observation well at $r = 500$ m and $t = 20$ day?*

Following Eq. (5.16), the drawdown is contributed by three constant rate pumping schedules, 700, 300, and −500 m³/day, respectively started at 20, 10, and 2 days before the current time. Hence the drawdown is

$$
\begin{aligned}
s \;=\; & \frac{1}{4\pi \times 2000 \text{ m}^2/\text{day}} \\
& \left[700 \text{ m}^3/\text{day} \times W\left(\frac{(500\,\text{m})^2 \times 0.00005}{4 \times 2000 \text{ m}^2/\text{day} \times 20\,\text{day}} \right) \right. \\
& +300 \text{ m}^3/\text{day} \times W\left(\frac{(500\,\text{m})^2 \times 0.00005}{4 \times 2000 \text{ m}^2/\text{day} \times 10\,\text{day}} \right) \\
& \left. -500 \text{ m}^3/\text{day} \times W\left(\frac{(500\,\text{m})^2 \times 0.00005}{4 \times 2000 \text{ m}^2/\text{day} \times 2\,\text{day}} \right) \right] \\
=\; & 0.214 \text{ m}
\end{aligned}
$$

Sometimes we can approximate a continuous, variable rate pumping as a finite number of discrete step changes with pumping rate $Q_1, Q_2, Q_3, \ldots$ In this case, it is more convenient to use constant time increment Δt. At a time $t = n\Delta t$, Eq. (5.16) becomes

$$s(r, n\Delta t) = \sum_{i=1}^{n} \frac{Q_i - Q_{i-1}}{4\pi T} W \left(\frac{r^2 S}{4T(n-i+1)\Delta t} \right)$$

$$= \sum_{i=1}^{n} \frac{Q_i}{4\pi T} \left[W \left(\frac{r^2 S}{4T(n-i+1)\Delta t} \right) - W \left(\frac{r^2 S}{4T(n-i)\Delta t} \right) \right]$$

$$= \sum_{i=1}^{n} \frac{Q_i}{4\pi T} w(n-i) \qquad (5.17)$$

In the $w(j, \Delta t)$ is defined as

$$w(j) = W \left(\frac{r^2 S}{4T(j+1)\Delta t} \right) - W \left(\frac{r^2 S}{4Tj\Delta t} \right), \quad j = 0, 1, \ldots, n-1$$

$$(5.18)$$

Equation (5.17) shows that for the calculation of drawdown at a single time, a large number of evaluations of well function are needed. It also indicates that the availability of the drawdown at the previous time step, $t = (n-1)\Delta t$, does not help in the evaluation of drawdown at $t = n\Delta t$. The whole history must be computed. By the use of constant time step, however, we find that many well function evaluations are repeated. Equations (5.17) and (5.18) show that for solutions up to $t = n\Delta t$, the function values $w(0), w(1), \ldots, w(n-1)$ can be pre-computed and stored to reduce the computation time.

5.3 Superposition in Space

The pumping well solutions not only can be superposed in time, but also in space. For example, we can simulate in an open field, i.e. a field of infinite horizontal extent, the drawdown at (x, y, t) caused by a group of m wells located at (x_i, y_i), with constant discharge Q_i

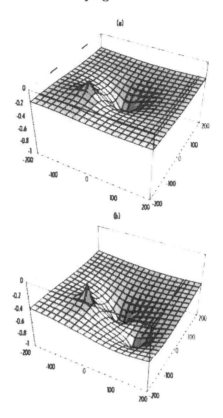

FIGURE 5.7. Piezometric head distribution due to two pumping and one recharge well, at (a) $t = 9$ day, and (b) $t = 10$ day.

and pumping initiation time t_i, as

$$s(x, y, t) = \sum_{i=1}^{m} \frac{Q_i}{4\pi T} W \left(\frac{r_i^2 S}{4T(t - t_i)} \right) \qquad (5.19)$$

where

$$r_i = \sqrt{(x - x_i)^2 + (y - y_i)^2} \qquad (5.20)$$

Example: *Given three pumping wells located at* (0 m, 0 m), (100 m, −100 m), (−75 m, −75 m), *pumped at constant rates of* $Q_w = 1000$,

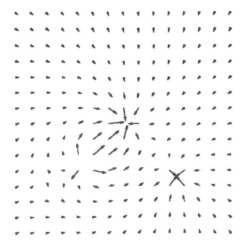

FIGURE 5.8. Velocity plot of two pumping wells and one recharge well.

600, −600 m³/day, *with starting time of* 0, 10 *and* 0 day, *plot the piezometric head distribution. The aquifer parameters are* $T = 1000$ *m²/day and* $S = 0.0003$.

The various drawdown solutions have been programmed into the *Mathematica* package *WField.m* (see Appendix C, and Cheng and Ouazar[30]). We can use them to produce the two dimensional contour plot, or the three-dimensional surface plot. Assuming the Theis solution, we present the surface plots of drawdown in the range (±200 m, ±200 m), at two different times, $t = 9$ and 11 day, as Figure 5.7(a) and (b), respectively. We note in particular that one of the wells is a recharge well with negative Q_w. Also, the second well starts at the 10th day; hence it does not appear in the first diagram.

Not only the drawdown fields can be superimposed; the same principle works for the flux fields as well. The flux can be obtained by applying the specific discharge formula to Eq. (5.19). For example, Eq. (3.94) gives the radial flux of the Theis solution. Plotting the resulting vector field, we can obtain the flow pattern. Figure 5.8 shows the velocity field after 30 days of pumping.

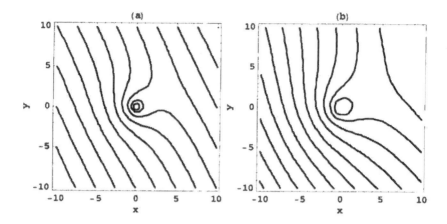

FIGURE 5.9. Piezometric head distribution of a well in a uniform flow field: (a) small time (b) large time.

Other groundwater flow fields can be combined with the well field. For example, for a uniform groundwater flow of specific discharge $\mathbf{U} = U_x \mathbf{i} + U_y \mathbf{j}$, the piezometric head field is

$$h = -\frac{1}{K}\left(U_x\, x + U_y\, y\right) \tag{5.21}$$

The pumping well can be placed in such a field and the drawdown can be superimposed. Figure 5.9 shows a uniform flow field making a 30° angle with the x-axis. A pumping well is located at the origin. Figure 5.9(a) presents the head distribution at a small time after pumping, and Figure 5.9(b) at a large time.

5.4   Method of Images

Wells are sometimes located near constant head boundaries such as streams, canals, and lakes. Other times wells may be situated near tight faults, rocky barriers bounding sedimentary basins, man-made dikes, or cutoff walls. If the pumping period is long enough, the cone of depression will reach to and interact with these boundaries. A treatment of these boundaries in arbitrary geometry requires a

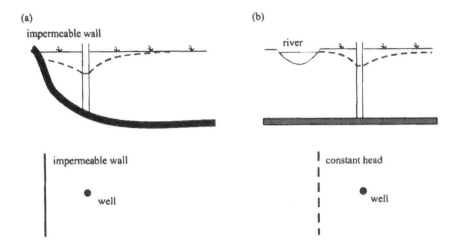

FIGURE 5.10. Well near an impermeable wall and a river, by the method of images.

numerical method. However, there exist a few problems with simple geometries such that their solution can be obtained by the principle of superposition. This technique is known as the *method of images.*[56]

Consider a pumping well situated in front of a straight and infinitely long impermeable boundary (see Figure 5.10(a)). The solution can be created by considering the wall as a mirror such that an image well exists on the other side. This artificially created well will compete for water with the real well. Due to symmetry, half of the water is claimed by the image well. This half of the domain is exactly what is not available for the real well due to presence of the impermeable wall. The correct solution is then produced without the need of formally solving the problem. In Figure 5.11 we demonstrate the position of the image well and the resultant streamlines. The streamline pattern clearly shows the competition of water, which acts just like a wall cutting off the water supply to the real well.

For a well located near a straight and infinitely long river of constant stage (Figure 5.10(b)), the solution can also be created using an image well, using the nearer of the river banks as the mirror surface. The image well, however, is of negative discharge, i.e., with the sign of the discharge reversed. Due to the anti-symmetry, there is

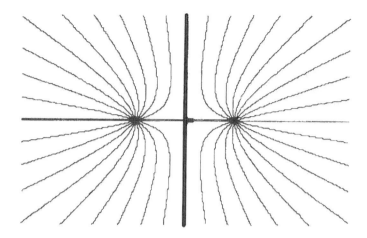

FIGURE 5.11. Streamlines for a well near an impermeable wall.

no drawdown along the symmetry line, satisfying the constant head condition at the river bank. The position of the image well and the resultant streamlines are presented in Figure 5.12. We observe that the streamlines are perpendicular to the river, suggesting that the river is supplying water to the pumping well.

Example: *A well is situated in between a river and an impermeable rock formation that are parallel to each other (Figure 5.13). The well is 120 m away from the river and 170 m from the impermeable wall. What are the image well positions and discharge strengths that can simulate the solution of this problem?*

Assume that the well is located at $x = 0$, the river at -120 m, and the impermeable wall at 170 m. There is a negative image well at $x = -240$ m mirrored from the river, and a positive image at 340 m mirrored from the wall. Each image, however, can create an image from another reflecting surface, just like the images created by two parallel mirrors. Hence the negative image at -240 m creates another negative one at 580 m. The positive image at 340 m, on the other hand, is mapped to a negative one at -580 m Whether the well discharge will flip in sign or not is determined by the type of

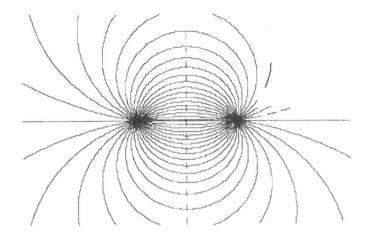

FIGURE 5.12. Streamlines for a well near a constant head boundary.

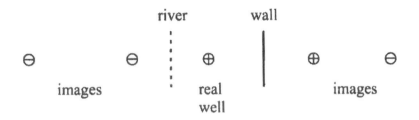

FIGURE 5.13. Multiple image of a well between a river and an impermeable wall. ($\oplus$: positive discharge; $\ominus$: negative discharge.)

reflecting surface: a wall image retains its sign, and a river image changes.

The successive positions and discharge signs of the images are summarized in Table 5.1. In principle, this process continues until an infinite number of images are created, or until the subsequent images overlap. In reality, however, the influence of the successive image wells on the drawdown near the real well diminishes rapidly as the distance gets large. In that case, only a few of the image wells are sufficient for the simulation.

TABLE 5.1. Real and image well locations.

Real/Image	R	I	I	I	I	I	I	I
Discharge	+	−	+	−	−	+	−	+
Location (m)	0	−240	340	580	−580	−820	920	1160

5.5 Pathlines

For the purpose of contaminant transport, it is of interest to trace the *pathlines* of the groundwater flow. The pathlines of the flow can be approximately determined by releasing a number of "particles" at specified locations and then by tracking their trajectories. At each of the particle location, the velocity of groundwater flow can be calculated using the drawdown and other flow field formulae. Given a small time increment Δt, the x- and y-displacements, Δx and Δy, can be calculated. The particle is then moved to the new position. The process continues. With this technique we can roughly assess whether and when a contaminant particle will reach a certain location.

This particle tracking process is sometimes reversed. Given a final destination, a negative Δt is used to backtrack the trajectory. The can help to determine the origin of contaminant that has been detected at certain locations.

It should be cautioned, however, that the above technique is only a crude approximation of the true contaminant transport process, as it considers only one transport mechanism, the advection. Other important transport mechanisms including dispersion, adsorption, etc., are ignored. Hence this technique does not yield the information of contaminant concentration. Nevertheless, it is still a powerful tool for investigating groundwater pollutant transport problems.

As an illustration, we can write the velocity field induced by a pumping well given by the Theis solution Eq. (3.90) as follows:

$$v_x = -\frac{Q_w}{2\pi b\phi}\frac{x - x_o}{(x - x_o)^2 + (y - y_o)^2}$$
$$\exp\left[-\frac{\left[(x - x_o)^2 + (y - y_o)^2\right]S}{4T(t - t_o)}\right] \qquad (5.22)$$

$$v_y = -\frac{Q_w}{2\pi b\phi} \frac{y - y_o}{(x - x_o)^2 + (y - y_o)^2}$$

$$\exp\left[-\frac{[(x - x_o)^2 + (y - y_o)^2]\,S}{4T(t - t_o)}\right] \quad (5.23)$$

where (x_o, y_o) is the well location, t_o is the pumping initiation time, b is the aquifer thickness, and ϕ is the porosity. We notice that porosity is needed to convert the specific discharge to average velocity, as demonstrated in Eq. (2.4).

For a pumping well in a Hantush-Jacob leaky aquifer, we can utilize the Laplace transform expression, Eq. (4.31), to obtain the velocity field:

$$\tilde{v}_x = -\frac{Q_w}{2\pi b\phi p} \sqrt{\frac{pS}{T} + \frac{1}{\lambda^2}} \frac{x - x_o}{\sqrt{(x - x_o)^2 + (y - y_o)^2}}$$

$$K_1\left[\sqrt{\left(\frac{pS}{T} + \frac{1}{\lambda^2}\right)[(x - x_o)^2 + (y - y_o)^2]}\right] \quad (5.24)$$

$$\tilde{v}_y = -\frac{Q_w}{2\pi b\phi p} \sqrt{\frac{pS}{T} + \frac{1}{\lambda^2}} \frac{y - y_o}{\sqrt{(x - x_o)^2 + (y - y_o)^2}}$$

$$K_1\left[\sqrt{\left(\frac{pS}{T} + \frac{1}{\lambda^2}\right)[(x - x_o)^2 + (y - y_o)^2]}\right] \quad (5.25)$$

Numerical inverse is then applied to find the solution in time.

Example: *Three pumping wells located at* (0 m, 0 m), (100 m, −100 m), (−75 m, −75 m) *are pumped at constant rates of* $Q_w = 1000, 600,$ *and* −600 m³/day. *The aquifer parameters are* $T = 1000$ m²/day, $S = 0.0003$, $b = 10$ m, *and* $\phi = 0.2$. *Demonstrate the pathlines of a few particles.*

Three particles with initial positions (−90, −60), (−50, −75), and (−50, −80) (in meter) are released. The time increment used is 1 day. Assuming the Theis aquifer and the velocity field given by Eqs. (5.22)

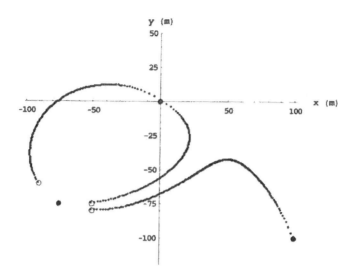

FIGURE 5.14. Pathlines of three particles released in a well field consisting of one recharge well at $(-75, -75)$, and two pumping wells at $(0, 0)$ and $(100, -100)$ (in meter).

and (5.23), the resultant trajectories are plotted in Figure 5.14. The initial locations are marked in open circles. The three wells are shown in filled circles. The successive locations of the particles with 1 day increment are marked in dots. The well located at $(-75, -75)$ is a recharging well; hence particles are repelled from it. The two other wells are pumping wells, so the particles are captured by them. The time of travel for the three particles at $(-90, -60)$, $(-50, -75)$, and $(-50, -80)$ are respectively 122, 100, and 223 days.

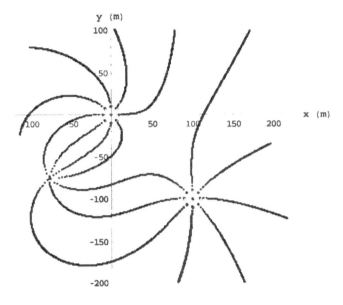

FIGURE 5.15. Capture zone of pumping wells.

Example: *For the same well field given in the preceding example, find the capture zones of the two pumping wells.*

To accomplish this, eight particles are deployed around each of the two pumping wells located at $(0, 0)$ and $(100, -100)$. With the flow field reversed, these particles are traced using 1-day increments. The results are shown in Figure 5.15. The pathline pattern delineates the zones within which particles are captured by the wells.

Chapter 6
MULTILAYERED AQUIFER THEORY

6.1 Aquifer-Aquitard Systems

Underground formations are usually layered due to their geomorphologic origins. For water supply purposes, these layers are often grouped into units of aquifers and aquitards. Figure 6.1 gives the schematic sketch of an aquifer-aquitard system.

The grouping of layers is based on their hydraulic properties, not on their geological ages. Several layers can be lumped to form a single aquifer or aquitard, if their hydraulic properties fall within a certain range. The equivalent horizontal and vertical hydraulic conductivities should be used, depending on the flow direction (see Sec. 2.6). The contrast of hydraulic conductivities between adjacent aquifer and aquitard should be at least five fold and preferably ten or more fold. The final system consists of alternating layers of aquifers and aquitards. Figure 6.2 illustrates the conceptualization of such a multilayered aquifer-aquitard system.

For convenience, these layers are shown to be horizontal and with constant thickness. Constant thickness is not a required condition and both the aquifer thickness b and the aquitard thickness b' can be functions of x and y. The slope and curvature of the layer interface, however, should be small. Large scale trending with small degree of inclination can be tolerated, as long as the effect of elevation on piezometric head is properly taken into account. The key conditions are that:

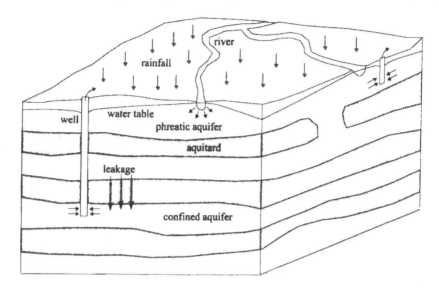

FIGURE 6.1. Schematic sketch of an aquifer-aquitard system.

1. the streamlines in the aquifer layers must be nearly horizontal for the hydrostatic assumption to hold true (see Secs. 3.1 and 3.2); and

2. the aquifer/aquitard hydraulic conductivity contrast must be large enough for the vertical aquitard flow assumption to be valid (see Secs. 2.7 and 4.1).

Consequently, for mathematical modeling purposes, aquifers become two-dimensional horizontal planes separated by aquitards. Their vertical dimension is collapsed and absorbed into the definitions of aquifer properties such as transmissivity and storativity. The aquitards, however, retain their vertical dimension in the mathematical model.

Example: *Figure 6.3 illustrates a slightly inclined confined aquifer with an outcrop where a recharge is available. For convenience, the aquifer is projected onto a horizontal plane $\overline{AB}$. For the head difference, the actual aquifer elevation should be referred, as shown in*

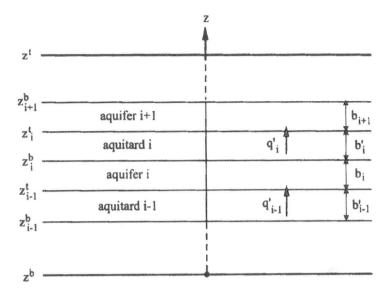

FIGURE 6.2. Definition sketch of a multilayered aquifer-aquitard system.

the piezometric head location in dashed line. Given the piezometric head difference Δh, what is the aquifer discharge?

According to Darcy's law, the discharge (per unit width of aquifer) is directly proportional to the thickness of the layer, b, and inversely proportional to the distance traveled, ΔL,

$$Q = Kb\frac{\Delta h}{\Delta L}$$

The thickness b as indicated in Figure 6.3 is correctly shown to be perpendicular to the flow direction. However, for a convenient approximation, we may replace it by the difference of elevation between the aquifer top and bottom,

$$b' = z^t - z^b = b\sec\theta$$

assuming a constant inclination angle θ. Obviously, b' overestimates b, hence producing a slightly larger discharge prediction. The correct distance traveled is shown as ΔL. As an approximation, we can use

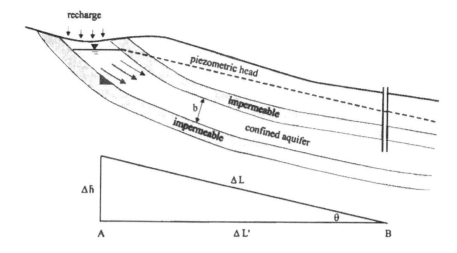

FIGURE 6.3. A slightly inclined confined aquifer.

its projection

$$\Delta L' = \Delta L \cos\theta$$

The combined effect of using b' and $\Delta L'$ leads to an error of

$$\varepsilon = \sec^2\theta - 1 \approx \theta^2 + \frac{2}{3}\theta^4$$

where θ is in radian unit. For $\theta = 10° = 0.17$ rad, the error is around 3%.

If the problem geometry is as simple as the one shown in Figure 6.3, correct references to b and ΔL should be made. However, for more complex two-dimensional planar geometry with variable inclination, it may not be feasible to keep track of the correct quantities. Their projections b' and $\Delta L'$ will be used instead. If the inclination is restricted to no more than 10°, the error is tolerable, particularly in view of the other possible errors resulting from parameter uncertainties of the system (see Chapter 11).

6.2   Neuman-Witherspoon Formulation

Figure 6.2 illustrates a groundwater system consisting of alternating layers of aquifers and aquitards. Consider the flow in aquifer layer i with leakage from the top and bottom aquitards, layer i and $i-1$. The hydraulic assumption as discussed in Sec. 3.1 leads to the following two-dimensional governing equation for aquifer i:[87,88]

$$\frac{\partial}{\partial x}\left(T_i^x \frac{\partial s_i}{\partial x}\right) + \frac{\partial}{\partial y}\left(T_i^y \frac{\partial s_i}{\partial y}\right)$$

$$- q'_{i-1}\big|_{z=z_{i-1}^t} + q'_i\big|_{z=z_i^b} - R_i = S_i \frac{\partial s_i}{\partial t} \qquad (6.1)$$

In the above, $s_i(x,y,t)$ is the drawdown in aquifer i, T_i^x and T_i^y are components of anisotropic aquifer transmissivity in the principal directions x and y, S_i is the aquifer storativity, z_{i-1}^t and z_i^b are the coordinates at the top and the bottom of the aquifer (Figure 6.2), q'_{i-1} and q'_i are vertical specific discharge in aquitard $i-1$ and i, respectively, and R_i accounts for all forms of recharge (or extraction with a negative sign) that are not included in the flux exchange terms.

The aquitard specific discharge can be evaluated from aquitard drawdown according to Darcy's law,

$$q'_i = K'_i \frac{\partial s'_i}{\partial z} \qquad (6.2)$$

where $s'_i(x,y,z,t)$ is the drawdown in aquitard i, and K'_i the aquitard hydraulic conductivity. Substituting the above into Eq. (6.1) we obtain

$$\frac{\partial}{\partial x}\left(T_i^x \frac{\partial s_i}{\partial x}\right) + \frac{\partial}{\partial y}\left(T_i^y \frac{\partial s_i}{\partial y}\right) - K'_{i-1}\frac{\partial s'_{i-1}}{\partial z}\bigg|_{z=z_{i-1}^t}$$

$$+ K'_i \frac{\partial s'_i}{\partial z}\bigg|_{z=z_i^b} - R_i = S_i \frac{\partial s_i}{\partial t}; \qquad i = 1,\dots,n \qquad (6.3)$$

which couples the aquifer drawdown with that of aquitard, for n layers of aquifer.

The aquitard drawdown, on the other hand, obeys the one-dimensional equation

$$\frac{\partial}{\partial z} K_i' \frac{\partial s_i'}{\partial z} = S_{s(i)}' \frac{\partial s_i'}{\partial t}; \qquad i = 0, \ldots, n \qquad (6.4)$$

where $S_{s(i)}'$ is the aquitard specific storage. Since it is generally difficult to resolve the vertical variations of hydraulic conductivity and specific storage of the aquitards, the parameters may be replaced by their equivalent constant values. In that case, Eq. (6.4) can be rewritten as

$$b_i' K_i' \frac{\partial^2 s_i'}{\partial z^2} = S_i' \frac{\partial s_i'}{\partial t}; \qquad i = 0, \ldots, n \qquad (6.5)$$

in which $S_i' = b_i' S_{s(i)}'$ is the aquitard storage, and $b_i' = z_i^t - z_i^b$ is the aquitard thickness. Combined with proper initial, boundary, and interfacial conditions, Eq. (6.3) with $i = 1, \ldots, n$ and Eq. (6.5) with $i = 0, \ldots, n$ form a sufficient solution system for the $2n + 1$ aquifer and aquitard drawdown quantities. Since the entire system is coupled, these equations must be solved simultaneously.

If the uppermost aquifer is a phreatic aquifer, the flow equation is generally nonlinear:

$$\frac{\partial}{\partial x} \left(T_n^x \frac{\partial s_n^*}{\partial x} \right) + \frac{\partial}{\partial y} \left(T_n^y \frac{\partial s_n^*}{\partial y} \right)$$
$$- K_{n-1}' \frac{\partial s_{n-1}'}{\partial z} \bigg|_{z=z_{n-1}^t} - R_n = S_n^* \frac{\partial s_n^*}{\partial t} \qquad (6.6)$$

where $s^* = s - s^2/2H$ is the modified drawdown, H the unperturbed phreatic aquifer thickness, $T^x = K_x H$ and $T^y = K_y H$ are the transmissivities, $S^* = S_y/\sqrt{1 - 2(s^*/H)}$ is the modified storativity, and S_y the specific yield, see Sec. 3.2. For small drawdown in the phreatic layer, the above equation can be linearized by assuming $s^* \approx s$, and $S^* \approx S_y$. Equation (6.6) is therefore linear and is compatible with the rest of the equations in (6.3).

The above system of governing equations needs to be supplemented by initial and boundary conditions. For the aquitards, conditions are needed at the top and bottom of each layer. When the top and bottom are in contact with aquifers, the continuity condition of piezometric head requires that

$$s'_i(x, y, z^t_i, t) = s_{i+1}(x, y, t)$$
$$s'_i(x, y, z^b_i, t) = s_i(x, y, t) \tag{6.7}$$

If the uppermost or the lowermost layer is an aquitard, which in turn is overlain or underlain by an impermeable layer, the zero flux condition should be prescribed:

$$\left.\frac{\partial s'_0}{\partial z}\right|_{z=z^b_0} = 0$$

$$\left.\frac{\partial s'_n}{\partial z}\right|_{z=z^t_n} = 0 \tag{6.8}$$

If the bounding aquitards are in contact with a highly permeable, and infinite water supply aquifer, the zero drawdown condition applies:

$$s'_0(x, y, z^b_0, t) = 0$$
$$s'_n(x, y, z^t_n, t) = 0 \tag{6.9}$$

We note that while conditions in Eqs. (6.8) and (6.9) are explicit, those in Eq. (6.7) are implicit, as aquifer drawdown needs to be solved as a part of the solution.

For the aquifer layers, boundary conditions are needed on boundaries enclosing the two-dimensional horizontal plane. On each part of the boundary, either the piezometric head, or the specific discharge needs to be prescribed, but not both. We also note that on at least one part of the boundary, the piezometric head must be given. These boundary conditions are then known as *well-posed*, and we have the *existence* and *uniqueness* of the solution.

Typical boundary conditions include piezometric head observed in wells, from groundwater outcrops, and at contact with large body of water, such as stream, lake, or ocean. Conditions of zero flux can be found along impermeable faults, bedrocks, and on the ridge of groundwater divide.

Since the governing equations are transient, initial conditions are needed everywhere in the aquitards and aquifers. We have formulated

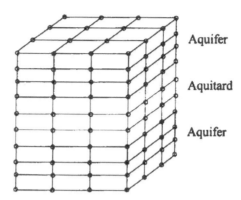

FIGURE 6.4. A three-dimensional solution mesh.

the problem in terms of drawdown, i.e., the perturbation from an initial equilibrated flow field. If the initial state is an established steady state flow field, the drawdown is zero everywhere. Hence the initial conditions are

$$s_i(x, y, 0) = s_i'(x, y, z, 0) = 0 \qquad (6.10)$$

The system is then complete for an analytical or a numerical solution.

Although the emphasis of this book is not on numerical solution, we shall give a limited scope discussion below. More detail on groundwater modeling can be found in a textbook.[2] Numerical solution involves the discretization of the solution geometry into a finite number of nodes, on which drawdown values are assigned. Figure 6.4 gives a schematic sketch of a three-dimensional grid for a two-aquifer-one-aquitard system. There exist a number of numerical techniques for solving discrete system like this. The more popular ones include the *Finite Element Method* (FEM)[98] and the *Finite Difference Method* (FDM).[70] For example, for groundwater flow in three-dimensional geometry, one of the highly popular computer programs is the finite-difference code *MODFLOW*[81] developed by the *U.S. Geological Survey*.

For the *Neuman-Witherspoon multilayered aquifer system* developed in this section the numerical solution geometry is simplified. With the application of hydraulic assumption, the aquifer vertical

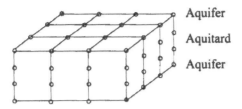

Aquifer

Aquitard

Aquifer

FIGURE 6.5. Solution mesh for the Neuman-Witherspoon multilay-ered aquifer system.

dimension is collapsed and the solution mesh becomes horizontal planes, as seen in Figure 6.5. The aquitards are modeled as a series of one-dimensional vertical strings connecting the aquifer nodes above and below. Although there is no lateral connection among aquitard nodes, the resultant mesh nevertheless has a three-dimensional appearance, as observed in Figure 6.5. We expect some improvement in numerical efficiency in this model as compared to the full three-dimensional system. Finite element solutions have been developed based on the Neuman-Witherspoon aquifer-aquitard model.[3,4,32,92]

6.3   Herrera Integro-Differential Formulation

If we use the equivalent aquitard properties in the vertical direction, such that the parameters are not functions of z (although they can still be functions of x and y), the aquitard equation is given by Eq. (6.5). Since Eq. (6.5) is one-dimensional in space and is of constant coefficient, its analytical solution can be easily found. (For aquitards that are not homogeneous in the vertical direction, i.e. K' is a function of z, analytical solution can also be found by utilizing the eigenfunction approach.[62] However, we shall restrict to the simpler case of using equivalent homogeneous aquifer properties.) This was realized by Herrera.[63,64]

We first re-state the aquitard problem as follows. The governing equation is

$$b_i' K_i' \frac{\partial^2 s_i'}{\partial z^2} = S_i' \frac{\partial s_i'}{\partial t} \tag{6.11}$$

with the initial and boundary conditions

$$
\begin{aligned}
s_i'(x, y, z_i^t, t) &= s_{i+1}(x, y, t) \\
s_i'(x, y, z_i^b, t) &= s_i(x, y, t) \\
s_i'(x, y, z, 0) &= 0
\end{aligned} \tag{6.12}
$$

For the moment we shall assume that $s_i(x, y, t)$ and $s_{i+1}(x, y, t)$ are known functions. This allows us to find the solution of the system, Eqs. (6.11) and (6.12). The mathematical technique for solving such problem has been discussed in Sec. 4.3. In principle, we shall first solve two standard problems to obtain the influence functions. In the first problem, the governing equation is Eq. (6.11), but the boundary conditions are modified to

$$
\begin{aligned}
s_i'(z_i^t, t) &= 0 \\
s_i'(z_i^b, t) &= 1
\end{aligned} \tag{6.13}
$$

with the same null initial condition. Such a problem has already been solved in Sec. 4.3. The solution Eq. (4.8) is reproduced here with a proper adjustment of reference coordinate as

$$
\mathcal{F}_i^{(1)}(x, y, z, t) = 1 - \frac{z - z_i^b}{b_i'} - \frac{2}{\pi} \sum_{m=1}^{\infty} \frac{1}{m} \exp\left(-\frac{m^2 \pi^2 K_i' t}{S_i' b_i'}\right)
$$
$$
\times \sin \frac{m\pi \left(z - z_i^b\right)}{b_i'} \tag{6.14}
$$

This solution will be denoted as the influence function $\mathcal{F}^{(1)}(x, y, z, t)$. In the second problem, we switch the boundary conditions at the top and the bottom such that

$$
\begin{aligned}
s_i'(z_i^t, t) &= 1 \\
s_i'(z_i^b, t) &= 0
\end{aligned} \tag{6.15}
$$

This is clearly the same problem but in a different direction. Its solution is obtained by replacing z in Eq. (6.14) with $z_i^t + z_i^b - z$. After expanding the sine function, we obtain

$$\mathcal{F}_i^{(2)}(x,y,z,t) = \frac{z - z_i^b}{b_i'} + \frac{2}{\pi} \sum_{m=1}^{\infty} \frac{(-1)^m}{m} \exp\left(-\frac{m^2\pi^2 K_i' t}{S_i' b_i'}\right)$$
$$\times \sin \frac{m\pi\left(z - z_i^b\right)}{b_i'} \qquad (6.16)$$

This gives another form of the influence function.

Based on the Duhamel principle of superposition, the aquitard drawdown that satisfies the set of conditions in Eq. (6.12) is

$$s_i'(x,y,z,t) = \int_0^t \mathcal{F}_i^{(1)}(x,y,z,\tau)\frac{\partial s_i(x,y,t-\tau)}{\partial t}\, d\tau$$
$$+ \int_0^t \mathcal{F}_i^{(2)}(x,y,z,\tau)\frac{\partial s_{i+1}(x,y,t-\tau)}{\partial t}\, d\tau \qquad (6.17)$$

In the aquifer equation (6.3), we need the derivative of s' to form the flux term, hence

$$K_i'\frac{\partial s_i'}{\partial z}\bigg|_{z=z_i^b} = \int_0^t K_i'\frac{\partial \mathcal{F}_i^{(1)}(x,y,z,\tau)}{\partial z}\bigg|_{z=z_i^b}\frac{\partial s_i(x,y,t-\tau)}{\partial t}\, d\tau$$
$$+ \int_0^t K_i'\frac{\partial \mathcal{F}_i^{(2)}(x,y,z,\tau)}{\partial z}\bigg|_{z=z_i^b}\frac{\partial s_{i+1}(x,y,t-\tau)}{\partial t}\, d\tau$$
$$= -\frac{K_i'}{b_i'}\int_0^t f_i(x,y,\tau)\frac{\partial s_i(x,y,t-\tau)}{\partial t}\, d\tau$$
$$+ \frac{K_i'}{b_i'}\int_0^t h_i(x,y,\tau)\frac{\partial s_{i+1}(x,y,t-\tau)}{\partial t}\, d\tau \qquad (6.18)$$

where we have defined

$$f_i(x,y,t) = -b_i'\frac{\partial \mathcal{F}_i^{(1)}(x,y,z,t)}{\partial z}\bigg|_{z=z_i^b}$$
$$= b_i'\frac{\partial \mathcal{F}_i^{(2)}(x,y,z,t)}{\partial z}\bigg|_{z=z_i^t} \qquad (6.19)$$

$$h_i(x,y,t) = b_i' \frac{\partial \mathcal{F}_i^{(2)}(x,y,z,t)}{\partial z}\bigg|_{z=z_i^b}$$

$$= -b_i' \frac{\partial \mathcal{F}_i^{(1)}(x,y,z,t)}{\partial z}\bigg|_{z=z_i^t} \tag{6.20}$$

We note that the second equalities in Eqs. (6.19) and (6.20) are obtained from the symmetric property of the functions. Differentiating Eqs. (6.14) and (6.16) we find

$$f_i(x,y,t) = 1 + 2\sum_{m=1}^{\infty} \exp\left(-\frac{m^2\pi^2 K_i' t}{S_i' b_i'}\right) \tag{6.21}$$

$$h_i(x,y,t) = 1 + 2\sum_{m=1}^{\infty} (-1)^m \exp\left(-\frac{m^2\pi^2 K_i' t}{S_i' b_i'}\right) \tag{6.22}$$

Following the terminology of Herrera,[65] $f_i(x,y,t)$ is called the *memory function* because it accounts for the history effect of drawdown at the current layer, and $h_i(x,y,t)$ the *influence function* for it is associated with the drawdown of adjacent layers.

By the same token, we can handle the other aquitard leakage term in Eq. (6.3) as

$$K_{i-1}' \frac{\partial s_{i-1}'}{\partial z}\bigg|_{z=z_{i-1}^t} =$$

$$\int_0^t K_{i-1}' \frac{\partial \mathcal{F}_{i-1}^{(1)}(x,y,z,\tau)}{\partial z}\bigg|_{z=z_{i-1}^t} \frac{\partial s_{i-1}(x,y,t-\tau)}{\partial t}\, d\tau$$

$$+ \int_0^t K_{i-1}' \frac{\partial \mathcal{F}_{i-1}^{(2)}(x,y,z,\tau)}{\partial z}\bigg|_{z=z_{i-1}^t} \frac{\partial s_i(x,y,t-\tau)}{\partial t}\, d\tau$$

$$= -\frac{K_{i-1}'}{b_{i-1}'} \int_0^t h_{i-1}(x,y,\tau) \frac{\partial s_{i-1}(x,y,t-\tau)}{\partial t}\, d\tau$$

$$+ \frac{K_{i-1}'}{b_{i-1}'} \int_0^t f_{i-1}(x,y,\tau) \frac{\partial s_i(x,y,t-\tau)}{\partial t}\, d\tau \tag{6.23}$$

Finally, with the substitution of Eqs. (6.18) and (6.23), Eq. (6.3) becomes

$$
\frac{\partial}{\partial x}\left(T_i^x \frac{\partial s_i}{\partial x}\right) + \frac{\partial}{\partial y}\left(T_i^y \frac{\partial s_i}{\partial y}\right)
$$

$$
+\frac{K'_{i-1}}{b'_{i-1}}\left[\int_0^t h_{i-1}(x,y,\tau)\frac{\partial s_{i-1}(x,y,t-\tau)}{\partial t}\,d\tau\right.
$$

$$
\left.-\int_0^t f_{i-1}(x,y,\tau)\frac{\partial s_i(x,y,t-\tau)}{\partial t}\,d\tau\right]
$$

$$
-\frac{K'_i}{b'_i}\left[\int_0^t f_i(x,y,\tau)\frac{\partial s_i(x,y,t-\tau)}{\partial t}\,d\tau\right.
$$

$$
\left.-\int_0^t h_i(x,y,\tau)\frac{\partial s_{i+1}(x,y,t-\tau)}{\partial t}\,d\tau\right]
$$

$$
-R_i = S_i\frac{\partial s_i}{\partial t}; \qquad i = 1,\ldots,n \tag{6.24}
$$

Equations of this type, which mix differential and integral expressions, are known as *integro-differential equations*.

At the uppermost and lowermost aquifers, Eq. (6.24) needs to be specialized for a number of occasions. When Eq. (6.24) is applied to aquifer 1 and n, it involves aquifer 0 and $n+1$, which do not exist. These terms are simply dropped. The same applies to terms involving aquitard layers that do not exist. When the outer aquifers are flanked by aquitards, boundary conditions at the outer boundaries of aquitards are needed. If the aquitard is in contact with a zero drawdown aquifer, no special care is required. If the outer boundary of the aquitard is confined by an impermeable layer, a different memory function is needed. The problem is the same as the one solved in Eq. (4.11). Adjusting to the present coordinate system, we define the influence function for the last layer as

$$
\mathcal{G}_n^{(1)}(x,y,z,t) = 1 - \frac{4}{\pi}\sum_{m=1}^{\infty}\frac{1}{2m-1}\exp\left[-\frac{(2m-1)^2\pi^2 K'_n t}{4S'_n b'_n}\right]
$$

$$
\times \sin\frac{(2m-1)\pi\left(z-z_n^b\right)}{2b'_n} \tag{6.25}
$$

For the first layer, we have

$$
\mathcal{G}_0^{(2)}(x,y,z,t) = 1 + \frac{4}{\pi}\sum_{m=1}^{\infty}\frac{(-1)^m}{2m-1}\exp\left[-\frac{(2m-1)^2\pi^2 K_0' t}{4S_0' b_0'}\right]
$$

$$
\times \cos\frac{(2m-1)\pi\left(z-z_0^b\right)}{2b_0'} \tag{6.26}
$$

If the drawdown in the aquifers are given, the aquitard drawdown is found as

$$
s_0'(x,y,z,t) = \int_0^t \mathcal{G}_0^{(2)}(x,y,z,\tau)\frac{\partial s_1(x,y,t-\tau)}{\partial t}\,d\tau \tag{6.27}
$$

$$
s_n'(x,y,z,t) = \int_0^t \mathcal{G}_n^{(1)}(x,y,z,\tau)\frac{\partial s_n(x,y,t-\tau)}{\partial t}\,d\tau \tag{6.28}
$$

For the leakage term in Eq. (6.3), we need the memory function, as obtained from Eqs. (6.27) and (6.28),

$$
g_i(x,y,t) = 2\sum_{m=1}^{\infty}\exp\left[-\frac{(2m-1)^2\pi^2 K_i' t}{4S_i' b_i'}\right]; \qquad i=0 \text{ or } n \tag{6.29}
$$

Hence at the first and the last aquifer layer, g_0 and g_n respectively replace f_0 and f_n in Eq. (6.24). For instance, the first aquifer equation becomes

$$
\frac{\partial}{\partial x}\left(T_1^x\frac{\partial s_1}{\partial x}\right) + \frac{\partial}{\partial y}\left(T_1^y\frac{\partial s_1}{\partial y}\right)
$$

$$
-\frac{K_0'}{b_0'}\int_0^t g_0(x,y,\tau)\frac{\partial s_1(x,y,t-\tau)}{\partial t}\,d\tau
$$

$$
-\frac{K_1'}{b_1'}\left[\int_0^t f_1(x,y,\tau)\frac{\partial s_1(x,y,t-\tau)}{\partial t}\,d\tau\right.
$$

$$
\left.-\int_0^t h_1(x,y,\tau)\frac{\partial s_2(x,y,t-\tau)}{\partial t}\,d\tau\right] - R_1 = S_1\frac{\partial s_1}{\partial t} \tag{6.30}
$$

The equation for the last aquifer can be similarly presented. This completes the Herrera formulation of multilayered aquifer theory.

What is the significance of the above result? We note that the solution system is the n aquifer equations presented as Eq. (6.24),

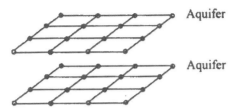

FIGURE 6.6. Solution mesh for the Herrera multilayered aquifer system.

together with the known, auxiliary functions, Eqs. (6.21), (6.22), and (6.29). The n aquifer drawdown variables are solved without the need of coupling with the aquitard drawdown! Once the aquifer drawdown is solved, the aquitard drawdown is found from the convolutional integral shown as Eqs. (6.17), (6.27), and (6.28).

Since most practical problems are likely to be solved by a numerical method, it is of interest to examine a typical solution mesh. Figure 6.6 shows the mesh for a two-aquifer-one-aquitard system corresponding to the Herrera formulation. This can be compared to Figure 6.5 for the Neuman-Witherspoon formulation. It is most interesting to observe that the aquitard discretization, which retains the three-dimensional appearance in the Neuman-Witherspoon formulation, has disappeared in the Herrera scheme. A significant saving in computational cost is expected.[67] Numerical solution of this integro-differential system has been attempted using the finite element method[66] and the finite difference method.[77, 100]

This method however is not without its deficiency. Since most numerical methods solve partial differential equations, special algorithms need to be developed to handle the simultaneous presence of integral and differential equations. These integrals are in the form of convolutional integrals, also known as *history integrals*. As the name suggests, to obtain the solution at a given time, the whole history of evolution up to that time is needed. This is compared to a standard time-stepping scheme, in which only the information of the prior time step is needed to obtain the solution at the current time. The

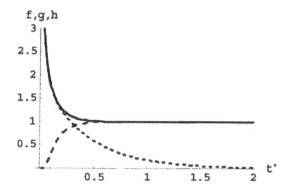

FIGURE 6.7. Influence and memory functions. (f: solid line; g: long dashed line; h: short dashed line.)

need for storage of the history in the integro-differential equation scheme may become burdensome when the solution time gets large.

Example: *Examine the behavior of the influence and memory functions.*

These functions, defined as Eqs. (6.21), (6.22) and (6.29), can be plotted as functions of the dimensionless time

$$t^* = \frac{K't}{S'b'}$$

Figure 6.7 shows the result. We notice the limits:

$$\begin{aligned}
h &\to 0, & f, g &\to \infty, & f &\approx g, & &\text{as } t \to 0 \\
f, h &\to 1, & g &\to 0, & & & &\text{as } t \to \infty
\end{aligned}$$

6.4   Laplace Transform Formulation

We have demonstrated along the way that the use of the Laplace transform in aquifer theories can significantly simplify the mathematics and the evaluation of the solution. The key to this success is due to the availability of an easy-to-use numerical Laplace inverse algorithm. The Stehfest algorithm has been employed for this purpose (see Sec. 3.10). This Laplace transform technique has also been applied to the Herrera formulation of multilayered aquifer theory with good success.[22, 25]

The use of the Laplace transform in the Herrera formulation provides double benefits. Not only it removes the time variable in the governing equations, but also it eliminates the cumbersome convolutional integrals. This is based on the well known Laplace transform property given by the convolutional theorem (see Appendix A)

$$\mathcal{L}\left\{\int_0^t f_1(t - \tau)\, f_2(t)\, d\tau\right\} = \tilde{f}_1(p)\, \tilde{f}_2(p) \qquad (6.31)$$

We also utilize this formula

$$\mathcal{L}\left\{\frac{\partial f(t)}{\partial t}\right\} = p\tilde{f}(p) - f(0) \qquad (6.32)$$

Applying the Laplace transform to Eq. (6.24), we find:

$$\frac{\partial}{\partial x}\left(T_i^x \frac{\partial \tilde{s}_i}{\partial x}\right) + \frac{\partial}{\partial y}\left(T_i^y \frac{\partial \tilde{s}_i}{\partial y}\right) + \frac{pK'_{i-1}}{b'_{i-1}}\left(\tilde{h}_{i-1}\, \tilde{s}_{i-1} - \tilde{f}_{i-1}\, \tilde{s}_i\right)$$
$$-\frac{pK'_i}{b'_i}\left(\tilde{f}_i\, \tilde{s}_i - \tilde{h}_i\, \tilde{s}_{i+1}\right) - \tilde{R}_i = p\, S_i\, \tilde{s}_i; \quad i = 1, \ldots, n \qquad (6.33)$$

The above is a system of coupled, second order, linear partial differential equations of elliptic type. We clearly observe that the convolutional integrals no longer exist.

To use Eq. (6.33), we need the Laplace transform of the memory and influence functions, $\tilde{f}$, $\tilde{g}$, and $\tilde{h}$. Rather than performing term-by-term transformation based on the series expressions, Eqs. (6.21), (6.22) and (6.29), which are inefficient for numerical evaluation, we re-solved these problems in the Laplace transform space. Referring to the two problems defined in Figure 4.2, we apply the Laplace transform to the governing equation (4.5)

$$\frac{d^2 \tilde{s}'}{dz^2} - \frac{S'p}{b'K'}\, \tilde{s}' = 0 \qquad (6.34)$$

The basic solution of the above is

$$\tilde{s}' = A \sinh \sqrt{\frac{S'p}{b'K'}}\, z + B \cosh \sqrt{\frac{S'p}{b'K'}}\, z \qquad (6.35)$$

After determining the constant coefficients A and B for the various boundary conditions, we obtain the Laplace transform counterparts of Eqs. (6.14), (6.16), (6.25) and (6.26), as

$$\tilde{\mathcal{F}}_i^{(1)}(x, y, z, p) = \frac{1}{p} \operatorname{csch}\left(\sqrt{\frac{S_i' b_i' p}{K_i'}}\right) \sinh\left(\sqrt{\frac{S_i' b_i' p}{K_i'}} \frac{z_i^t - z}{b_i'}\right) \qquad (6.36)$$

$$\tilde{\mathcal{F}}_i^{(2)}(x, y, z, p) = \frac{1}{p} \operatorname{csch}\left(\sqrt{\frac{S_i' b_i' p}{K_i'}}\right) \sinh\left(\sqrt{\frac{S_i' b_i' p}{K_i'}} \frac{z - z_i^b}{b_i'}\right) \qquad (6.37)$$

$$\tilde{\mathcal{G}}_n^{(1)}(x, y, z, p) = \frac{1}{p} \operatorname{sech}\left(\sqrt{\frac{S_n' b_n' p}{K_n'}}\right) \cosh\left(\sqrt{\frac{S_n' b_n' p}{K_n'}} \frac{z_n^t - z}{b_n'}\right) \qquad (6.38)$$

$$\tilde{\mathcal{G}}_0^{(2)}(x, y, z, p) = \frac{1}{p} \operatorname{sech}\left(\sqrt{\frac{S_0' b_0' p}{K_0'}}\right) \cosh\left(\sqrt{\frac{S_0' b_0' p}{K_0'}} \frac{z - z_0^b}{b_0'}\right) \qquad (6.39)$$

Based on the above, the memory and influence functions needed in Eq. (6.33) are provided as

$$\tilde{f}_i(x, y, p) = \sqrt{\frac{S_i' b_i'}{K_i' p}} \coth\left(\sqrt{\frac{S_i' b_i' p}{K_i'}}\right) \qquad (6.40)$$

$$\tilde{h}_i(x, y, p) = \sqrt{\frac{S_i' b_i'}{K_i' p}} \operatorname{csch}\left(\sqrt{\frac{S_i' b_i' p}{K_i'}}\right) \qquad (6.41)$$

$$\tilde{g}_i(x, y, p) = \sqrt{\frac{S_i' b_i'}{K_i' p}} \tanh\left(\sqrt{\frac{S_i' b_i' p}{K_i'}}\right); \quad i = 1 \text{ or } n \quad (6.42)$$

We note that $\tilde{g}_i$ should replace $\tilde{f}_i$ in cases where the uppermost and lowermost aquitards are bound by impermeable layers. Once

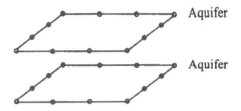

FIGURE 6.8. A boundary element method solution mesh.

the aquifer drawdown is solved, the aquitard drawdown can be found form these algebraic equations:

$$\tilde{s}'_i(x,y,z,p) = p\,\tilde{s}_i\,\widetilde{\mathcal{F}}^{(1)}_i(x,y,z,p)$$
$$+p\,\tilde{s}_{i+1}\,\widetilde{\mathcal{F}}^{(2)}_i(x,y,z,p) \qquad (6.43)$$

$$\tilde{s}'_0(x,y,z,p) = p\,\tilde{s}_1\,\widetilde{\mathcal{G}}^{(2)}_0(x,y,z,p) \qquad (6.44)$$

$$\tilde{s}'_n(x,y,z,p) = p\,\tilde{s}_n\,\widetilde{\mathcal{G}}^{(1)}_n(x,y,z,p) \qquad (6.45)$$

This completes the formulation.

Finally, we examine the numerical modeling issues. Since the Laplace transform formulation is based on the Herrera integro-differential equations, the solution mesh is the same as that represented in Figure 6.6. To find the drawdown solution at a given time, the system shown in Eq. (6.33) is solved for half a dozen to a dozen times, each time with a different Laplace transform parameter p value, according to the number of terms used in the Stehfest inverse formula, Eq. (3.127). Time stepping is not needed. Solution at any time, large or small, can be found without the knowledge of solution at a prior time step.

This scheme has been used in a finite difference solution of a multilayered aquifer system.[22] For the cases that the aquifers and aquitards are homogeneous, it has been demonstrated that a special numerical technique, known as the Boundary Element Method (BEM), can further simplify the numerical solution system[26] such that the resultant solution mesh has the appearance as the one shown in Figure 6.8.

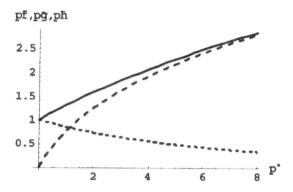

FIGURE 6.9. Influence and memory functions. ($p\tilde{f}$: solid line; $p\tilde{g}$: long dashed line; $p\tilde{h}$: short dashed line.)

Example: *Examine the behavior of influence and memory functions in the Laplace transform domain.*

These functions given as Eqs. (6.40)–(6.42) have the following limiting properties:

$$\lim_{p\to0} p\,\tilde{f}(p) \;=\; \lim_{t\to\infty} f(t)$$

$$\lim_{p\to\infty} p\,\tilde{f}(p) \;=\; \lim_{t\to0} f(t)$$

and similarly for $\tilde{g}(p)$ and $\tilde{h}(p)$. Hence it is more convenient to present them as $p\tilde{f}(p)$, $p\tilde{g}(p)$, and $p\tilde{h}(p)$. In Figure 6.9 these functions are plotted versus the dimensionless quantity

$$p^* = \frac{S'b'p}{K'}$$

These results can be compared to their time domain counterparts in Figure 6.7.

Chapter 7

PUMPING WELL IN MULTILAYERED AQUIFER SYSTEMS

7.1 One-Aquifer System

In Chapter 6 we presented the multilayered aquifer theory in its general form. For a heterogeneous aquifer system in arbitrary geometry, a numerical method is required for its solution. In this chapter, we shall examine the special problem of pumping extraction from a homogeneous, isotropic, multilayered aquifer system with infinite horizontal extent. Due to these simplifying assumptions, analytical solutions are possible.[25,61,87,88,106] In this chapter, the analytical solutions are presented using the Laplace transform scheme in Sec. 6.4.

With the assumption of homogeneity and isotropy, the governing equation (6.33) is simplified to

$$\nabla^2 \tilde{s}_i + \frac{pK'_{i-1}}{T_i b'_{i-1}} \left(\tilde{h}_{i-1}\, \tilde{s}_{i-1} - \tilde{f}_{i-1}\, \tilde{s}_i \right) - \frac{pK'_i}{T_i b'_i} \left(\tilde{f}_i\, \tilde{s}_i - \tilde{h}_i\, \tilde{s}_{i+1} \right)$$
$$-\frac{pS_i}{T_i}\, \tilde{s}_i = \frac{\tilde{R}_i}{T_i}; \qquad i = 1, \dots, n \tag{7.1}$$

The functional dependence of the influence and memory functions becomes: $\tilde{f}(p)$, $\tilde{g}(p)$, $\tilde{h}(p)$, $\widetilde{\mathcal{F}}^{(1)}(z,p)$, $\widetilde{\mathcal{F}}^{(2)}(z,p)$, $\widetilde{\mathcal{G}}^{(1)}(z,p)$, $\widetilde{\mathcal{G}}^{(2)}(z,p)$. They are no longer dependent on the horizontal spatial coordinates (x, y) due to the homogeneity assumption.

Figure 7.1 depicts a one-aquifer-two-aquitard system. For convenience, we first assume that the top and the bottom of the system are confined by impermeable layers. The governing equation for this

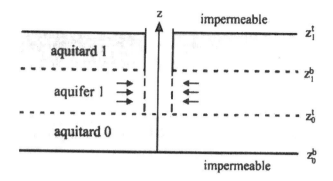

FIGURE 7.1. A one-aquifer-two-aquitard system.

system is

$$\nabla^2 \tilde{s}_1 - \frac{pK_0'}{T_1 b_0'}\tilde{g}_0\,\tilde{s}_1 - \frac{pK_1'}{T_1 b_1'}\tilde{g}_1\,\tilde{s}_1 - \frac{pS_1}{T_1}\tilde{s}_1 = -\frac{Q_1}{pT_1}\delta(\mathbf{x}-\mathbf{x}') \quad (7.2)$$

In the above, there are a few steps worth commenting. First, we notice that terms involving the nonexistent aquifers, $\tilde{s}_0$ and $\tilde{s}_2$, are dropped. Second, the function $\tilde{f}_i$ has been replaced by $\tilde{g}_i$ due to the impermeable boundary condition. Also, we have introduced a delta function (see Sec. 3.5) on the right hand side, in place of the recharge term, to simulate a pumping well with a constant (step rise) discharge Q_1 and located at $\mathbf{x}'$. For this three-layer system, there exists a single equation because the aquitard drawdown has been eliminated by the exact solution. We now seek the Green's function of Eq. (7.2).

We can rewrite Eq. (7.2) into the standard form of a modified Helmholtz equation (see Eq. (3.83))

$$\nabla^2 \tilde{s}_1 - k^2 \tilde{s}_1 = -\frac{Q_1}{pT_1}\delta(\mathbf{x}-\mathbf{x}') \quad (7.3)$$

where

$$k^2 = \frac{p}{T_1}\left(\frac{K_0'}{b_0'}\tilde{g}_0 + \frac{K_1'}{b_1'}\tilde{g}_1 + S_1\right) \quad (7.4)$$

Solution of this system is by now familiar,

$$\tilde{s}_1 = \frac{Q_1}{2\pi p T_1}\,K_0(kr) \quad (7.5)$$

where $r = |\mathbf{x} - \mathbf{x}'|$ is the radial distance from the well. We note that this solution is the same as that for the Theis solution, Eq. (3.75), the Jacob-Lohman solution Eq. (3.122), the Hantush-Jacob solution Eq. (4.31), and the Hantush solution Eq. (4.58). The difference lies only in the definition of k and the discharge rate. We shall later learn that k is the *eigenvalue* of the partial differential equation system Eq. (7.1).

We are now ready to examine a few other special cases. First, if we change the boundary condition at the outmost boundary from impermeable to constant head (zero drawdown), the solution is obtained by replacing $\tilde{g}_i$ with $\tilde{f}_i$; hence

$$k^2 = \frac{p}{T_1} \left(\frac{K_0'}{b_0'} \tilde{f}_0 + \frac{K_1'}{b_1'} \tilde{f}_1 + S_1 \right) \qquad (7.6)$$

Next, if one of the aquitard, say, the lower aquitard, does not exist, and the lower boundary of the aquifer is an impermeable layer, we shall drop the corresponding aquitard term in Eq. (7.4) or (7.6), denoted by the index 0, to obtain

$$k^2 = \frac{p}{T_1} \left(\frac{K_1'}{b_1'} \tilde{f}_1 + S_1 \right) \qquad (7.7)$$

We notice that this is exactly the Hantush solution (cf. Eq. (4.59)) of case 1, and also of case 2 if $\tilde{f}_1$ is replaced by $\tilde{g}_1$.

If we further assume that the aquitard storage S_1' is zero, meaning that there is no delay effect in the aquitard, we find that the function $\tilde{f}_1$ has the limit

$$\lim_{S_1' \to 0} \tilde{f}_1(p) = \frac{1}{p} \qquad (7.8)$$

based on the relation

$$\lim_{\eta \to 0} \eta \coth \eta = 1 \qquad (7.9)$$

This leads to

$$k^2 = \frac{p}{T_1} \left(\frac{K_1'}{p b_1'} + S_1 \right) \qquad (7.10)$$

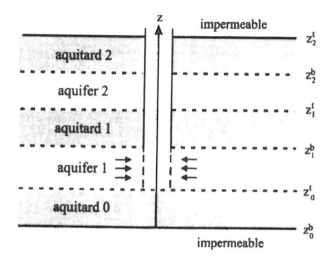

FIGURE 7.2. A two-aquifer-three-aquitard system.

which is the Hantush-Jacob solution (cf. Eq. (4.31)).

Finally, when both aquitards disappear, we drop all the aquitard terms and find

$$k^2 = \frac{pS_1}{T_1} \tag{7.11}$$

This is the Theis solution. Hence all the single aquifer solutions (with up to two flanking aquitards) are unified.

7.2   Two-Aquifer System

Figure 7.2 shows a two-aquifer-three-aquitard system with a pumping well located in the first aquifer. The governing equations, according to Eq. (7.1), are

$$\nabla^2 \tilde{s}_{11} - a_{11}\tilde{s}_{11} - a_{12}\tilde{s}_{21} = -\frac{Q_1}{pT_1}\delta(\mathbf{x} - \mathbf{x}') \tag{7.12}$$

$$\nabla^2 \tilde{s}_{21} - a_{21}\tilde{s}_{11} - a_{22}\tilde{s}_{21} = 0 \tag{7.13}$$

where we have defined the following coefficients for shorthand:

$$a_{11} = \frac{pK_0'\tilde{g}_0}{b_0'T_1} + \frac{pK_1'\tilde{f}_1}{b_1'T_1} + \frac{pS_1}{T_1}$$

$$a_{22} = \frac{pK_1'\tilde{f}_1}{b_1'T_2} + \frac{pK_2'\tilde{g}_2}{b_2'T_2} + \frac{pS_2}{T_2}$$

$$a_{12} = -\frac{pK_1'\tilde{h}_1}{b_1'T_1}$$

$$a_{21} = -\frac{pK_1'\tilde{h}_1}{b_1'T_2} \tag{7.14}$$

We note that we have used double subscripts to denote drawdown $\tilde{s}_{ij}$. The convention is needed because the first subscript is used to indicate the aquifer layer to which the drawdown is referred, and the second subscript shows the aquifer layer in which the pumping takes place. In other words, $\tilde{s}_{21}$ is the drawdown in aquifer 2 due to a pumping well in aquifer 1.

In Eq. (7.14) we note that $\tilde{g}_0$ and $\tilde{g}_2$ can be replaced by $\tilde{f}_0$ and $\tilde{f}_2$, respectively, if the constant head boundary condition, instead of impermeable wall, is prescribed, as

$$a_{11} = \frac{pK_0'\tilde{f}_0}{b_0'T_1} + \frac{pK_1'\tilde{f}_1}{b_1'T_1} + \frac{pS_1}{T_1}; \quad \text{if } s_0'(z_0^b,t) = 0$$

$$a_{22} = \frac{pK_1'\tilde{f}_1}{b_1'T_2} + \frac{pK_2'\tilde{f}_2}{b_2'T_2} + \frac{pS_2}{T_2}; \quad \text{if } s_2'(z_0^t,t) = 0 \tag{7.15}$$

Furthermore, if the upper and lower aquitards do not exist, the coefficients are modified to

$$a_{11} = \frac{pK_1'\tilde{f}_1}{b_1'T_1} + \frac{pS_1}{T_1}; \quad \text{without lower aquitard}$$

$$a_{22} = \frac{pK_1'\tilde{f}_1}{b_1'T_2} + \frac{pS_2}{T_2}; \quad \text{without upper aquitard} \tag{7.16}$$

Otherwise the solution system remains unchanged.

Equations (7.12) and (7.13) are two coupled elliptic equations. To find their solution, we seek to decouple them. This can be achieved by raising the order of the PDEs. For example, $\tilde{s}_{11}$ can be solved from Eq. (7.13) in terms of $\tilde{s}_{21}$, and substituted into Eq. (7.12). This gives

$$\nabla^4 \tilde{s}_{21} - (a_{22} + a_{11})\nabla^2 \tilde{s}_{21} + (a_{11}a_{22} - a_{12}a_{21})\tilde{s}_{21}$$
$$= -\frac{Q_1}{pT_1} a_{21} \delta(\mathbf{x} - \mathbf{x}') \tag{7.17}$$

where $\nabla^4 = \nabla^2 \nabla^2$ is the *biharmonic operator*. Similarly we obtain

$$\nabla^4 \tilde{s}_{11} - (a_{22} + a_{11})\nabla^2 \tilde{s}_{11} + (a_{11}a_{22} - a_{12}a_{21})\tilde{s}_{11}$$
$$= -\frac{Q_1}{pT_1} \nabla^2 \delta(\mathbf{x} - \mathbf{x}') + \frac{Q_1}{pT_1} a_{22} \delta(\mathbf{x} - \mathbf{x}') \tag{7.18}$$

We notice that these partial differential operators are linear and can be manipulated (addition, multiplication, etc.) based on algebraic rules. We hence factor them into the following:

$$(\nabla^2 - k_1^2)(\nabla^2 - k_2^2)\tilde{s}_{21} = -\frac{Q_1}{pT_1} a_{21} \delta(\mathbf{x} - \mathbf{x}') \tag{7.19}$$

$$(\nabla^2 - k_1^2)(\nabla^2 - k_2^2)\tilde{s}_{11} = -\frac{Q_1}{pT_1}(\nabla^2 - a_{22}) \delta(\mathbf{x} - \mathbf{x}') \tag{7.20}$$

where

$$k_1^2 = \frac{a_{11} + a_{22} - \sqrt{(a_{11} - a_{22})^2 + 4a_{12}a_{21}}}{2} \tag{7.21}$$

$$k_2^2 = \frac{a_{11} + a_{22} + \sqrt{(a_{11} - a_{22})^2 + 4a_{12}a_{21}}}{2} \tag{7.22}$$

By writing Eqs. (7.21) and (7.22) in terms of k_1^2 and k_2^2 we have assumed that these expressions are real and positive. It is necessary to check this assumption. First we note that all physical parameters must be positive. From Eq. (7.14) it is easily seen that a_{11} and a_{22} are positive, whereas a_{12} and a_{21} are negative. The quantities inside the radicals of Eqs. (7.21) and (7.22) are positive; hence k_1^2 and k_2^2 are real. While k_2^2 is now obviously positive, for k_1^2 to be positive we need the condition $(a_{11} + a_{22})^2 > (a_{11} - a_{22})^2 + 4a_{12}a_{21}$. The reader can easily prove this relation using the definitions in Eq. (7.14).

To find the solution of Eqs. (7.19) and (7.20), we first examine the standard form

$$(\nabla^2 - k_1^2)(\nabla^2 - k_2^2)\phi_2 = -2\pi\delta(\mathbf{x} - \mathbf{x}') \qquad (7.23)$$

The above can be rewritten as

$$(\nabla^2 - k_2^2)\phi_2 = \phi_1 \qquad (7.24)$$

and

$$(\nabla^2 - k_1^2)\phi_1 = -2\pi\delta(\mathbf{x} - \mathbf{x}') \qquad (7.25)$$

It is clear that

$$\phi_1 = K_0(k_1 r) \qquad (7.26)$$

as demonstrated in Eqs. (3.83) and (3.84). Equation (7.24) becomes

$$(\nabla^2 - k_2^2)\phi_2 = K_0(k_1 r) \qquad (7.27)$$

The above equation can be solved in a rigorous manner using Hankel transform.[25] Here, however, we notice from Eq. (7.23) that the solution must be symmetrical to k_1 and k_2. We simply guess that

$$\phi_2 = A\,K_0(k_1 r) + B\,K_0(k_2 r) \qquad (7.28)$$

Substituting the above into Eq. (7.27) and making sure that the delta functions thus generated cancel with each other, we obtain the coefficients A and B such that

$$\phi_2 = \frac{K_0(k_1 r)}{k_1^2 - k_2^2} + \frac{K_0(k_2 r)}{k_2^2 - k_1^2} \qquad (7.29)$$

With the above, it is clear that

$$\tilde{s}_{21} = \frac{Q_1}{2\pi p T_1}\, a_{21}\, \phi_2 \qquad (7.30)$$

$$\tilde{s}_{11} = \frac{Q_1}{2\pi p T_1}(\nabla^2 - a_{22})\, \phi_2 \qquad (7.31)$$

With the substitution of ϕ_2 into Eqs. (7.30) and (7.31), and the relation

$$\nabla^2 \phi_2 = K_0(k_1 r) + k_2^2 \phi_2 \tag{7.32}$$

based on Eq. (7.27), the solution becomes

$$\tilde{s}_{11} = \frac{Q_1}{2\pi p T_1} \left[\frac{(k_1^2 - a_{22})K_0(k_1 r)}{k_1^2 - k_2^2} + \frac{(k_2^2 - a_{22})K_0(k_2 r)}{k_2^2 - k_1^2} \right] \tag{7.33}$$

$$\tilde{s}_{21} = \frac{Q_1}{2\pi p T_1} a_{21} \left[\frac{K_0(k_1 r)}{k_1^2 - k_2^2} + \frac{K_0(k_2 r)}{k_2^2 - k_1^2} \right] \tag{7.34}$$

Based on symmetry, the drawdown in aquifers 1 and 2, due to the a pumping well located in aquifer 2, can be obtained by switching subscripts 1 and 2 in the above:.

$$\tilde{s}_{12} = \frac{Q_2}{2\pi p T_2} a_{12} \left[\frac{K_0(k_1 r)}{k_1^2 - k_2^2} + \frac{K_0(k_2 r)}{k_2^2 - k_1^2} \right] \tag{7.35}$$

$$\tilde{s}_{22} = \frac{Q_2}{2\pi p T_2} \left[\frac{(k_1^2 - a_{11})K_0(k_1 r)}{k_1^2 - k_2^2} + \frac{(k_2^2 - a_{11})K_0(k_2 r)}{k_2^2 - k_1^2} \right] \tag{7.36}$$

This completes the derivation of aquifer drawdown due to a pumping well in a two-aquifer system.

7.3   Multi-Aquifer System

Before moving on to further special cases, we shall examine the most general case of an n-aquifer system. To facilitate the presentation, we need to adopt the tensor notation. Particularly, we notice that the drawdown in the i-th aquifer due to pumping in the j-th aquifer, $\tilde{s}_{ij}$, is already a notation of the second order tensor. Utilizing the Einstein summation convention, Eq. (7.1) can be rewritten into this concise form:

$$\nabla^2 \tilde{s}_{ij} - a_{ik}\tilde{s}_{kj} = \tilde{b}_{ij}; \qquad i, j, k = 1, 2, \ldots, n \tag{7.37}$$

Following the convention, the repeated indices are summed. The right hand side in Eq. (7.37) represents a pumping well in the j-th layer,

$$\tilde{b}_{ij} = -\frac{Q_j}{pT_j}\delta_{ij}\delta(\mathbf{x} - \mathbf{x}') \tag{7.38}$$

in which δ_{ij} is the *Kronecker delta*, which has the property of

$$
\begin{aligned}
\delta_{ij} &= 1, && \text{if } i = j \\
\delta_{ij} &= 0, && \text{if } i \neq j
\end{aligned} \tag{7.39}
$$

In Eq. (7.38), however, no summation is applied for Q_j and T_j. Equation (7.38) therefore means that for the i-th equation, its right hand side is

$$
\begin{aligned}
\tilde{b}_{ij} &= -\frac{Q_j}{pT_j}\delta(\mathbf{x} - \mathbf{x}'), && \text{if } i = j \\
\tilde{b}_{ij} &= 0, && \text{if } i \neq j
\end{aligned} \tag{7.40}
$$

The coefficients a_{ik} in Eq. (7.37) are

$$a_{ii} = \frac{pK'_{i-1}\tilde{f}_{i-1}}{b'_{i-1}T_i} + \frac{pK'_i\tilde{f}_i}{b'_iT_i} + \frac{pS_i}{T_i}; \qquad 1 < i < n$$

$$a_{11} = \frac{pK'_0\tilde{g}_0}{b'_0T_1} + \frac{pK'_1\tilde{f}_1}{b'_1T_1} + \frac{pS_1}{T_1}$$

$$a_{nn} = \frac{pK'_{n-1}\tilde{f}_{n-1}}{b'_{n-1}T_n} + \frac{pK'_n\tilde{g}_n}{b'_nT_n} + \frac{pS_n}{T_n}$$

$$a_{i,i-1} = -\frac{pK'_{i-1}\tilde{h}_{i-1}}{b'_{i-1}T_i}; \qquad 1 < i \leq n$$

$$a_{i,i+1} = -\frac{pK'_i\tilde{h}_i}{b'_iT_i}; \qquad 1 \leq i < n$$

$$a_{ik} = 0; \quad 1 \leq i, k \leq n; \qquad k \neq i-1, i, i+1 \tag{7.41}$$

By the above definition we notice that a maximum of three coefficients a_{ik} exist in a single equation. This is because each aquifer is directly related only to the aquifer above, the aquifer below, and itself.

It is important to notice that the coefficient a_{ik} is non-symmetrical, i.e. $a_{ik} \neq a_{ki}$ generally.

Similar to Eqs. (7.15) and (7.16), the coefficients a_{11} and a_{nn} need to be modified if the condition of either constant head, or no outermost aquitard layer, exists. We now seek to solve the coupled system of n equations, Eq. (7.37) with $i = 1, \ldots, n$, for the n unknown drawdown values $\tilde{s}_{ij}$, for a fixed j.

Following the clue in the preceding section, we shall treat the Laplacian operator as an algebraic entity that can be subjected to elementary operations such as addition and multiplication.[69] In addition, we shall utilize the matrix operations for the solution.

This linear system of equations, Eq. (7.37), can be written into a matrix form

$$\mathbf{B}\,\mathbf{s} = \mathbf{r} \tag{7.42}$$

where $\mathbf{B}$ is a an $n \times n$ "coefficient" matrix of the form

$$\mathbf{B} = \begin{bmatrix} \nabla^2 - a_{11} & -a_{12} & 0 & 0 & \cdots & 0 \\ -a_{21} & \nabla^2 - a_{22} & -a_{23} & 0 & \cdots & 0 \\ 0 & -a_{32} & \nabla^2 - a_{33} & -a_{34} & \cdots & 0 \\ \vdots & \vdots & \vdots & \vdots & \ddots & \vdots \\ 0 & 0 & 0 & 0 & \cdots & \nabla^2 - a_{nn} \end{bmatrix} \tag{7.43}$$

We notice the *tri-diagonal* nature of the above matrix. The s matrix contains the unknowns to be solved, which is normally a column matrix. However, in an attempt to solve all $\tilde{s}_{ij}$, with a pumping well located in aquifer j, with $j = 1, \ldots, n$, s becomes an $n \times n$ matrix comprising of the drawdown elements

$$\mathbf{s} = \begin{bmatrix} \tilde{s}_{11} & \tilde{s}_{12} & \cdots & \tilde{s}_{1n} \\ \tilde{s}_{21} & \tilde{s}_{22} & \cdots & \tilde{s}_{2n} \\ \vdots & \vdots & \ddots & \vdots \\ \tilde{s}_{n1} & \tilde{s}_{n1} & \cdots & \tilde{s}_{nn} \end{bmatrix} \tag{7.44}$$

The corresponding right hand side is also an $n \times n$ matrix, but of diagonal form

$$\mathbf{r} = -\frac{1}{p}\delta(\mathbf{x} - \mathbf{x}')\begin{bmatrix} Q_1/T_1 & 0 & \cdots & 0 \\ 0 & Q_2/T_2 & \cdots & 0 \\ \vdots & \vdots & \ddots & \vdots \\ 0 & 0 & \cdots & Q_n/T_n \end{bmatrix} \qquad (7.45)$$

Equation (7.42) can be solved by *Cramer's rule*. For a fixed j, the drawdown in the i-th aquifer is

$$\tilde{s}_{ij} = -\frac{Q_j}{pT_j}\delta(\mathbf{x} - \mathbf{x}')\frac{1}{|\mathbf{B}|}.$$

$$\begin{vmatrix} \nabla^2 - a_{11} & -a_{12} & 0 & 0 & \cdots & 0 & \cdots & 0 \\ -a_{21} & \nabla^2 - a_{22} & -a_{23} & 0 & \cdots & 0 & \cdots & 0 \\ 0 & -a_{32} & \nabla^2 - a_{33} & -a_{34} & \cdots & 0 & \cdots & 0 \\ \vdots & \vdots & \vdots & \vdots & \ddots & \vdots & \vdots & \vdots \\ \cdots & \cdots & \cdots & \cdots & \cdots & 0 & \cdots & \cdots \\ \cdots & \cdots & \cdots & \cdots & \cdots & 1 & \cdots & \cdots \\ \cdots & \cdots & \cdots & \cdots & \cdots & 0 & \cdots & \cdots \\ \vdots & \vdots & \vdots & \vdots & \vdots & \vdots & \ddots & \vdots \\ 0 & 0 & 0 & 0 & \cdots & 0 & \cdots & \nabla^2 - a_{nn} \end{vmatrix}$$

$$\qquad (7.46)$$

where $|\cdot|$ means the determinant of a matrix. The determinant shown just above is formed by replacing the i-th column of $|\mathbf{B}|$ by the j-th column of the right-hand-side matrix $\mathbf{r}$, following Cramer's rule. Hence the elements of column i are all zero, except for the j-th element, which is 1, with the term $-Q_j\delta(\mathbf{x}-\mathbf{x}')/pT_j$ extracted outside. By the rule of reduction of determinants, this allows us to cancel the j-th row and the i-th column of the above determinant, with a sign correction $(-1)^{i+j}$. The resultant is exactly the *cofactor* of $|\mathbf{B}|$, denoted as $|\mathbf{B}^{ji}|$. We hence have achieved the solution in shorthand as

$$\tilde{s}_{ij} = -\frac{Q_j}{pT_j}\delta(\mathbf{x} - \mathbf{x}')\frac{|\mathbf{B}^{ji}|}{|\mathbf{B}|} \qquad (7.47)$$

Or, rather we write

$$|\mathbf{B}|\,\tilde{s}_{ij} = -\frac{Q_j}{pT_j}\delta(\mathbf{x} - \mathbf{x}')\,|\mathbf{B}^{ji}| \tag{7.48}$$

Here we would like to expand the determinant $|\mathbf{B}|$. But rather than conducting a straightforward expansion, we seek a more organized form. It follows from Eq. (7.43) that $|\mathbf{B}|$ may be written in the form of an *eigensystem*

$$|\mathbf{B}| = |\nabla^2\mathbf{I} - \mathbf{A}| \tag{7.49}$$

where $\mathbf{I}$ is the *identity matrix*,

$$\mathbf{I} = \begin{bmatrix} 1 & 0 & \cdots & 0 \\ 0 & 1 & \cdots & 0 \\ \vdots & \vdots & \ddots & \vdots \\ 0 & 0 & \cdots & 1 \end{bmatrix} \tag{7.50}$$

and $\mathbf{A}$ is the coefficient matrix $[a_{ik}]$

$$\mathbf{A} = \begin{bmatrix} a_{11} & a_{12} & 0 & 0 & \cdots & 0 \\ a_{21} & a_{22} & a_{23} & 0 & \cdots & 0 \\ 0 & a_{32} & a_{33} & a_{34} & \cdots & 0 \\ \vdots & \vdots & \vdots & \vdots & \ddots & \vdots \\ 0 & 0 & 0 & 0 & \cdots & a_{nn} \end{bmatrix} \tag{7.51}$$

In Eq. (7.49) we view the Laplacian operator ∇^2 as a scalar quantity, say λ. Just as the system $|\mathbf{B}| = |\lambda\mathbf{I} - \mathbf{A}|$, which can be factored into an *eigenvalue* system $(\lambda-\lambda_1)(\lambda-\lambda_2)\cdots(\lambda-\lambda_n)$, $|\mathbf{B}|$ can be expressed as

$$|\mathbf{B}| = (\nabla^2 - k_1^2)(\nabla^2 - k_2^2)\cdots(\nabla^2 - k_n^2) \tag{7.52}$$

where $k_i^2 = \lambda_i$ are the eigenvalues of the matrix $\mathbf{A}$. (Readers who are not familiar with eigenvalue systems can consult a mathematics textbook.[52])

For $n = 2$ we have demonstrated that the eigenvalues k_1^2 and k_2^2 in Eqs. (7.19) and (7.20) are real and positive. For $n \geq 3$, we shall forgo the mathematical proof. But we argue on physical ground that if a physically acceptable set of parameters is entered, all the eigenvalues must be real and positive. This is because complex eigenvalues produce solutions that have no physical meaning, and negative eigenvalues generate wave-type solutions that are not characteristic of diffusion phenomenon.

Substituting Eq. (7.52) into Eq. (7.48), we obtain

$$(\nabla^2 - k_1^2)(\nabla^2 - k_2^2) \cdots (\nabla^2 - k_n^2)\tilde{s}_{ij} = -\frac{Q_j}{pT_j}\delta(\mathbf{x} - \mathbf{x}') \left| \mathbf{B}^{ji} \right| \quad (7.53)$$

The partial differential operator on the left-hand side is a *poly-metaharmonic operator*.[28] We need to find its free-space Green's function in standard form

$$(\nabla^2 - k_1^2)(\nabla^2 - k_2^2) \cdots (\nabla^2 - k_n^2)\phi_n = -2\pi\delta(\mathbf{x} - \mathbf{x}') \quad (7.54)$$

Comparing Eq. (7.54) with Eqs. (7.23) to (7.25), we find that we can decompose the above into a recurrence relation

$$(\nabla^2 - k_n^2)\phi_n = \phi_{n-1}$$
$$(\nabla^2 - k_{n-1}^2)\phi_{n-1} = \phi_{n-2}$$
$$\vdots$$
$$(\nabla^2 - k_1^2)\phi_1 = -2\pi\delta(\mathbf{x} - \mathbf{x}') \quad (7.55)$$

As we have already found ϕ_2 from ϕ_1 in the preceding section, we can follow suit to successively find ϕ_3, ϕ_4, etc. Without going into the detail of derivation, it turns out that the solutions are simply summations of modified Bessel functions K_0, as

$$\phi_k = \sum_{i=1}^{k} \frac{K_0(k_i r)}{\prod_{j=1, j \neq i}^{n} \left(k_i^2 - k_j^2 \right)}; \qquad j = 2, \ldots, n \quad (7.56)$$

where Π is the product function. To fix ideas, we can try for $k = 3$,

$$\phi_3 = \frac{K_0(k_1 r)}{\left(k_1^2 - k_2^2 \right)\left(k_1^2 - k_3^2 \right)} + \frac{K_0(k_2 r)}{\left(k_2^2 - k_1^2 \right)\left(k_2^2 - k_3^2 \right)}$$
$$+ \frac{K_0(k_3 r)}{\left(k_3^2 - k_1^2 \right)\left(k_3^2 - k_2^2 \right)} \quad (7.57)$$

and $k = 4$,

$$\phi_4 = \frac{K_0(k_1 r)}{\left(k_1^2 - k_2^2\right)\left(k_1^2 - k_3^2\right)\left(k_1^2 - k_4^2\right)}$$
$$+ \frac{K_0(k_2 r)}{\left(k_2^2 - k_1^2\right)\left(k_2^2 - k_3^2\right)\left(k_2^2 - k_4^2\right)}$$
$$+ \frac{K_0(k_3 r)}{\left(k_3^2 - k_1^2\right)\left(k_3^2 - k_2^2\right)\left(k_3^2 - k_4^2\right)}$$
$$+ \frac{K_0(k_4 r)}{\left(k_4^2 - k_1^2\right)\left(k_4^2 - k_2^2\right)\left(k_4^2 - k_3^2\right)} \tag{7.58}$$

With the fundamental solution of Eq. (7.54) given by Eq. (7.56), the solution of Eq. (7.53).is simply

$$\tilde{s}_{ij} = \frac{Q_j}{2\pi p T_j} \left| \mathbf{B}^{ji} \right| \phi_n \tag{7.59}$$

We hence have completed the drawdown solution of a multilayered aquifer system with a pumping well located in an arbitrary aquifer. The above procedure seems abstract. But we can carry it out for specific cases as demonstrated in the following section.

7.4   Three-Aquifer System

Although the complete solution has been constructed in the preceding section, it is desirable to explicitly present the special cases for application purposes. In this section, the solution of the three-aquifer system, which includes up to seven layers (with four aquitards), is obtained as a reduction from the general solution.

The coefficients a_{ij} are explicitly listed following Eq. (7.41):

$$a_{11} = \frac{pK_0'\tilde{g}_0}{b_0'T_1} + \frac{pK_1'\tilde{f}_1}{b_1'T_1} + \frac{pS_1}{T_1}$$

$$a_{22} = \frac{pK_1'\tilde{f}_1}{b_1'T_2} + \frac{pK_2'\tilde{f}_2}{b_2'T_2} + \frac{pS_2}{T_2}$$

$$a_{33} = \frac{pK_2'\tilde{f}_2}{b_2'T_3} + \frac{pK_3'\tilde{g}_3}{b_3'T_3} + \frac{pS_3}{T_3}$$

$$a_{12} = -\frac{pK_1'\tilde{h}_1}{b_1'T_1}$$

$$a_{21} = -\frac{pK_1'\tilde{h}_1}{b_1'T_2}$$

$$a_{23} = -\frac{pK_2'\tilde{h}_2}{b_2'T_2}$$

$$a_{32} = -\frac{pK_2'\tilde{h}_2}{b_2'T_3}$$

$$a_{13} = a_{31} = 0 \tag{7.60}$$

For $\tilde{s}_{11}$, the solution according to Eq. (7.59) is

$$\tilde{s}_{11} = \frac{Q_1}{2\pi p T_1} \begin{vmatrix} \nabla^2 - a_{22} & -a_{23} \\ -a_{32} & \nabla^2 - a_{33} \end{vmatrix} \phi_3 \tag{7.61}$$

We note that the cofactor is obtained by eliminating the first row and the first column of

$$|\mathbf{B}| = \begin{vmatrix} \nabla^2 - a_{11} & -a_{12} & 0 \\ -a_{21} & \nabla^2 - a_{22} & -a_{23} \\ 0 & -a_{32} & \nabla^2 - a_{33} \end{vmatrix} \tag{7.62}$$

We can expand the determinant in Eq. (7.61) into

$$
\begin{aligned}
\tilde{s}_{11} &= \frac{Q_1}{2\pi p T_1} \left[(\nabla^2 - a_{22})(\nabla^2 - a_{33}) - a_{32}a_{23} \right] \phi_3 \\
&= \frac{Q_1}{2\pi p T_1} \left[\nabla^4 - (a_{22} + a_{33})\nabla^2 + a_{22}a_{33} - a_{32}a_{23} \right] \phi_3 \\
&= \frac{Q_1}{2\pi p T_1} \{ \phi_1 + (k_2^2 + k_3^2 - a_{22} - a_{33})\phi_2 \\
&\quad + \left[(k_3^2 - a_{22})(k_3^2 - a_{33}) - a_{32}a_{23} \right] \phi_3 \}
\end{aligned}
\tag{7.63}
$$

In the above we have utilized the recurrence relation for ϕ_i as shown in Eq. (7.55), which can be expressed in this form

$$\nabla^2 \phi_i = k_i^2 \phi_i + \phi_{i-1} \tag{7.64}$$

and further on

$$\nabla^4 \phi_i = k_i^4 \phi_i + \left(k_i^2 + k_{i-1}^2 \right) \phi_{i-1} + \phi_{i-2} \tag{7.65}$$

Although not used here, for four- and five-aquifer systems, we also need

$$\nabla^6 \phi_i = k_i^6 \phi_i + \left(k_i^4 + k_i^2 k_{i-1}^2 + k_{i-1}^4\right) \phi_{i-1} +$$
$$\left(k_i^2 + k_{i-1}^2 + k_{i-2}^2\right) \phi_{i-2} + \phi_{i-3} \qquad (7.66)$$

and

$$\nabla^8 \phi_i = k_i^8 \phi_i + \left(k_i^6 + k_i^4 k_{i-1}^2 + k_i^2 k_{i-1}^4 + k_{i-1}^6\right) \phi_{i-1}$$
$$+ \left(k_i^4 + k_{i-1}^4 + k_{i-2}^4 + k_i^2 k_{i-1}^2 + k_i^2 k_{i-2}^2 + k_{i-1}^2 k_{i-2}^2\right) \phi_{i-2}$$
$$+ \left(k_i^2 + k_{i-1}^2 + k_{i-2}^2 + k_{i-3}^2\right) \phi_{i-3} + \phi_{i-4} \qquad (7.67)$$

Higher order ones can be constructed by observing the pattern of the above relations.

Since the solutions of ϕ_1, $\phi_2, \ldots$, have already been found, Eq. (7.63) gives the drawdown in aquifer 1 due to pumping in aquifer 1, in a three-aquifer system. Other drawdown quantities are found in similar manner, and are expressed in terms of ϕ_i as follows:

$$\tilde{s}_{21} = \frac{Q_1}{2\pi p T_1} a_{21} \left[\phi_2 + (k_3^2 - a_{33})\phi_3\right] \qquad (7.68)$$

$$\tilde{s}_{31} = \frac{Q_1}{2\pi p T_1} a_{32} a_{21} \phi_3 \qquad (7.69)$$

$$\tilde{s}_{12} = \frac{Q_2}{2\pi p T_2} a_{12} \left[\phi_2 + (k_3^2 - a_{33})\phi_3\right] \qquad (7.70)$$

$$\tilde{s}_{22} = \frac{Q_2}{2\pi p T_2} \left[\phi_1 + (k_2^2 + k_3^2 - a_{11} - a_{33})\phi_2 \right.$$
$$\left. + (k_3^2 - a_{11})(k_3^2 - a_{33})\phi_3\right] \qquad (7.71)$$

$$\tilde{s}_{32} = \frac{Q_2}{2\pi p T_2} a_{32} \left[\phi_2 + (k_3^2 - a_{11})\phi_3\right] \qquad (7.72)$$

$$\tilde{s}_{13} = \frac{Q_3}{2\pi p T_3} a_{12} a_{23} \phi_3 \qquad (7.73)$$

$$\tilde{s}_{23} = \frac{Q_3}{2\pi p T_3} a_{23} \left[\phi_2 + (k_3^2 - a_{11})\phi_3\right] \qquad (7.74)$$

$$\tilde{s}_{33} = \frac{Q_3}{2\pi p T_3} \left\{\phi_1 + (k_2^2 + k_3^2 - a_{11} - a_{22})\phi_2 \right.$$
$$\left. + \left[(k_3^2 - a_{11})(k_3^2 - a_{22}) - a_{12} a_{21}\right] \phi_3\right\} \qquad (7.75)$$

This completes the explicit solution of the three-aquifer system.

Explicit solution for the four-aquifer system is shown in Cheng and Morohunfola.[25] Solutions for up to seven aquifers can be found in Morohunfola.[85]

Chapter 8

APPLICATION OF MULTILAYERED AQUIFER SOLUTIONS

8.1 One-Aquifer Problems

In this chapter, analytical solutions developed in Chapter 7 are used to examine some practical problems. Although these solutions are analytical, they need to be evaluated. Numerical tools based on *Fortran* and *Mathematica* are be constructed. For one- and two-aquifer problems, a *Mathematica* macro package *mwfield.m* is provided in Appendix C. For three-aquifer problems, a *Fortran* program *multi.f* is presented in Appendix B.

We first analyze the one-aquifer-two-aquitard system as shown in Figure 7.1. The solution is given by Eqs. (7.4) and (7.5). The same solution incorporates the one-aquifer-one-aquitard and one-aquifer only cases as well. This is achieved by using $K'_0 = 0$ and/or $K'_1 = 0$ (but $b'_0 \neq 0$ and $b'_1 \neq 0$) in data input. These aquitards are then automatically eliminated. Otherwise no change in the *Mathematica* macro and *Fortran* program is needed.

Example: *A pumping well with discharge $Q_w = 200$ m^3/day is located in an aquifer with $T = 200$ m^2/day, and $S = 0.00005$. The aquifer is sandwiched by a lower aquitard with $K' = 0.002$ m/day, $S' = 0.0005$, $b' = 30$ m, and an upper aquitard with $K' = 0.0002$ m/day, $S' = 0.001$, $b' = 30$ m. What is the drawdown history in the aquifer at a radial distance of 500 m from the pumping well?*

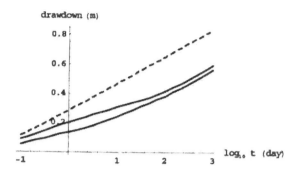

FIGURE 8.1. Drawdown versus logarithmic time. (Lower solid line: both aquitards present; upper solid line: only upper aquitard present; dashed line: both aquitards absent (Theis solution).)

The *Mathematica* function S1[Q, Kp0, Sp0, bp0, T1, S1, Kp1, Sp1, bp1, x, y, t] provided in the *mwfield.m* package allows the drawdown versus time curve to be plotted using commands such as Plot[S1[200, 0.002, 0.0005, 30, 200, 0.00005, 0.0002, 0.001, 30, 500, 0, t], {t, 0, 100}], where the numbers in the parentheses correspond to the parameters given above.

To enhance the small time behavior, we present the result as drawdown s versus logarithmic time $\log_{10} t$. This is shown in Figure 8.1 as the lower solid line. For comparison, we eliminate the lower aquitard by setting $K_0' = 0$ in the above data set. The corresponding drawdown is shown as the upper solid line. If both aquitards are eliminated, we recover the Theis solution. This is shown as a dashed line in Figure 8.1. We observe that the drawdown is much reduced for the cases with aquitards. This is expected due to the extra water supplied by the aquitards during the pumping.

We notice that at large times these curves become asymptotically a straight line. For the Theis solution, this is predicted by the Cooper-Jacob approximation, Eq. (3.99), which is rewritten here in base 10 logarithm:

$$s \approx \frac{2.30 \, Q_1}{4\pi T_1} \log_{10} \frac{2.25 T_1 t}{r^2 S_1}$$

For the one-aquifer-two-aquitard solution, we can investigate as follows. For large time behavior, we can expand the solution Eqs. (7.4) and (7.5) in Taylor series for small p using *Mathematica* and obtain:

$$\tilde{s} \approx \frac{Q_w}{4\pi T_1} \frac{1}{p} \left[\ln p + 2\gamma + \ln \sqrt{\frac{r^2 \left(S_1 + S_0' + S_1'\right)}{4T_1}} \right]$$

Here we recall that $\gamma = 0.5772\ldots$ is the Euler number. Using the inversion table, Table A-1, and after some algebraic manipulation, we obtain

$$s \approx \frac{2.30\, Q_1}{4\pi T_1} \log_{10} \frac{2.25 T_1 t}{r^2 \left(S_1 + S_0' + S_1'\right)}$$

Hence the Theis and the one-aquifer-two-aquitard curves have the same asymptotic slope of

$$\frac{2.30 Q_1}{4\pi T_1} = 0.183$$

and they are

$$\frac{2.30 Q_1}{4\pi T_1} \log_{10} \left(1 + \frac{S_0'}{S_1} + \frac{S_1'}{S_1}\right) = 0.273 \text{ m}$$

apart.

Example: *Given the same one-aquifer system in the preceding example, but without the lower aquitard, investigate the effect of aquitard hydraulic conductivity on the aquifer drawdown.*

In the above example we have particularly chosen relatively large storage for aquitards. Consider the case of a clayey layer with relatively large storage, yet small hydraulic conductivity, overlaying an aquifer. Can the aquitard be ignored and the Theis solution be applied because the hydraulic conductivity of clay is very small?

Assuming the same pumping rate and the observation distance as the previous example, we plot in Figure 8.2 the aquifer drawdown versus logarithmic time as the lowest solid curve. We then

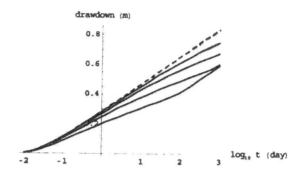

FIGURE 8.2. Aquifer drawdown for a one-aquifer-one-aquitard system, with different aquitard hydraulic conductivities. (Solid curves from bottom up: $K_1' = 2 \times 10^{-4}$, 2×10^{-5}, 2×10^{-6}, and 2×10^{-7} m/day; dashed curve: Theis solution, $K_1' = 0$.)

continuously decrease the aquitard hydraulic conductivity K_1' from 2×10^{-4} to 2×10^{-5}, 2×10^{-6}, and 2×10^{-7} m/day and plot the drawdown curves in the same figure. Also presented is the Theis solution (dashed line), which corresponds to the case $K_1' = 0$. The maximum time plotted in the figure is $t = 1000$ days. We observe that as the aquitard hydraulic conductivity decreases, the drawdown approaches that of the Theis case. However, even for the case of $K_1' = 2 \times 10^{-7}$ m/day, which can be classified as impermeable (see Figure 2.3), the two curves still depart after only 10 days of pumping. Hence for long term water supply purposes, the presence of a low hydraulic conductivity, but large storativity layer should not be ignored.

Example: *For the same one-aquifer-two-aquitard system considered in the first example of this section, present the aquitard drawdown.*

The aquitard drawdown can be calculated from Eqs. (6.36)–(6.39) and (6.43)–(6.45). For example, the drawdown in the upper aquitard is found as

$$\tilde{s}_1'(x, y, z, p) = p\tilde{s}_1(x, y, p)\widetilde{\mathcal{G}}_1^{(1)}(z, p)$$

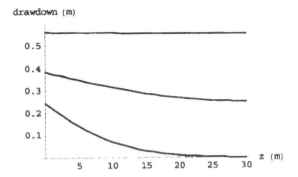

FIGURE 8.3. Drawdown in the upper aquitard of the one-aquifer-two-aquitard problem. (Lower curve: $t = 10$ day; middle curve: $t = 100$ day; upper curve: $t = 1000$ day)

where

$$\widetilde{\mathcal{G}}_1^{(1)}(z,p) = \frac{1}{p} \operatorname{sech}\left(\sqrt{\frac{S_1' b_1' p}{K_1'}}\right) \cosh\left(\sqrt{\frac{S_1' b_1' p}{K_1'}} \frac{z_1^t - z}{b_1'}\right)$$

This is programmed as the function $\mathtt{Sp1[\cdots]}$ in the *Mathematica* macro. Figure 8.3 plots the aquitard drawdown versus the depth z at three different times, $t = 10$, 100, and 1000 day. We note that $z = 0$ corresponds to the aquitard bottom, and $z = 30$ m the top. It is observed that at 1000 day, the aquitard drawdown is practically uniform throughout.

8.2 Two-Aquifer Problems

The geometry of a two-aquifer-three-aquitard system is shown in Figure 7.2. The *Mathematica* macro package *mwfield.m* provides functions $\mathtt{Sij[\cdots]}$ for the evaluation of drawdown in the i-th aquifer due to a pump in the j-th aquifer. A few examples utilizing this tool are presented below.

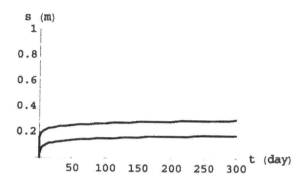

FIGURE 8.4. Drawdown in aquifer 1 (upper curve) and 2 (lower curve) due to pumping in aquifer 1.

Example: *A two-aquifer-one-aquitard system has the following properties:* $T_1 = 3000$ m^2/day, $S_1 = 0.0001$, $K_1' = 0.02$ m/day, $S_1' = 0.0008$, $b_1' = 30$ m, $T_2 = 500$ m^2/day, $S_2 = 0.00005$. *Plot the drawdown as a function of time caused by a well pumped at a constant rate of* 1000 m^3/day.

First we assume that the well is located in aquifer 1, which has the larger transmissivity. To find the drawdown in aquifer 1, at a distance $r = 100$ m, the function calls S11[1000, 0, 0, 1, 3000, 0.0001, 0.02, 0.0008, 30, 500, 0.00005, 0, 0, 1, 100, 0, t] and S21[···] are used. In the above functions, the parameters enter in the sequence of Q, K_0', S_0', b_0', T_1, S_1, K_1', S_1', b_1', T_2, S_2, K_2', S_2', and b_2'. We notice that for aquitard 0 and 2 the parameter values $K_0' = K_2' = 0$ are used, and b_0' and b_2' are set to an arbitrary non-zero constant (or else zero division will result in the program). This automatically "turns off" these two aquitards. The last three arguments in the function are set as $x = 100$ m, $y = 0$ m, and t a variable. The drawdown versus time is plotted in Figure 8.4 for the two aquifers for the first 300 days.

Next, we place the well in aquifer 2 and plot the drawdown using S22[···] and S12[···]. The result is presented in Figure 8.5. We notice that the drawdown in the pumped aquifer is larger in this

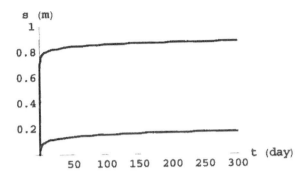

FIGURE 8.5. Drawdown in aquifer 1 (upper curve) and 2 (lower curve) due to pumping in aquifer 2.

case due to the smaller transmissivity of aquifer 2. We also note with curiosity that the drawdown in the unpumped aquifer for the two cases, shown as the lower curves in Figures 8.4 and 8.5, are identical. This is not a coincidence as we observe from Eqs. (7.34) and (7.35), with coefficients defined in Eq. (7.14), that

$$s_{12} = s_{21}$$

if the same pumping rate is used. This reciprocal relation also exists in general, i.e.,

$$s_{ij} = s_{ji}$$

for an n-layer system.

Example: *For the same two-aquifer-one-aquitard system as the above example, plot the aquifer drawdown as a function of distance at different times.*

This is achieved by using function calls in the form S11[···, x, 0, 50], etc. Assume that the well is located in aquifer 1 with the same pumping rate as the above example. We plot in Figure 8.6 the drawdown versus distance in aquifer 1 for the first 2000 m at three different times, $t = 10$, 50, and 200 day. The shape of the drawdown

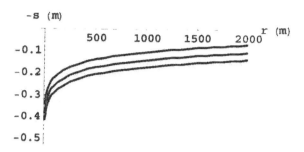

FIGURE 8.6. Drawdown versus distance in aquifer 1, due to pumping in aquifer 1. (From top curve down: $t = 10$, 50, and 200 day)

curve is inverted by plotting its negative value to show the cone of depression.

In Figure 8.7, we present the drawdown in aquifer 2, the unpumped aquifer. We observe that the drawdown is rather uniform in the range plotted. Particularly, the drawdown is not singular toward the well, i.e. $r \to 0$. We can check this behavior from the solution Eq. (7.34). Following the asymptotic behavior of Bessel function as shown in Eq. (3.71), we find

$$\lim_{r \to 0} \tilde{s}_{21} = \frac{Q_1}{2\pi p T_1} a_{21} \frac{\ln(k_2/k_1)}{k_1^2 - k_2^2}$$

which is indeed non-singular.

Example: *A two-aquifer-one-aquitard system has the following the properties: $T_1 = 2000$ m²/day, $T_2 = 1000$ m²/day, $S_1 = 0.0001$, $S_2 = 0.00003$, $K_1' = 0.1$ m/day, $S_1' = 0.0008$, and $b_1' = 20$ m. Two pumping wells, one located in aquifer 1 at the spatial coordinates $(0\,\mathrm{m},\ 0\,\mathrm{m})$ and pumped at the rate of 500 m³/day, and the second in aquifer 2 at $(2000\,\mathrm{m},\ 2000\,\mathrm{m})$ with the rate 800 m³/day. If the two pumps started at the same time, what is the drawdown in the aquifers?*

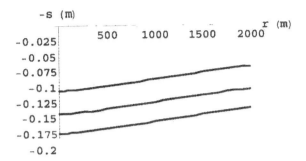

FIGURE 8.7. Drawdown versus distance in aquifer 2, due to pumping in aquifer 1. (From top curve down: $t = 10$, 50, and 200 day)

Multiple pumping wells can be simulated by adding the drawdown functions at different locations and times. In the present case, the drawdown in aquifer 1 due to the two pumps after 200 day of pumping can be obtained as S11[500,$\cdots$, x, y, 200, 0, 0, 0] + S12[800,$\cdots$, x, y, 200, 2000, 2000, 0]. The plot of the drawdown surface is presented in Figure 8.8 for both aquifers.

Example: *A well penetrates a two-aquifer-one-aquitard system. It is screened in the two aquifers such that water is extracted from both of them. For a total discharge Q, what is its partition of discharge in the two aquifers?*

Assume that the discharge from aquifer 1 and 2 are respectively Q_1 and Q_2. The total discharge is

$$Q = Q_1 + Q_2$$

The condition that determines the partition is based on the assumption that the drawdown at the well bore is the same for both aquifers at all times:

$$s_1(r_w, t) = s_2(r_w, t)$$

In fact, to be consistent with pumping well solutions, this condition will be imposed not at $r = r_w$, but at $r \to 0$. Taking the limit of

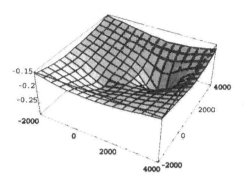

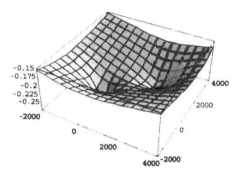

FIGURE 8.8. Drawdown in a two-aquifer-one-aquitard system with pumping wells in both aquifers, at $t = 200$ day. Upper diagram: drawdown in aquifer 1; lower diagram: aquifer 2.

Eqs. (7.33)–(7.36), we find

$$\lim_{r \to 0} (\tilde{s}_{11} + \tilde{s}_{12}) \sim -\frac{Q_1}{2\pi p T_1} \ln r$$

$$\lim_{r \to 0} (\tilde{s}_{21} + \tilde{s}_{22}) \sim -\frac{Q_2}{2\pi p T_2} \ln r$$

The requirement that the drawdown must be equal produces the relation

$$\frac{Q_1}{T_1} = \frac{Q_2}{T_2}$$

We hence find

$$Q_1 = \frac{T_1 Q}{T_1 + T_2}$$

$$Q_2 = \frac{T_2 Q}{T_1 + T_2}$$

The drawdown in aquifer 1 and 2 are then found as the summation of the two pumping solutions:

$$s_1 = s_{11} + s_{12}$$

$$s_2 = s_{21} + s_{22}$$

8.3   Three-Aquifer Problems

For three-aquifer-four-aquitard systems, a *Fortran* program *multi.for* is constructed and listed in Appendix C. The structure of the program closely follows the solution presented in Sec. 7.4. It is however worthwhile to discuss the ways of finding eigenvalues of matrix $\mathbf{A}$, as defined in Eq. (7.51). For an n-layer system, eigenvalues are generally found by numerical means. *Fortran* subroutines for this purpose can be found in numerical analysis textbooks, such as the *Numerical Recipes*.[101] Public domain subroutine packages such as the *EISPACK*[108] are also available. For *Mathematica*, the function `Eigenvalues[ ]` can perform this task. For one- and two-aquifer systems, however, the eigenvalues k_i^2 have been explicitly presented in Secs. 7.1 and 7.2; hence no numerical procedure is necessary. This can be extended to three-aquifer systems, as demonstrated in the following.

The eigenvalues λ_i for the three-aquifer system is found from the following determinant

$$|\mathbf{A} - \lambda \mathbf{I}| = 0 \qquad (8.1)$$

where I is the *identity matrix*. By noting the tri-diagonal nature of A, the above can be expressed as:

$$
\begin{vmatrix}
a_{11} - \lambda & a_{12} & 0 \\
a_{21} & a_{22} - \lambda & a_{23} \\
0 & a_{32} & a_{33} - \lambda
\end{vmatrix} = 0 \tag{8.2}
$$

The determinant can be expanded to give the polynomial

$$
\lambda^3 + c_2 \lambda^2 + c_1 \lambda + c_0 = 0 \tag{8.3}
$$

in which

$$
\begin{aligned}
c_0 &= a_{11}a_{23}a_{32} + a_{12}a_{21}a_{33} - a_{11}a_{22}a_{33} \\
c_1 &= a_{11}a_{22} + a_{11}a_{33} + a_{22}a_{33} - a_{12}a_{21} - a_{23}a_{32} \\
c_2 &= -a_{11} - a_{22} - a_{33}
\end{aligned} \tag{8.4}
$$

The three eigenvalues needed for the solution, $k_i^2 = \lambda_i$, where $i = 1, 2, 3$, are the roots of Eq. (8.3). They can be solved as

$$
\lambda_1 = -\frac{c_2}{3} - \frac{2^{1/3}\left(3c_1 - c_2^2\right)}{3D} + \frac{2^{2/3}}{6}D \tag{8.5}
$$

$$
\lambda_2 = -\frac{c_2}{3} + \frac{2^{1/3}\left(1 - \sqrt{3}\,i\right)\left(3c_1 - c_2^2\right)}{6D}
$$

$$
- \frac{2^{2/3}}{12}\left(1 + \sqrt{3}\,i\right)D \tag{8.6}
$$

$$
\lambda_3 = -\frac{c_2}{3} + \frac{2^{1/3}\left(1 + \sqrt{3}\,i\right)\left(3c_1 - c_2^2\right)}{6D}
$$

$$
- \frac{2^{2/3}}{12}\left(1 - \sqrt{3}\,i\right)D \tag{8.7}
$$

where $i = \sqrt{-1}$ and

$$
D = \left[-27c_0 + 9c_1c_2 - 2c_2^3 \right.
$$

$$
\left. + \sqrt{4\left(3c_1 - c_2^2\right) + \left(27c_0 - 9c_1c_2 + 2c_2^3\right)^2} \right]^{1/3} \tag{8.8}
$$

Although the roots are expressed in the form of complex variables, they are all real and positive, once evaluated.

The program *multi.for* can be used to calculate drawdown at any given radial distance and time.

Example: *Prepare input data for a three-aquifer-four-aquitard system.*

The program *multi.for* requires a data file *multi.dat* that list all the aquifer parameters. A sample file is given below:

```
0.005    0.0002    50.    K'(0), S'(0), b'(0)
3000.    0.0001           T(1),  S(1)
0.02     0.0008    30.    K'(1), S'(1), b'(1)
500.     0.00005          T(2),  S(2)
0.05     0.0005    20.    K'(2), S'(2), b'(2)
300.     0.00002          T(3),  S(3)
0.0008   0.00008   45.    K'(3), S'(3), b'(3)
```

In the above, the units used are m for length and day for time. The text at the end of each line annotates the parameters. Its presence does not interfere with the data input. Hence the data structure is self-explanatory. Users can edit the above data file to suit their needs.

Once started, the user will be prompted by program:

```
Pumping rate for aquifer 1, 2, 3 =>
```

Take, for example, a well that is screened in aquifer 2 and 3, but sealed off in aquifer 1. For a pumping rate of 1000 m³/day, the partition based on transmissivity values leads to: $Q_2 = 625$ m³/day and $Q_3 = 375$ m³/day. The user then enters

```
0 625 375
```

Next, the program will prompt for input of distance r and time t. If we enter 1000 (in meter) and 100 (in day), the drawdown for the three aquifer layers is calculated:

```
s(1) = 0.112115430386386
s(2) = 0.191444130923056
s(3) = 0.199708345412525
```

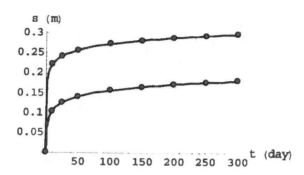

FIGURE 8.9. Drawdown in aquifer 1 (upper curve) and 2 (lower curve) for a two-aquifer-one-aquitard system (solid lines: *Mathematica* solution; dots: *Fortran* solution).

Example: *Use the same program to simulate a two-aquifer-one-aquitard system.*

The same program can be used for systems with fewer aquifers and aquitards. The flow exchange among aquifers can be cut off by assigning zero hydraulic conductivity to the intervening aquitard. For example, the data set below:

0.	0.0002	50.	K'(0), S'(0), b'(0)
3000.	0.0001		T(1), S(1)
0.02	0.0008	30.	K'(1), S'(1), b'(1)
500.	0.00005		T(2), S(2)
0.	0.0005	20.	K'(2), S'(2), b'(2)
300.	0.00002		T(3), S(3)
0.0008	0.00008	45.	K'(3), S'(3), b'(3)

shows that aquifer 1, aquitard 1, and aquifer 2 form a two-aquifer-one-aquitard system. Also, aquifer 3 and aquitard 3 become a one-aquifer-one-aquitard system.

For comparison, we simulate the two-aquifer-one-aquitard case using both the *Fortran* program the *Mathematica* macro. The drawdown history at a distance of 100 m due to a pumping rate of

$Q_1 = 1000$ m^3/day is presented in Figure 8.9 for aquifer 1 and 2. The results are in excellent agreement.

Chapter 9
SENSITIVITY ANALYSIS

9.1 Parameter Sensitivity

Sensitivity analysis is the study of the changes of a system's response due to disturbances in the various factors that the system is dependent upon. It is a useful tool in the management and planning of water resources systems. For application in aquifer systems, the disturbances can be associated with the lack of knowledge of the initial conditions, boundary conditions, forcing functions, boundary geometry, physical coefficients, etc. The input data concerning these factors may not be reliably determined; or they are future events that cannot be predicted. Under these circumstances, sensitivity analysis can assist a manager to assess the impact due to these uncertainties.

While there exist different types of sensitivity factors, in this chapter we focus on *parameter sensitivity*, namely a measure of the system sensitivity to changes in physical parameters. For aquifer systems, the physical parameters are transmissivity, storativity, hydraulic conductivity, thickness, etc. If the input parameters have an error of a certain magnitude, it is of interest to find out the magnitude of output error. If the assessed error is too large to be acceptable, sensitivity analysis may be used to determine the degree of refinement needed in input data to achieve the desirable accuracy of the output prediction.

Sensitivity analysis is also closely related to *parameter determination*. Often it is difficult to determine the physical parameters of a formation by direct measurements. Indirect means, such as observing aquifer response due to disturbances by pumping activities, are employed. Mathematically, an *inverse problem* is solved. If the response we choose to observe is sensitive to a certain parameter, then that parameter stands a good chance to be accurately determined. On the other hand, if the response is insensitive to the variation of a parameters, then the mathematical problem can becomes ill posed in the presence of data noise. The parameter cannot be accurately, or even uniquely determined.

The stochastic response of a system is also related to its sensitivity. Given the input data uncertainty in terms of statistical measures such as the standard deviation of transmissivity estimate, the output uncertainty such as drawdown standard deviation is to be determined. A standard method for such an analysis is the perturbation technique. The perturbation method is again tied to parameter sensitivity.

The parameter determination problems are discussed in Chapter 10 and the stochastic problems are presented in Chapter 11.

9.2   Sensitivity of Theis Solution

We shall first examine the drawdown given by the Theis solution, Eq. (3.91), which is repeated here for convenience:

$$s = \frac{Q_w}{4\pi T} W(u) \tag{9.1}$$

where

$$u = \frac{r^2 S}{4Tt} \tag{9.2}$$

The drawdown is obviously directly proportional to Q_w. Hence only the sensitivity of the two aquifer parameters S and T needs to be investigated. A numerical example is given below as a demonstration.

Example: *An aquifer has its properties estimated to be: $T = 300$ m^2/day and $S = 0.0001$. Investigate the sensitivity of drawdown due*

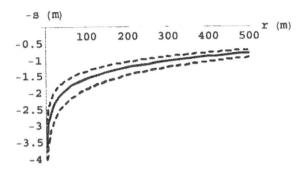

FIGURE 9.1. Drawdown sensitivity due to ±20% perturbation of transmissivity. Solid line: no perturbation; upper dashed line: +20% perturbation; lower dashed line: −20% perturbation.

to a pumping of $Q_w = 900$ m^3/day, with respect to transmissivity and storativity, at $t = 1$ day.

The drawdown at $t = 1$ day is evaluated based on the given parameters and plotted as $-s$ vs. r in Figure 9.1, as the solid line. We assume that the transmissivity has a ±20% uncertainty, given by the limits $T = 240$ and 360 m^2/day. The drawdown is re-calculated using these values and plotted as the dashed lines in Figure 9.1. We notice that within the distance plotted (500 m), an increase in transmissivity causes an decrease in drawdown, and vice versa. The drawdown perturbation diminishes with distance.

We next perturb the storativity by 20% while keeping the transmissivity constant. The perturbed result is shown in dashed lines in Figure 9.2. The result demonstrates that an increase in storativity causes a decrease in drawdown, and vice versa. The magnitude of drawdown change is significantly smaller than that of the transmissivity case, indicating that the drawdown is more sensitive to a variation in transmissivity, than to storativity.

In sensitivity analysis, the perturbations are meant to be small. In that case, we can approximate the perturbed drawdown by Taylor series expansion. Take for example, the drawdown corresponding to

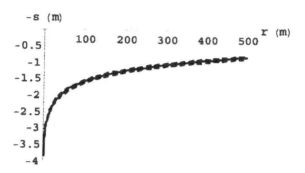

FIGURE 9.2. Drawdown sensitivity due to $\pm 20\%$ perturbation of storativity. Solid line: no perturbation; upper dashed line: $+20\%$ perturbation; lower dashed line: -20% perturbation.

a perturbation in transmissivity, $s(r, t; S, T + \Delta T)$, which can be approximated as

$$s(r, t; S, T + \Delta T) = s(r, t; S, T) + \frac{\partial s(r, t; S, T)}{\partial T} \Delta T + \ldots \quad (9.3)$$

For the case of storativity, we find

$$s(r, t; S + \Delta S, T) = s(r, t; S, T) + \frac{\partial s(r, t; S, T)}{\partial S} \Delta S + \ldots \quad (9.4)$$

For small ΔT and ΔS, the perturbed drawdown is approximately given by the second terms on the right hand sides of Eqs. (9.3) and (9.4). We hence recognize the significance of the first derivatives, $\partial s / \partial T$ and $\partial s / \partial S$, which characterize the system sensitivity to the corresponding parameter. These derivatives are known as the *sensitivity coefficients*. Since the drawdown is given in analytical form, we can explicitly express the sensitivity coefficients as:

$$\frac{\partial s}{\partial T} = \frac{Q_w}{4\pi T^2} \left[e^{-u} - W(u) \right] \quad (9.5)$$

$$\frac{\partial s}{\partial S} = -\frac{Q_w}{4\pi ST} e^{-u} \quad (9.6)$$

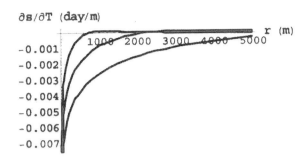

FIGURE 9.3. Sensitivity coefficient $\partial s/\partial T$ versus r. Upper curve: at $t = 0.1$ day; middle curve: 1 day; lower curve: 10 day.

Example: *For the same aquifer properties and pumping rate as the above example, examine the sensitivity coefficients.*

In Figure 9.3 the sensitivity coefficient $\partial s/\partial T$ is plotted as a function of distance for three values of time, $t = 0.1$, 1 and 10 day. We observe that the sensitivity coefficient is large and negative for small values of r. Its magnitude drops and crosses zero at a certain distance, and is of negligible magnitude beyond that. The distance at which the crossing takes place increases as time increases.

In Figure 9.4 the sensitivity coefficient $\partial s/\partial S$ is presented. We also find that the sensitivity is larger at smaller distance, and it increases with time.

It is a common practice to present sensitivity coefficients in their dimensional form, as those in Figures 9.3 and 9.4. The result however is dependent on the particular set of parameter values chosen. These sensitivity coefficients have different dimensions; hence they may give the wrong impression of the importance of the coefficients. We have already observed in the first example in this section that the drawdown is more sensitive to the transmissivity changes than to storativity. The magnitude of $\partial s/\partial T$ and $\partial s/\partial S$ shown in Figures

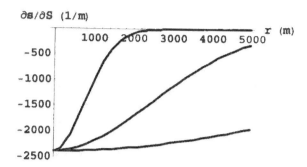

FIGURE 9.4. Sensitivity coefficient $\partial s/\partial S$ versus r. Upper curve: at $t = 0.1$ day; middle curve: 1 day; lower curve: 10 day.

9.3 and 9.4 does not reflect this fact. But they should not be directly compared. It is better to present the result in dimensionless form.

We can normalize Eqs. (9.5) and (9.6) to become

$$\frac{T}{s}\frac{\partial s}{\partial T} = -1 + \frac{e^{-u}}{W(u)} \tag{9.7}$$

$$\frac{S}{s}\frac{\partial s}{\partial S} = -\frac{e^{-u}}{W(u)} \tag{9.8}$$

These *dimensionless sensitivity coefficients* can now be compared. The above functional relationships are condensed as they are functions of one dimensionless variable, u, only. They can each be presented as a universal curve for all values of dimensional parameters.

In Figure 9.5, we plot Eqs. (9.7) and (9.8) for $10^{-6} < u < 10^{1}$, which is the typical range of u for the Theis type curve (Figure 3.13). Similar to the Theis type curve, logarithmic scale is used to display u to enhance the behavior at small u. In reading these curves, we should recall that $u \sim r^2$ or $u \sim t^{-1}$.

We can compare the transmissivity sensitivity curve in Figure 9.5 (solid line) with those in Figure 9.3. In Figure 9.3 we observe that $\partial s/\partial T$ is unbounded as $r \to 0$, indicating an infinite sensitivity. This observation is misleading because it is created by the unboundedness of drawdown s as $r \to 0$. The relative sensitivity, i.e. the change normalized by the drawdown itself, is given by the dimensionless

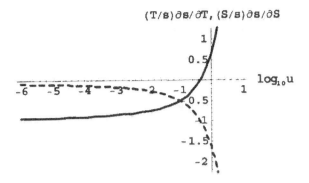

FIGURE 9.5. Relative sensitivity coefficients versus u. Solid line: $(T/s)\partial s/\partial T$; dashed line: $(S/s)\partial s/\partial S$.

sensitivity coefficient. Figure 9.5 shows that as $r \to 0$ $(u \to 0)$, $(T/s)(\partial s/\partial T)$ approaches -1. In other words,

$$\frac{\partial s}{\partial T} \to -\frac{s}{T}; \qquad \text{as } r \to 0 \tag{9.9}$$

On the other hand, as $r \to \infty$ $(u \to \infty)$, Figure 9.5 shows that the relative sensitivity goes to infinity. While this is true, it is not of practical interest because the drawdown approaches zero when $r \to \infty$. In this case, Figure 9.3 gives the more useful information because it indicates that the absolute sensitivity goes to zero as $r \to \infty$.

Similarly, we can compare Figures 9.4 and 9.5 for sensitivity concerning storativity. As $r \to 0$, the relative sensitivity goes to zero, but the absolute sensitivity approaches a constant:

$$\frac{\partial s}{\partial S} \to -\frac{Q_w}{4\pi ST}; \qquad \text{as } r \to 0 \tag{9.10}$$

As $r \to \infty$, we find $\partial s/\partial S \to 0$ and $(S/s)(\partial s/\partial S) \to -\infty$. But neither is of practical interested due to the diminishing drawdown.

If we discard the limiting ranges of $r \to 0$ and $r \to \infty$, we observe from Figure 9.5 that the magnitude of sensitivity coefficient for transmissivity is much greater than that for storativity for the most part of the curve. The curves however cross at $u = 0.102$, where they are of equal magnitude.

Also, we note that the transmissivity sensitivity coefficient crosses zero at $u = 0.435$ and becomes positive. Beyond this point, a reversal in trend, that is, an increase in transmissivity causes an increase in drawdown, is observed. Around this point, the sensitivity is small or is zero. As indicated by McElwee,[80] this seems to be a suitable choice for the radius of influence. This distance can be found by solving

$$u = \frac{r^2 S}{4Tt} = 0.435 \tag{9.11}$$

Hence, we find the radius of influence R as

$$R = 1.32 \left(\frac{Tt}{S} \right)^{1/2} \tag{9.12}$$

This can be compared to Eq. (3.52).

9.3   Sensitivity of Hantush-Jacob Solution

The Hantush-Jacob solution in Eqs. (4.33)–(4.36) is rewritten here for convenience,

$$s = \frac{Q_w}{4\pi T} W(u, \beta) \tag{9.13}$$

where

$$W(u, \beta) = \int_u^\infty \frac{1}{u} \exp\left(-u - \frac{\beta^2}{4u} \right) du \tag{9.14}$$

and

$$u = \frac{r^2 S}{4Tt} \tag{9.15}$$

$$\beta = \sqrt{\frac{L'}{T}} r \tag{9.16}$$

In the above we have introduced a *leakance* defined as (see also Sec. 4.4)

$$L' = \frac{K'}{b'} \tag{9.17}$$

We note in the above solution that the two aquitard parameters K' and b' always appear in the combination indicated in Eq. (9.17). In a parameter determination problem using aquifer drawdown as data, it is not possible to separately identify these two parameters. We hence shall investigate the sensitivity of aquifer drawdown subject to perturbation of the three parameters only, T, S, and L'.

The sensitivity coefficients can be obtained by differentiating Eq. (9.13) with respect to these three parameters. In their normalized form, we have

$$\frac{T}{s}\frac{\partial s}{\partial T} = -1 + \frac{\exp\left(-u - \frac{\beta^2}{4u}\right)}{W(u,\beta)}$$
$$+ \frac{\frac{\beta^2}{4}\int_u^\infty \frac{1}{u^2}\exp\left(-u - \frac{\beta^2}{4u}\right)\,du}{W(u,\beta)} \tag{9.18}$$

$$\frac{S}{s}\frac{\partial s}{\partial S} = -\frac{\exp\left(-u - \frac{\beta^2}{4u}\right)}{W(u,\beta)} \tag{9.19}$$

$$\frac{L'}{s}\frac{\partial s}{\partial L'} = -\frac{\frac{\beta^2}{4}\int_u^\infty \frac{1}{u^2}\exp\left(-u - \frac{\beta^2}{4u}\right)\,du}{W(u,\beta)} \tag{9.20}$$

However, to avoid the cumbersome evaluation of the integrals, we shall present these results using the Laplace transform approach.

The Laplace transform drawdown for the Hantush-Jacob solution is given by Eq. (4.31)

$$\tilde{s} = \frac{Q_w}{2\pi pT}\,K_0(kr) \tag{9.21}$$

The above expression in fact incorporates all the one-aquifer system cases. This includes the Hantush-Jacob case in which (Eq. (4.61))

$$k(T,S,L') = \sqrt{\frac{pS}{T} + \frac{L'}{T}} \tag{9.22}$$

the Hantush case (Eq. (4.59))

$$k(T,S,L',S') = \sqrt{\frac{pS}{T} + \frac{L'}{T}p\tilde{f}(p)} \tag{9.23}$$

and the one-aquifer-two-aquitard case (Eqs. (7.4) or (7.6))

$$k(T, S, L_0', S_0', L_1', S_1') = \sqrt{\frac{pS_1}{T_1} + \frac{L_0'}{T_1} p\tilde{g}_0 + \frac{L_1'}{T_1} p\tilde{g}_1} \qquad (9.24)$$

where

$$p\tilde{f}(p) = \sqrt{\frac{pS'}{L'}} \coth \sqrt{\frac{pS'}{L'}} \qquad (9.25)$$

$$p\tilde{g}(p) = \sqrt{\frac{pS'}{L'}} \tanh \sqrt{\frac{pS'}{L'}} \qquad (9.26)$$

In the above we have explicitly shown the functional dependence of k on the physical parameters.

To find the sensitivity coefficients, we differentiate Eq. (9.21),

$$\frac{\partial \tilde{s}}{\partial T} = -\frac{Q_w}{2\pi pT} \left[\frac{1}{T} K_0(kr) + r\frac{\partial k}{\partial T} K_1(kr) \right] \qquad (9.27)$$

$$\frac{\partial \tilde{s}}{\partial C} = -\frac{Q_w r}{2\pi pT} \frac{\partial k}{\partial C} K_1(kr) \qquad (9.28)$$

where the symbol C stands for one of the following parameters, S, L_0', S_0', L_1', S_1'. The normalized form can be found as

$$\frac{T}{s}\frac{\partial s}{\partial T} = \frac{T\mathcal{L}^{-1}\left\{ \dfrac{\partial \tilde{s}}{\partial T} \right\}}{\mathcal{L}^{-1}\{\tilde{s}\}}$$

$$= -1 - Tr\frac{\mathcal{L}^{-1}\left\{ \dfrac{\partial k}{\partial T}\dfrac{K_1(kr)}{p} \right\}}{\mathcal{L}^{-1}\left\{ \dfrac{K_0(kr)}{p} \right\}} \qquad (9.29)$$

$$\frac{C}{s}\frac{\partial s}{\partial C} = \frac{C\mathcal{L}^{-1}\left\{ \dfrac{\partial \tilde{s}}{\partial C} \right\}}{\mathcal{L}^{-1}\{\tilde{s}\}}$$

$$= -Cr\frac{\mathcal{L}^{-1}\left\{ \dfrac{\partial k}{\partial C}\dfrac{K_1(kr)}{p} \right\}}{\mathcal{L}^{-1}\left\{ \dfrac{K_0(kr)}{p} \right\}} \qquad (9.30)$$

Particularly, for the case of the Hantush-Jacob solution, the above becomes

$$\frac{T}{s}\frac{\partial s}{\partial T} = -1 + \frac{r}{2}\frac{\mathcal{L}^{-1}\left\{\dfrac{k\,K_1\,(k\,r)}{p}\right\}}{\mathcal{L}^{-1}\left\{\dfrac{K_0\,(k\,r)}{p}\right\}} \tag{9.31}$$

$$\frac{S}{s}\frac{\partial s}{\partial S} = -\frac{Sr}{2T}\frac{\mathcal{L}^{-1}\left\{\dfrac{K_1\,(k\,r)}{k}\right\}}{\mathcal{L}^{-1}\left\{\dfrac{K_0\,(k\,r)}{p}\right\}} \tag{9.32}$$

$$\frac{L'}{s}\frac{\partial s}{\partial L} = -\frac{L'r}{2T}\frac{\mathcal{L}^{-1}\left\{\dfrac{K_1\,(k\,r)}{pk}\right\}}{\mathcal{L}^{-1}\left\{\dfrac{K_0\,(k\,r)}{p}\right\}} \tag{9.33}$$

with k given by Eq. (9.22). The right-hand-sides of Eq. (9.31)–(9.33) can be further expressed in terms of dimensionless variables, u and β. To achieve this, the Laplace inversion needs to be performed in dimensionless parameters. Instead of using (t, p) as the transformation pair, we use (t^*, p^*) (see also Sec. 3.10), where

$$t^* = \frac{1}{u} = \frac{4Tt}{r^2S} \tag{9.34}$$

$$p^* = \frac{r^2Sp}{4T} \tag{9.35}$$

We hence can form the counterparts of Eqs. (9.18)–(9.20) as

$$\frac{T}{s}\frac{\partial s}{\partial T} = -1$$

$$+\frac{1}{2}\frac{\mathcal{L}^{-1}\left\{\dfrac{\sqrt{4p^* + \beta^2}\,K_1\left(\sqrt{4p^* + \beta^2}\right)}{p^*}\right\}}{\mathcal{L}^{-1}\left\{\dfrac{K_0\left(\sqrt{4p^* + \beta^2}\right)}{p^*}\right\}} \tag{9.36}$$

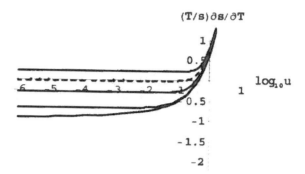

FIGURE 9.6. Sensitivity coefficient $(T/s)\partial s/\partial T$ versus u. Solid lines, from the bottom up: $\beta = 0$, 0.3, 1, and 2; dashed line: $\beta = 1.55$.

$$\frac{S}{s}\frac{\partial s}{\partial S} = -2\frac{\mathcal{L}^{-1}\left\{\dfrac{K_1\left(\sqrt{4p^* + \beta^2}\right)}{\sqrt{4p^* + \beta^2}}\right\}}{\mathcal{L}^{-1}\left\{\dfrac{K_0\left(\sqrt{4p^* + \beta^2}\right)}{p^*}\right\}} \qquad (9.37)$$

$$\frac{L'}{s}\frac{\partial s}{\partial L'} = -\frac{\beta^2}{2}\frac{\mathcal{L}^{-1}\left\{\dfrac{K_1\left(\sqrt{4p^* + \beta^2}\right)}{p^*\sqrt{4p^* + \beta^2}}\right\}}{\mathcal{L}^{-1}\left\{\dfrac{K_0\left(\sqrt{4p^* + \beta^2}\right)}{p^*}\right\}} \qquad (9.38)$$

Similar to the well function, the above expressions are functions of the dimensionless parameters u and β. Hence the sensitivity of the Hantush-Jacob solution can be summarized into three charts, shown as Figures 9.6 to 9.8.

In Figure 9.6 the sensitivity coefficients for transmissivity is shown for five values of β. In interpreting these results, we recall that large u represents large r or small t, and vice versa. First we observe that the bottom curve, corresponding to the case $\beta = 0$, is identical to the solid line in Figure 9.5, for $\beta = 0$ corresponds to the Theis solution (see also Figure 4.12). As β increases, the sensitivity curve shifts

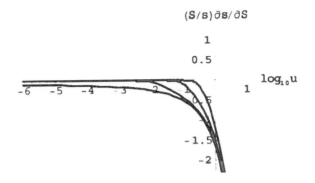

FIGURE 9.7. Sensitivity coefficient $(S/s)\partial s/\partial S$ versus u. Solid lines, from the bottom up: $\beta = 0$, 0.3, 1, and 2.

upward. For the case of $\beta = 2$, the sensitivity coefficient becomes entirely positive. In that range, which generally corresponds to cases with large aquitard leakance values, an increase in transmissivity increases drawdown. The crossing that marks the reversal of trend takes place around $\beta = 1.55$. As shown in dashed line in Figure 9.6, the sensitivity for this case is nearly zero for the most part of u. This reversal of trend is caused by two opposing effects. As observed from the $1/T$ term in front of the drawdown solution, an increase of transmissivity reduces the drawdown. On the other hand, the leakage received through the aquitard, characterized by the parameter β, becomes smaller as T becomes larger. Smaller leakage implies larger drawdown, as evident from the Hantush-Jacob type curves, Figure 4.11.

We next examine the drawdown sensitivity to aquifer storativity. Figure 9.7 displays the dimensionless sensitivity coefficient for four β values. We observe that the sensitivity is generally smaller than that caused by transmissivity. The bottom curve, corresponding to the case $\beta = 0$, is identical to the dashed curve in Figure 9.5, as expected. We also find that when the leakance is large (β large), the sensitivity caused by storativity variation decreases.

Finally, the effect of leakance is investigated. Figure 9.8 plots the normalized sensitivity coefficient for four β values. First, we observe that all the curves are in the negative range, meaning that an increase

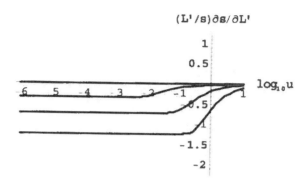

FIGURE 9.8. Sensitivity coefficient $(L'/s)\partial s/\partial L'$ versus u. Solid lines, from the top down: $\beta = 0$, 0.3, 1, and 2.

in leakance causes a decrease in drawdown. For a given set of aquifer parameters, the sensitivity is approximately a constant for nearly all ranges of distance and time. The magnitude of the sensitivity is observed to be quite large, comparable to that of the transmissivity effect. As the leakance decreases, however, the sensitivity decreases. For the case $\beta = 0$, meaning no leakage, the curve coincides with the abscissa.

In the above we have given a general discussion of drawdown sensitivity to various aquifer and aquitard parameters. For the two cases discussed, the Theis solution and the Hantush-Jacob solution, the results are presented in dimensionless forms. This reduction of parameters allows the entire sensitivity behavior to be condensed into a few charts, shown as Figures 9.5–9.8. The analysis can be extended to multilayered systems,[71] but the analysis is more complex.

Chapter 10
PARAMETER DETERMINATION

10.1 Inverse Problem

In previous chapters, we have devoted our attention mainly to *forward problems* (or *direct problems*, *forecast problems*). We assume that the mathematical *system* together with its *parameters* that govern a physical phenomenon are known. Our task is to solve the system by analytical or numerical means to produce a *forecast*. For example, the Theis solution is a mathematical system that predicts drawdown in aquifers. The drawdown values are controlled by the physical parameters defining the system, such as transmissivity, storativity, and pumping rate. Given the full set of parameter values, a drawdown forecast over space and time can be made.

Inverse problems (or *identification problems*, *calibration problems*), on the other hand, are aimed at determining unknown parameters, or even the system itself, based on observed outcomes of the system. In groundwater flow, drawdown data can be collected in wells over time and space. Given the operation condition such as the pumping rate, and the boundary conditions of known head and flow rate, we seek to determine the aquifer transmissivity and storativity. On other occasions, the location of a boundary, such as the existence of an impermeable fault, is not known, and is to be searched. It is also possible that we need to identify not only aquifer parameters, but also the system itself. For example, is the aquifer confined or leaky? Is a certain set of governing equations appropriate for the aquifer system? This is known as *system identification*.

These problems described above are generally known as inverse problems. Particularly, if the quantities to be determined are physical parameters, it is called a *parameter determination* (or *identification*) *problem*. In this chapter, we focus on aquifer parameter determination problems only.

In real-world applications, the inverse problem typically precedes the forward problem, because the system and its parameters must first be identified before any forecast can be made. Inverse problems are more difficult to solve. In most inverse solution algorithms, a forward solution technique is required to continuously produce trial solutions based on optimized guesses. Hence an efficient forward solution capability must exist before the inverse solution becomes feasible.

Another reason that the inverse problem is difficult is its general ill posedness in terms of stability and uniqueness of mathematical solutions. In practical problems, the data collected are sometimes not sufficient in quantity, or not sensitive enough to identify certain parameters. Other times the observed data may contain large measurement errors as well as other kinds of errors. To minimize the impact of errors associated with a few data points, a great many more data points are collected. A theoretical solution that satisfies all these data cannot exist. Hence a different interpretation of the solution based on the minimization of discrepancy must be taken.

Even if the problem has a solution, there may exist a multiplicity of solutions making the parameters unidentifiable in the presence of errors. In other words, one may find a large number of plausible solutions, each of which is slightly better or worse than the other in terms of minimizing the discrepancy. If the differences among the solutions are within the noise level of data collected, there is no physical ground of selecting one solution over the other.

When the problem is identifiable, there can exist numerical difficulties in solving the system. Most numerical techniques rely on local gradients in search of a correct direction to make subsequent guesses. One can easily lands on a location where the local gradients send the search into a divergent direction, or to a convergent one corresponding to a local, rather than global, minimum. All of these issues add to the complexity of inverse problems. For groundwater flow, some of the implications are illustrated below.

Take, for example, the Theis solution, Eq. (3.91). Given the pumping rate and drawdown data, we seek to determine the two physical parameters T and S. From mathematical principles, only two drawdown data, taken at different times or radial distances, are needed. It is possible that these two data cannot simultaneously satisfy the Theis solution. For instance, the drawdown taken at a later time is smaller than that at an earlier time at the same location. Then there is no solution. However, if the aquifer system is indeed governed by the Theis solution, then these two data should uniquely define a set of parameters.

The fact that the system has a solution does not mean that it can be easily found. One possible solution procedure is to use the graphical Theis type curve method, as explained in Sec. 3.8. A second possibility utilizes the asymptotic behavior of the Theis solution. As we observe from the Cooper-Jacob approximation, Eq. (3.99), the transmissivity can be explicitly solved from two drawdown data using either Eq. (3.102) or (3.104). Storativity then emerges from Eq. (3.99). Hence we have achieved the parameter inverse.

The above procedure, although theoretically possible, is unacceptable in practice. As any field hydrogeologist knows, pumping test data are prone to errors. The common practice is to sample drawdown data for a period of time to gather a great many more data points than the minimum requirement of two. The mathematical problem then becomes *over-determined*. If all data are taken without error, then we have redundancy, as any pair of data will produce the same set of aquifer parameters. This is however not the case. Therefore, a mathematical solution in the ordinary sense does not exist. The definition of an inverse solution must be modified. In this case, it means that we seek to find the solution of "best-fit."

The traditional technique of finding the best-fit solution is by visual inspection using the graphical type-curve procedure described in Sec. 3.8. The type curve is moved such that data are evenly scattered around the theoretical curve. A large number of type curves exist for different aquifer assumptions. A few of them were examined in Chapters 3 and 4, and more can be found in books on this subject.[10,117,119] This solution procedure is however subjective. Two hydrogeologists will not come up with exactly the same answer given the same set of data, although the solutions should not be far away. To have the consistency of the solution, and to be able to have an automated procedure, it is desirable to rigorously redefine the solution. This is accomplished through the theory of *optimization*, which is discussed in the next section.

10.2 Nonlinear Least Square

In an over-determined system, the solution in the traditional sense does not exist. A condition based on the minimization of error can be introduced to redefine the problem to achieve the existence and uniqueness of the solution. This concept is illustrated below using the pumping test as an example.

Assume that we are given a set of discrete drawdown data observed at different location and time pairs (r_i, t_i),

$$\hat{s}(r_i, t_i) = \hat{s}_i; \qquad i = 1, \cdots, n \tag{10.1}$$

in a pumped aquifer. In practice, the location is typically fixed at an observation well, and the time varies. Sometimes, however, several observation wells are used. All these time and spatial data can be lumped together.

We first need to judge that the aquifer response obeys a certain pumping well drawdown theory, such as the Theis solution, the Hantush-Jacob solution, etc. The theoretical drawdown solution is given as $s(\mathbf{p}; r, t)$, in which $\mathbf{p} = (p_1, p_2, \ldots, p_m)$ is the parameter vector to be identified. For the Theis theory, we have $\mathbf{p} = (T, S)$; for the Hantush-Jacob theory, $\mathbf{p} = (T, S, L')$, etc.

To quantify the discrepancy between the theoretical and the measured drawdown, a *square error* is defined as

$$E(\mathbf{p}) = \sum_{i=1}^{n} [s(\mathbf{p}; r_i, t_i) - \hat{s}(r_i, t_i)]^2 \tag{10.2}$$

In the above, the individual error from each data point is squared and summed. The squaring ensures that the error accumulates, instead of canceling out. For the purpose of *optimization*, E is called an *objective function*. Our objective is to minimize E, corresponding to the discrepancy with the theoretical solution, with respect to the physical parameters $\mathbf{p}$, known as *design variables*.

We observe that once the objective function, and our decision to minimize it, is defined, the solution can become unique. However, there may exist multiple ways to define the objective function. In fact, we can propose a different square error to minimize. To be compatible with the traditional type-curve method, which is performed on a log-log plot using visual techniques, it may be more desirable to minimize the square error of logarithmic drawdown, $\log_{10} s$, instead of the drawdown itself. Hence it can be proposed:

$$E_\ell(\mathbf{p}) = \sum_{i=1}^{n} w_i \left[\log_{10} s(\mathbf{p}; r_i, t_i) - \log_{10} \hat{s}(r_i, t_i) \right]^2 \qquad (10.3)$$

In the above, we have further introduced the data weights, w_i, that allow user intervention. The weights can be adjusted to diminish the influence of certain data points, or to discard outliers in the data. Or we can enhance the weight of a certain group of the data, such as large time or small time data. This mimics the intuition that is practiced by an experienced hydrogeologist.

For $E(\mathbf{p})$ to be minimized with respect to a set of design variables, $p_j, j = 1, \ldots, m$, the following conditions must be satisfied:

$$\frac{\partial E}{\partial p_j} = 0; \qquad j = 1, \ldots, m \qquad (10.4)$$

For simplicity, we shall use Eq. (10.2) as the objective function. It can be differentiated to yield

$$\sum_{i=1}^{n} \left[s(\mathbf{p}; r_i, t_i) - \hat{s}(r_i, t_i) \right] \frac{\partial s(\mathbf{p}; r_i, t_i)}{\partial p_j} = 0; \qquad j = 1, \ldots, m \qquad (10.5)$$

These are m equations defining the m unknowns, p_j.

We note that the conditions in Eq. (10.5) only guarantee that E takes an extreme value, either a minimum, or a maximum. For E

to be a minimum, as required in the present solution, the *Hessian matrix*, defined as

$$
\mathbf{H} =
\begin{bmatrix}
\frac{\partial^2 E}{\partial p_1^2} & \frac{\partial^2 E}{\partial p_1 \partial p_2} & \cdots & \frac{\partial^2 E}{\partial p_1 \partial p_m} \\
\frac{\partial^2 E}{\partial p_2 \partial p_1} & \frac{\partial^2 E}{\partial p_2^2} & \cdots & \frac{\partial^2 E}{\partial p_2 \partial p_m} \\
\vdots & \vdots & \ddots & \vdots \\
\frac{\partial^2 E}{\partial p_m \partial p_1} & \frac{\partial^2 E}{\partial p_m \partial p_2} & \cdots & \frac{\partial^2 E}{\partial p_m^2}
\end{bmatrix}
\tag{10.6}
$$

must be *positive definite*. In other words, the determinants of the submatrices

$$
\mathbf{H}_i =
\begin{bmatrix}
\frac{\partial^2 E}{\partial p_1^2} & \frac{\partial^2 E}{\partial p_1 \partial p_2} & \cdots & \frac{\partial^2 E}{\partial p_1 \partial p_i} \\
\frac{\partial^2 E}{\partial p_2 \partial p_1} & \frac{\partial^2 E}{\partial p_2^2} & \cdots & \frac{\partial^2 E}{\partial p_2 \partial p_i} \\
\vdots & \vdots & \ddots & \vdots \\
\frac{\partial^2 E}{\partial p_i \partial p_1} & \frac{\partial^2 E}{\partial p_i \partial p_2} & \cdots & \frac{\partial^2 E}{\partial p_i^2}
\end{bmatrix} ; \quad i = 1, 2, \ldots, m
\tag{10.7}
$$

are all positive,

$$
|\mathbf{H}_i| > 0; \quad i = 1, 2, \ldots, m
\tag{10.8}
$$

In this case, a minimum is found.

Once the extremum is shown to be a minimum, we need to be aware that it may only be a *local minimum*, not a *global minimum*, as many local minima can exist. To reach the correct solution, we need to have the insight to make an initial guess close to the true solution. Otherwise we may solve the problem with a number of initial trials and select the best among the solutions.

10.3   Optimization Techniques

A large number of optimization methods exist that can minimize (or maximize) objective functions. We shall discuss only a few in our effort of aquifer parameter determination. Generally speaking, we can classify optimization algorithms into three categories according to the type of mathematical information required:[112]

1. *Search method*: It requires the evaluation of only the objective function.

2. *Gradient method*: It requires the objective function, as well as its gradients.

3. *Second order method*: Further to the above, the second derivatives are needed.

The search method is most straightforward. If the objective function is easy to evaluate, it can be calculated for a large number of selected parameter values with a tolerable amount of computer CPU time. The set that gives the minimum value of the objective function is considered the solution. If the number of dimensions of the parameter space is small, say one to two dimensions, and the search range is finite, we can practically perform a "complete" search for a given resolution. (A true complete search is not possible as the parameter space is continuous.) In other words, the objective function can be evaluated for a large number of points in the parameter space. It can be presented as a curve for the one-parameter case, and a contour map for the two-parameter case. The minimum can easily be identified. We shall demonstrate in the next section that it is possible to perform such a search for the Theis solution, which has two parameters.

If the objective function is costly to evaluate, the exhaustive search cannot be afforded, and more efficient algorithms need to be devised. In this case, methods like *bracketing, golden section search, Fibonacci section search*, etc. can be used in a one-dimensional search.[101] These search methods assume that there exists a simple minimum within the search brackets.

As the number of parameters increases, and the objective function map becomes more complex due to a large number of local minima, an advanced search technique is needed. Due to the impossibility of having a complete search, and the futility of gradient search which often settles into a local rather than global minimum, a new generation of probability and evolution based search methods has emerged. We shall mention only one method based on the idea of natural evolution of species, known as *Genetic Algorithms*.[47,68,94] The concept is based on the initial selection of a relatively small *population*. Each

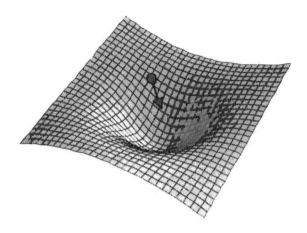

FIGURE 10.1. Method of steepest descent.

individual in the population represents a possible solution in the parameter space. The fitness of each individual is determined by the value of the objective function calculated based on that set of parameters. The natural evolutionary processes of *reproduction, crossover, mutation*, and *selection*, are applied using probability rules to evolve the new and better generations. At the same time of the selection based on fitness, some less fit individuals are allowed to survive. By maintaining a diverse population, genetic algorithms slow down the convergence process of gradient methods, which almost invariably settles into a local minimum and misses the global minimum. In a search space where there exists a large number of minima, the evolution based search algorithms claim to find much better near-optimal solution than any other method.

Next we discuss the gradient methods. If we are near the global minimum, and behavior of the objective function is not complex near that region, gradient methods are highly efficient in guiding the trial solution into convergence. One of the gradient methods is known as the *methods of steepest descent*. Assume an objective function defined by two parameters, p_1 and p_2. Figure 10.1 shows a surface plot of the objective function E near the minimum. The current trial solution is marked as a dot on the surface. We want to find the direction of next trial that will lead us to the bottom of the valley most rapidly. A most reasonable try is to utilize the negative gradient of the objective function

$$-\nabla E = -\frac{\partial E}{\partial p_1}\mathbf{i} - \frac{\partial E}{\partial p_2}\mathbf{j} \tag{10.9}$$

because this vector points to the steepest downhill direction, as illustrated in Figure 10.1. A certain empirical formula is needed to determine a proper step size that can guide the solution smoothly downhill to the bottom of the valley. For the ease of visualization, the above presentation is done in a two-parameter space. In general cases, the surface plot cannot be presented. The gradient and the descent into the minimum is conducted in an abstract sense.

Despite the simplicity of the concept, the method of steepest descent does not work well for highly distorted gradient field, i.e. for a valley that is long and narrow. For problems involving aquifer parameter determination, this happens to be the case. As we have learned in the preceding chapter on sensitivity analysis, the drawdown is much more sensitive to transmissivity than to storativity. Certain modification of the scheme is desirable. This is often accomplished by the *conjugate gradient method*. Detail of this and other methods can be found in books on numerical methods and optimization.[101,112]

Next, the second order methods, which use not only the gradient, but also the second derivatives to guide the search, are discussed. A popular method is the *BFGS (Broyden-Fletcher-Goldfarb-Shanno) method*, which is presented here and will be used for the parameter determination of the Hantush-Jacob aquifer.

Assume that in the process of successive trials in search of the optimum, we are at the n-th trial location given by parameter vector $\mathbf{p}_n$. The true solution is located at $\mathbf{p}$, which is a small distance away. We can approximate the objective function at $\mathbf{p}$ by a multiple-variable Taylor series expansion including up to the second order terms:

$$E(\mathbf{p}) \approx E(\mathbf{p}_n) + \Delta\mathbf{p}_n^{\mathrm{T}} \cdot \nabla E(\mathbf{p}_n) + \frac{1}{2}\Delta\mathbf{p}_n^{\mathrm{T}} \cdot \mathbf{H}(\mathbf{p}_n) \cdot \Delta\mathbf{p}_n \tag{10.10}$$

where $\mathbf{p}$, $\mathbf{p}_n$, and $\Delta\mathbf{p}_n$ are column matrices, the superscript $(.)^{\mathrm{T}}$ denotes matrix transpose, $\mathbf{H}$ is the Hessian matrix defined in Eq. (10.6), and

$$\Delta\mathbf{p}_n = \mathbf{p} - \mathbf{p}_n \qquad (10.11)$$

For readers who are not familiar with the matrix notation, Eq. (10.10) can be expressed for the two-variable case as

$$E(p_1, p_2) \approx E(p_{1(n)}, p_{2(n)})$$

$$+ \frac{\partial E(p_{1(n)}, p_{2(n)})}{\partial p_{1(n)}} \Delta p_{1(n)} + \frac{\partial E(p_{1(n)}, p_{2(n)})}{\partial p_{2(n)}} \Delta p_{2(n)}$$

$$+ \frac{1}{2} \left[\frac{\partial^2 E(p_{1(n)}, p_{2(n)})}{\partial p_{1(n)}^2} \Delta p_{1(n)}^2 + 2 \frac{\partial^2 E(p_{1(n)}, p_{2(n)})}{\partial p_{1(n)} \partial p_{2(n)}} \Delta p_{1(n)} \Delta p_{2(n)} \right.$$

$$\left. + \frac{\partial^2 E(p_1^{(n)}, p_2^{(n)})}{\partial p_{2(n)}^2} \Delta p_{2(n)}^2 \right] \qquad (10.12)$$

We next take the gradient of Eq. (10.10) with respect to $\mathbf{p}$, to obtain

$$\nabla E(\mathbf{p}) \approx \nabla E(\mathbf{p}_n) + \mathbf{H}(\mathbf{p}_n) \cdot \Delta\mathbf{p}_n \qquad (10.13)$$

In the above we note that $\mathbf{p}_n$ is regarded as constant when differentiating with respect to $\mathbf{p}$. The condition that $\mathbf{p}$ is a minimum is given by (see also Eq. (10.4))

$$\nabla E(\mathbf{p}) = 0 \qquad (10.14)$$

We can then solve Eq. (10.13) for $\Delta\mathbf{p}_n$,

$$\Delta\mathbf{p}_n = -\mathbf{H}^{-1}(\mathbf{p}_n) \cdot \nabla E(\mathbf{p}_n) \qquad (10.15)$$

where the superscript $(.)^{-1}$ denotes matrix inverse. $\Delta\mathbf{p}$ hence is the increment for the next trial. In other words, the iterative formula

$$\begin{aligned} \mathbf{p}_{n+1} &= \mathbf{p}_n + \Delta\mathbf{p}_n \\ &= \mathbf{p}_n - \mathbf{H}^{-1}(\mathbf{p}_n) \cdot \nabla E(\mathbf{p}_n) \end{aligned} \qquad (10.16)$$

can be used to update the parameters until they converge. This method is called *Newton's method*.

Although Newton's method converges very fast near the minimum, it has several disadvantages. One disadvantage is that it requires second derivatives of the objective function. When the second derivatives are not analytically available, numerically evaluating them at every iteration step can be inaccurate and time consuming. The major difficulty, however, is that when the trial solution p_n is still far away from the true minimum, the Hessian matrix may not be positive definite. In other words, the search sequence may not converge. Alternative methods, known as *quasi-Newton methods*, are often used instead.

In quasi-Newton methods, the inverse of Hessian matrix H^{-1} is replaced by a symmetric positive definite matrix G_n which evolves with iteration steps, and eventually is expected to approximate H^{-1} near the true minimum. Hence

$$p_{n+1} = p_n - G_n \cdot \nabla E(p_n) \tag{10.17}$$

Different selection of G_n leads to different quasi-Newton method. Particularly, the BFGS method uses[112]

$$G_1 = I \tag{10.18}$$

$$G_{n+1} = G_n + \left(1 + \frac{\Delta g_n^T \cdot G_n \cdot \Delta g_n}{\Delta p_n^T \cdot \Delta g_n}\right) \frac{\Delta p_n \cdot \Delta p_n^T}{\Delta p_n^T \cdot \Delta g_n}$$

$$- \frac{\Delta p_n \cdot \Delta g_n^T \cdot G_n + G_n \cdot \Delta g_n \cdot \Delta p_n^T}{\Delta p_n^T \cdot \Delta g_n} \tag{10.19}$$

where I is the identity matrix, and

$$\Delta g_n = \nabla E(p_{n+1}) - \nabla E(p_n) \tag{10.20}$$

This completes the definition of the iteration steps of the BFGS scheme.

We shall not go into further detail of the BFGS scheme, and neither other methods for optimization. As canned subroutines[101] will be used for the present purposes, an understanding of the concept is sufficient.

10.4 Theis Aquifer

For the Theis solution, the sum of square error between the theoretical prediction and the observed drawdown can be expressed as

$$E(T, S) = \sum_{i=1}^{n} \left[\frac{Q_w}{4\pi T} W \left(\frac{r_i^2 S}{4 T t_i} \right) - \hat{s}(r_i, t_i) \right]^2 \qquad (10.21)$$

based on Eq. (10.2). Given the observed drawdown $\hat{s}_i$ and pumping rate Q_w, the square error E is a function of T and S only. The objective is to search the pair of T and S values that minimizes the objective function E. Because the search space is only two-dimensional, and the function is easy to evaluate, it is possible to perform an enumeration of all possible solutions by specifying a certain resolution of discretization. The optimum can then be found. This is accomplished by plotting the values of E on the T-S plane as a two-dimensional contour plot, or a three-dimensional surface plot. The *Mathematica* functions *ContourPlot* and *Plot3D* are ideally suited for this purpose. Similarly, we can use a *Fortran* program for the evaluation of E by scanning a range of T and S values. A plotting software is needed for the graphical presentation. The location of the minimum E can be visually inspected. No sophisticated inverse technique is needed. With the computing power nowadays, this evaluation and plotting can be performed within seconds using a microcomputer.

Example: *The following drawdown data from a pumping test were reported in Walton,*[118]

t (min)	3	5	8	12	20	24	30	38	47	50
$\hat{s}$ (ft)	0.3	0.7	1.3	2.1	3.2	3.6	4.1	4.7	5.1	5.3

t (min)	60	70	80	90	100	130	160	200	260
$\hat{s}$ (ft)	5.7	6.1	6.3	6.7	7.0	7.5	8.3	8.5	9.2

t (min)	320	380	500
$\hat{s}$ (ft)	9.7	10.2	10.9

The pumping rate is $Q_w = 29.4$ ft^3/min, and the radial distance to observation well is $r = 824$ ft. Identify the aquifer parameters.

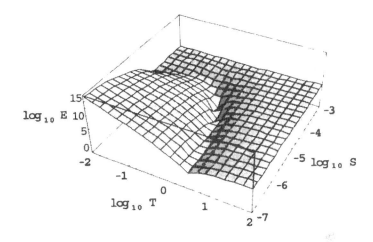

FIGURE 10.2. Plot of least square error on the T-S plane.

A *Mathematica* macro *WInv.m* is presented in Appendix C, which has programmed a function for contour plot (*ContourPlotLSqError*[...]) and one for surface plot (*Plot3DLSqError*[...]). A data file *PumpData.m* is needed for data input.

In Figure 10.2 we plot the least square error defined in Eq. (10.21) in logarithmic scale ($\log_{10} E$), versus $\log_{10} T$ and $\log_{10} S$. We have presented the result in a relatively wide range, $10^{-2} < T < 10^2$ (in ft^2/min) and $10^{-7} < S < 10^{-3}$. The presence of a minimum is clearly observed. Based on the visual observation, we narrow the range of search to $10^{-1} < T < 10^1$ and $10^{-6} < S < 10^{-4}$ and plot the result in contour lines in Figure 10.3. By zooming into the minimum a few more times, we settle for the values $T = 0.92$ ft^2/min and $S = 2.1 \times 10^{-5}$ as the solution. To check the resultant match, we plot the drawdown based on the Theis solution using the predicted transmissivity and storativity values, which is compared with the field observation, in Figure 10.4.

Although the above ad-hoc inverse procedure based on enumeration and visual inspection works well, it is still desirable to present

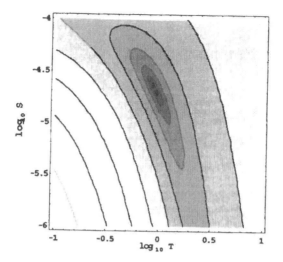

FIGURE 10.3. Contour plot identifying the minimum least square error in the T-S map.

below a formal inverse procedure that uses less human intervention, and is more suitable for finding a solution with high accuracy.

We return to the least square method as described in Sec. 10.2. The parameters T and S are to be solved from Eq. (10.5). For the present case, it becomes

$$\sum_{i=1}^{n} [s(T,S;r_i,t_i) - \hat{s}(r_i,t_i)] \frac{\partial s(T,S;r_i,t_i)}{\partial T} = 0 \quad (10.22)$$

$$\sum_{i=1}^{n} [s(T,S;r_i,t_i) - \hat{s}(r_i,t_i)] \frac{\partial s(T,S;r_i,t_i)}{\partial S} = 0 \quad (10.23)$$

The two sensitivity coefficients $\partial s/\partial T$ and $\partial s/\partial S$ in the above have been obtained in Eqs. (9.5) and (9.6). With the substitution and some simplification, Eqs. (10.22) and (10.23) become

$$\sum_{i=1}^{n} \left[\frac{Q_w}{4\pi T} W \left(\frac{r_i^2 S}{4Tt_i} \right) - \hat{s}(r_i,t_i) \right] \cdot$$
$$\left[\exp \left(-\frac{r_i^2 S}{4Tt_i} \right) - W \left(\frac{r_i^2 S}{4Tt_i} \right) \right] = 0 \quad (10.24)$$

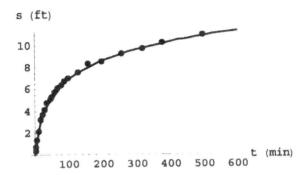

FIGURE 10.4. Predicted (solid line) and observed (dots) drawdown.

$$\sum_{i=1}^{n} \left[\frac{Q_w}{4\pi T} W \left(\frac{r_i^2 S}{4 T t_i} \right) - \hat{s}(r_i, t_i) \right] \exp \left(-\frac{r_i^2 S}{4 T t_i} \right) = 0 \qquad (10.25)$$

We can simplify Eq. (10.24) by subtracting Eq. (10.25) from it to obtain

$$\sum_{i=1}^{n} \left[\frac{Q_w}{4\pi T} W \left(\frac{r_i^2 S}{4 T t_i} \right) - \hat{s}(r_i, t_i) \right] W \left(\frac{r_i^2 S}{4 T t_i} \right) = 0 \qquad (10.26)$$

The pair of nonlinear equations (10.25) and (10.26) can be used to solve for T and S.

Rather than solving these two equations, we find it possible to further reduce the system. Equation (10.25) can be written as

$$\frac{Q_w}{4\pi T} \sum_{i=1}^{n} W \left(\frac{r_i^2 S}{4 T t_i} \right) \exp \left(-\frac{r_i^2 S}{4 T t_i} \right) = \sum_{i=1}^{n} \hat{s}(r_i, t_i) \exp \left(-\frac{r_i^2 S}{4 T t_i} \right) \qquad (10.27)$$

where we have extracted the factor $Q_w/4\pi T$ out of the summation sign. We then solve for

$$\frac{Q_w}{4\pi T} = \sum_{i=1}^{n} \hat{s}(r_i, t_i) \exp \left(-\frac{r_i^2 S}{4 T t_i} \right) \bigg/ \sum_{i=1}^{n} W \left(\frac{r_i^2 S}{4 T t_i} \right) \exp \left(-\frac{r_i^2 S}{4 T t_i} \right) \qquad (10.28)$$

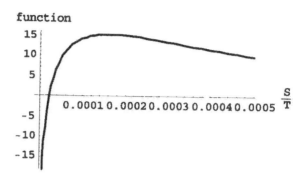

FIGURE 10.5. Plot of the nonlinear least square equation, Eq. (10.29).

The above equation can be used to eliminate the same term in Eq. (10.26) to obtain

$$\sum_{i=1}^{n} \left[\frac{\sum_{i=1}^{n} \hat{s}(r_i, t_i) \exp\left(-\frac{r_i^2 S}{4Tt_i}\right)}{\sum_{i=1}^{n} W\left(\frac{r_i^2 S}{4Tt_i}\right) \exp\left(-\frac{r_i^2 S}{4Tt_i}\right)} W\left(\frac{r_i^2 S}{4Tt_i}\right) - \hat{s}(r_i, t_i) \right]$$

$$\cdot W\left(\frac{r_i^2 S}{4Tt_i}\right) = 0 \qquad (10.29)$$

Equation (10.29) is a single nonlinear equation containing a single unknown S/T. A *Mathematica* function *FindRoot*, or a *Fortran* program based on a root finding algorithm, such as *Newton-Raphson method*,[101] can be used to find the root. Once S/T is found, it can be substituted into Eq. (10.28) to solve for T. S is then easily calculated.

Example: *For the same pumping test problem presented in the preceding example, find the aquifer parameters by the nonlinear least square method.*

Before finding the root of Eq. (10.29), we first attempt to visualize the solution by enumeration. Since Eq. (10.29) is a function of a single variable S/T, we can plot it as a curve shown in Figure 10.5. The solution is located at where the curve intersects the abscissa.

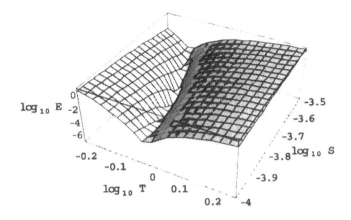

FIGURE 10.6. Least square error plotted on the T-S plane.

We can visually find the solution by zooming into the region a few times. This can be done with ease using a tool like *Mathematica*.

However, our goal is to have an automated procedure. Hence the *FindRoot* function of *Mathematica* is applied to in Eq. (10.29). We find $S/T = 2.28 \times 10^{-5}$. Substituting this value into Eq. (10.28), we obtain $T = 0.920$ ft^2/min. S is then calculated as 2.09×10^{-5}. This procedure is programmed into a single *Mathematica* function *ParInv*[...], as included in the *WInv.m* package (Appendix C).

Example: *A pumping test yields the following drawdown data:*[116]

t (min)	1	1.5	2	2.5	3	4	5	6	8
$\hat{s}$ (m)	0.2	0.27	0.3	0.34	0.37	0.41	0.45	0.48	0.53

t (min)	10	12	14	18	24	30	40	50
$\hat{s}$ (m)	0.57	0.6	0.63	0.67	0.72	0.76	0.81	0.85

t (min)	60	80	100	120	150	180	210	240
$\hat{s}$ (m)	0.9	0.93	0.96	1.0	1.04	1.07	1.1	1.12

The pumping rate is $Q_w = 1.736$ m^3/min, and the observation well is 60 m from the pumping well. Find the transmissivity and storativity.

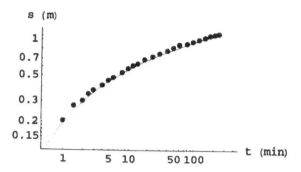

FIGURE 10.7. Match of observed and predicted drawdown.

The *Plot3DLSqError* function is used to produce the three-dimensional surface plot in Figure 10.6. We have zoomed in sufficiently to see the detail near the minimum. We observe that the error drops rather sharply in the transmissivity direction, while it varies slowly along the storativity direction. This behavior is consistent with the drawdown sensitivity analysis with respect to these two parameters, as discussed in Sec. 4.2.

The more accurate result is found using the *ParInv* function. We find $T = 0.790$ m^2/min, and $S = 1.93 \times 10^{-4}$. The final match is presented in Figure 10.7 in log-log scale to enhance the visualization of small time data.

10.5   Hantush-Jacob Solution

For pumping test in a Hantush-Jacob leaky aquifer, we adopt E_ℓ, as defined in Eq. (10.3) as the objective function to minimize:

$$E_\ell(T, S, \beta) = \sum_{i=1}^{n} [\log_{10} s(T, S, \beta; r, t_i) - \log_{10} \hat{s}(r, t_i)]^2 \quad (10.30)$$

where the weights have been taken as unity, and

$$s(T, S, \beta; r, t_i) = \frac{Q_w}{4\pi T} W(u_i, \beta) \quad (10.31)$$

and

$$u_i = \frac{r^2 S}{4T t_i} \tag{10.32}$$

There are two reasons that E_ℓ, rather than E in Eq. (10.2), is used. First, due to the presentation of type curves in log-log plots, the graphical matching procedure essentially minimizes the error of drawdown in logarithmic scale. Second, the gradient of E can be extremely large on parts of the parameter map. This can be demonstrated by observing the Theis case. Figures 10.2 and 10.3 present E over a range of parameters. The "terrain" becomes so steep that we have to suppress it by presenting the result in $\log_{10} E$. The function E_ℓ, on the other hand, is better behaved.

We note that in Eq. (10.30) we have chosen

$$\beta = \sqrt{\frac{K'}{b'T}}\, r \tag{10.33}$$

as the third parameter to determine, in order to be consistent with the conventional Hantush-Jacob type curve solution (Sec. 4.6). Because β incorporates the radial distance, only data from a single observation well can be used, which is the same for the type procedure. On the other hand, if a physical parameter such as the leakance

$$L' = \frac{K'}{b'} \tag{10.34}$$

is chosen, data from different observation wells can be lumped and used.

Due to the extra parameter involved, the least square minimization procedure as shown in Eq. (10.5) gives three nonlinear equations. Following a procedure similar to that in Eqs. (10.27)–(10.29), the three equations can be reduced to two. Although finding the root of a single nonlinear equation using a gradient method, such as the *bracketed Newton-Raphson method*, is relatively easy, finding roots of a system of nonlinear equations is not as reliable.[101] Methods based on optimization perform better in this case. Hence the BFGS method presented in Sec. 10.3 will be used for parameter determination.

Two pieces of information are needed in a quasi-Newton method: the analytical expressions of the objective function, and its derivatives. Before presenting these expressions, we note that the practical ranges of groundwater flow parameters cover several orders of magnitude. When presented in regular scale, the parameter space is very tight in small value ranges, where the objective function can behave in a singular manner. To have a smooth map in graphical presentation, such as Figures 10.2 and 10.3, we have presented the parameter space in logarithmic scale. This suggests that to have a stable, convergent search, the search space should be converted to logarithmic parameters.

We define three logarithmic parameters as

$$
\begin{aligned}
T_\ell &= \log_{10} T \\
S_\ell &= \log_{10} S \\
\beta_\ell &= \log_{10} \beta
\end{aligned}
\tag{10.35}
$$

The objective function according to Eqs. (10.30) and (10.31) is

$$
E_\ell(T_\ell, S_\ell, \beta_\ell) = \sum_{i=1}^{n} \left\{ \log_{10} \left[\frac{Q_w}{4\pi T} W\left(u_i, \beta\right) \right] - \log_{10} \hat{s}(r, t_i) \right\}^2
\tag{10.36}
$$

Its derivatives are

$$
\begin{aligned}
\frac{\partial E_\ell}{\partial T_\ell} &= -2 \sum_{i=1}^{n} \left\{ \log_{10} \left[\frac{Q_w}{4\pi T} W\left(u_i, \beta\right) \right] - \log_{10} \hat{s}(r, t_i) \right\} \\
&\quad \left[W\left(u_i, \beta\right) - \exp\left(-u_i - \frac{\beta^2}{4u_i} \right) \right] \Big/ W\left(u_i, \beta\right) \quad (10.37) \\
\frac{\partial E_\ell}{\partial S_\ell} &= -2 \sum_{i=1}^{n} \left\{ \log_{10} \left[\frac{Q_w}{4\pi T} W\left(u_i, \beta\right) \right] - \log_{10} \hat{s}(r, t_i) \right\} \\
&\quad \exp\left(-u_i - \frac{\beta^2}{4u_i} \right) \Big/ W\left(u_i, \beta\right) \quad (10.38)
\end{aligned}
$$

$$\frac{\partial E_\ell}{\partial \beta_\ell} = -2 \sum_{i=1}^{n} \left\{ \log_{10} \left[\frac{Q_w}{4\pi T} W(u_i, \beta) \right] - \log_{10} \hat{s}(r, t_i) \right\}$$
$$\frac{\beta^2}{2} D(u_i, \beta) \Big/ W(u_i, \beta) \qquad (10.39)$$

where

$$D(u_i, \beta) = \int_{u_i}^{\infty} \frac{1}{u_i^2} \exp\left(-u_i - \frac{\beta^2}{4u_i}\right) du_i \qquad (10.40)$$

Here we note that $W(u_i, \beta)$ is evaluated by the numerical Laplace inversion as (see Eq. (4.37)):

$$W(u_i, \beta) = W\left(\frac{1}{t_i^*}, \beta\right) = \mathcal{L}^{-1}\left\{ \frac{2}{p_i^*} K_0\left(\sqrt{4p_i^* + \beta^2}\right) \right\} \qquad (10.41)$$

and $D(u_i, \beta)$ can be found from

$$D(u_i, \beta) = D\left(\frac{1}{t_i^*}, \beta\right) = \mathcal{L}^{-1}\left\{ \frac{4K_1\left(\sqrt{4p_i^* + \beta^2}\right)}{p_i^* \sqrt{4p_i^* + \beta^2}} \right\} \qquad (10.42)$$

We thus have all the information needed to apply the BFGS quasi-Newton optimization scheme.

Example: *A pumping test conducted in a leaky aquifer yielded the following drawdown data:*[119]

t (min)	5	28	41	60	75	244	493	669
$\hat{s}$ (ft)	0.76	3.3	3.59	4.08	4.39	5.47	5.96	6.11

t (min)	958	1129	1185
$\hat{s}$ (ft)	6.27	6.4	6.42

The well discharge is $Q_w = 3.34$ ft³/min, and the radial distance of observation well to pumping well is 96 ft. Find the transmissivity and storativity.

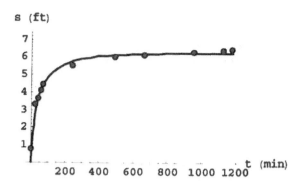

FIGURE 10.8. Matched drawdown for pumping test in a Hantush-Jacob leaky aquifer.

A *Fortran* program *InvHJ.for* has been written for this purpose (see Appendix B). The main part of the program is relatively straightforward. It involves the definition of two subroutines: *function func* and *subroutine dfunc*, for the evaluation of the objective function, Eq. (10.36), and its derivatives, Eqs. (10.37) to (10.39). The subsequent parameter determination by optimization is handled using a blackbox by calling the *dfpmin* subroutine found in *Numerical Recipes*.[101] The program requires a data file named *drawdown.dat* that contains the drawdown versus time data, and operation parameters such as pumping rate and observation well distance.

As with all optimization methods, an initial guess of the parameters is needed to seed the program. For example, if we enter $T = 0.02$ ft^2/min, $S = 0.00005$, $\beta = 0.05$ as the initial try, the program finds the optimized solution as $T = 0.142$ ft^2/min, $S = 0.000198$, $\beta = 0.218$, from which we can calculate $L' = 7.32 \times 10^{-7}$ min^{-1}. The result is basically the same as that found by graphical procedure in Walton:[119] $T = 0.14$ ft^2/min, $S = 0.0002$, $\beta = 0.22$. It takes 13 iterations for the program to converge with a relative tolerance of 0.01%. The root-mean-square error defined as $(E_\ell/n)^{1/2}$, where n is the number of data, is 0.015. The final match is visualized by plotting the observed drawdown together with the theoretical prediction, as shown in Figure 10.8.

Although the program works well, it should be commented that it has only a modest range of convergence for different initial trial values, despite all the careful considerations and parameter transformation. To have some idea, we use the final solution at $(T, S, \beta) = (0.14, 0.0002, 0.2)$ as the center point for initial trial, and vary one parameter at a time to test the range of convergence. We find the following ranges: $0.01 < T < 0.3$, $0.00003 < S < 0.0007$, and $0.06 < \beta < 0.4$. Outside the range, the solution either diverges, or converges to a wrong answer with its square error greater than the true minimum. This means that we either need to make a good initial guess based on the hydrogeologist's insight of the aquifer material, or should attempt multiple tries starting at different initial values. This latter approach is feasible because the program is extremely fast. One can create an automatic procedure that conducts the multiple tries.

Chapter 11
STOCHASTIC ANALYSIS

11.1   Uncertainty in Modeling

Traditionally, groundwater flow and transport problems are modeled as *deterministic processes*. In a deterministic process, all parameters in the model are assumed to be certain and known to a precision. But this is hardly true in real-world problems. Many elements of the model, such as the hydrological input, hydrogeological properties, extraction rate, etc., are random, uncertain, or unknown. For example, the aquifer hydraulic conductivity may have too rapid a spatial variation to be sampled in detail. The hydrological events cannot be controlled and thus are unpredictable. The lack of aquifer information due to the high cost of exploration also plays an important role in the uncertainty of modeling. The issues of randomness in groundwater, and the resulting uncertainty in modeling, should be closely examined.

Porous formations are by their nature random media. They display variations in physical properties in all kinds of length scales, up from the formation scale related to different geological materials, and down to the pore scale involving randomly shaped and distributed pores. It is normally of no practical interest to model the variability in the microscopic level. The statistical averaging process of continuum mechanics[11,13] is applied to take care of this problem. This leads to the deterministic governing partial differential equations of groundwater flow, as those presented in the preceding chapters. The random processes that we are interested in fact stem from these deterministic governing equations.

The formation is nevertheless heterogeneous beyond the microscopic scale. To model a heterogeneous formation based on these deterministic governing equations, it is necessary to know the physical properties of the formation to a fine enough resolution. In practical applications, such resolution is rarely attainable, due to the relative inaccessibility of the formation and the cost of exploration.

Modeling based on deterministic equations is presented in the form of initial and boundary value problems. If a mathematically well-posed set of conditions is supplied, a unique prediction is expected. These input conditions, however, are generally uncertain. Hydrological and hydraulic events such as precipitation and river stage, operation conditions such as pumping rate and duration, etc., are random or uncertain. This is particularly true if a prediction is to be made into the future. In other occasions, the precise location of an underground boundary, on which the boundary condition is to be prescribed, may not be known, creating another uncertain factor of the problem.

It seems that for water resources planning and management purposes, modeling groundwater as a deterministic process is not a proper approach. For water managers, one way to cope with uncertainty is through *scenario building*. Multiple scenarios, and particularly the *worst scenarios*, are created and investigated. Another approach to handle uncertainty is by sensitivity analysis. For example, if the pumping demand or the aquifer parameter estimate is off by a certain percentage, what is the sensitivity of drawdown corresponding to these changes? These scenario based approaches, although much more realistic than the deterministic approach, still lack an important element of addressing uncertainty: the *probability* of events. We know that these extreme events can happen, and the consequences of their happening. But what is the probability that they will happen?

The *stochastic analysis* is aimed at addressing the probability issue. The input information is presented either as a full *probability distribution*, or more realistically, a synopsis of the full probability in terms of *statistical moments*. The output will be similarly presented.

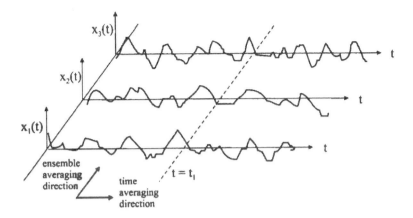

FIGURE 11.1. Three samples in the ensembles of random process $x(t)$.

In other words, water resources planners can obtain quantitative information such as the most probable outcome, with envelopes of confidence levels of a certain percentage.

In the sections to follow, we shall first review the fundamentals of random process in Secs. 11.2 and 11.3. They are however presented in a limited scope of being able to introduce the two elementary problems as illustrated in Secs. 11.4 and 11.6. For further issues of stochastic analysis, the reader is referred to a more advanced book.[38]

11.2   Random Processes and Ensemble Statistics

Suppose we have a *probability space* that contains *events* x_1, x_2, x_3, ... The *ensemble* collection of these events can be expressed as x, which is called a *random variable*. When these events are random functions of time, $x_1(t)$, $x_2(t)$, $x_3(t)$, ..., then $x(t)$ is called a *stochastic process*. Figure 11.1 illustrates three such *samples* (or *realizations*) in the ensemble. Each of these records is also called a *time series*. For groundwater applications, examples of random variable include precipitation records, river stage fluctuation, etc.

The stochastic processes need not be functions of time. They can be functions of a spatial variable, say z, where z is a measure of

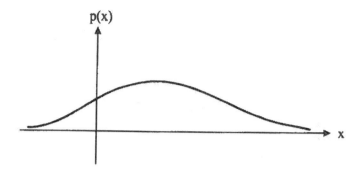

FIGURE 11.2. Probability density function of random variable $x(t_1)$.

distance. An example is the variation of hydraulic conductivity as a function of distance. For simplicity, however, we shall refer to these cases also as time series and the processes as stochastic processes.

These infinite number realizations of time series form a probability space. If we fix the time at $t = t_1$ (see Figure 11.1), the possible values of $x(t_1)$ can be expressed as a *probability density function* (pdf), denoted as $p(x)$, as shown in Figure 11.2.

Probability density function provides the complete information of the distribution of random variable. However, it is often desirable to condense the data by extracting the characteristics of the distribution using *statistical moments*. The single most important statistical measure is the *ensemble average*, or the *mean*, defined as

$$E\left[x(t)\right] = \bar{x}(t) = \lim_{n \to \infty} \frac{1}{n} \sum_{i=1}^{n} x_i(t) \tag{11.1}$$

where we have used the symbol E for *expectation*, or overbar $^-$, to denotes the process of ensemble averaging. We note that the averaging is taken at a fixed time and summed in the ensemble direction (see Figure 11.1). The result hence is generally a function of time. If after the ensemble averaging, the mean turns out to be not a function of time, then the process is called *stationary*. If the time variable is actually a spatial variable (say, the variation of hydraulic conductivity over distance), the process is sometimes called *statistically homogeneous*.

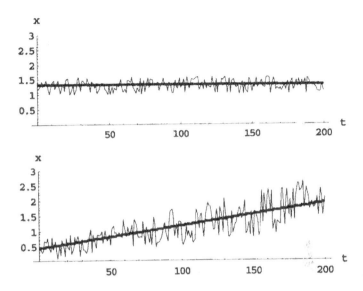

FIGURE 11.3. Time series with stationary mean and stationary standard deviation (upper diagram), and nonstationary mean and nonstationary standard deviation (lower diagram).

The next important statistical quantity is the *variance*

$$\sigma_x^2(t) = \overline{[x(t) - \bar{x}(t)]^2} = \overline{x^2}(t) - \bar{x}^2(t) \tag{11.2}$$

Here we recall that the overbar denotes an ensemble averaging operation like the one defined in Eq. (11.1). The square root of variance, $\sigma_x(t)$, is called *standard deviation*, which is a measure of the average magnitude of deviation from the mean. Figure 11.3 shows two time series, one with small standard deviation (upper diagram), and the other with large standard deviation (lower diagram). In the upper diagram, both the mean (in dark line) and the standard deviation (thin line) are stationary; whereas in the lower diagram, both are *nonstationary*, as they change with time.

Another important second order statistical measure is the *covariance*

$$
\begin{aligned}
c_{xy}(t_1, t_2) &= \overline{[x(t_1) - \bar{x}(t_1)][y(t_2) - \bar{y}(t_2)]} \\
&= \overline{x(t_1)y(t_2)} - \bar{x}(t_1)\,\bar{y}(t_2)
\end{aligned}
\tag{11.3}
$$

where t_1 and t_2 indicate that the samples are taken at fixed but different times for x and y. The covariance is a measure of correlation between two random variables. Consider two random variables $x(t_1)$ and $y(t_2)$. When x takes a certain value, say a value greater than or less than its mean $\bar{x}$, we are interested in the behavior of y. If y is somewhat influenced by x such that its deviation from the mean has a high probability to be positive or negative, following the trend of x, then a non-zero covariance is registered. If y has the same trend as x to be above and below the mean, the covariance is positive. If y has the opposite trend, a negative covariance is found. On the other hand, if y bears no relation to x, and has an equal chance to be above and below its mean, whatever the x value is, then the terms in the ensemble average tend to cancel out and return a zero value for the covariance.

A similar quantity, known as the *correlation*, is defined as

$$r_{xy}(t_1, t_2) = \overline{x(t_1)y(t_2)} \qquad (11.4)$$

For time series, it is of interest to find the correlation of the same quantity observed at two different times. This is called *autocovariance* and is given by

$$
\begin{aligned}
c_{xx}(t_1, t_2) &= \overline{[x(t_1) - \bar{x}(t_1)][x(t_2) - \bar{x}(t_2)]} \\
&= \overline{x(t_1)x(t_2)} - \overline{x(t_1)}\,\overline{x(t_2)}
\end{aligned}
\qquad (11.5)
$$

If an autocovariance is not a function of time itself, but a function of the lag time separating the two measurements,

$$c_{xx}(t_1, t_2) = c_{xx}(t_1 - t_2) = c_{xx}(\tau) \qquad (11.6)$$

where $\tau = t_2 - t_1$, then the process is called stationary. When t represents distance rather than time, we call this process statistically homogeneous.

A random process is *strongly stationary* if all orders of its statistical moment are stationary. In practical applications, due to our limitations in the measurement and utilization of random data, it is often sufficient to use only the first and the second order statistics. If the stationary condition exists in the first two orders, namely the mean and the standard deviation are independent of time, and the autocovariance is a function of lag time only, the random process is called *weakly stationary*.

When the two random variables of a covariance are referred to at the same time, we use the following notation

$$c_{xy}(t, t) = \sigma_{xy}(t) \tag{11.7}$$

Furthermore, if the two variables are the same variable, we find the variance

$$\sigma_{xx}(t) = \sigma_x^2(t) \tag{11.8}$$

as defined in Eq. (11.2).

The covariance is a dimensional quantity; hence it is difficult to interpret its magnitude. For a measure that shows the relative magnitude of a correlation, a *cross correlation coefficient* is defined by normalizing the covariance:

$$R_{xy}(t_1, t_2) = \frac{c_{xy}(t_1, t_2)}{\sqrt{\sigma_x^2(t_1)\,\sigma_y^2(t_2)}} \tag{11.9}$$

Similarly, we define the *autocorrelation coefficient* as

$$R_{xx}(t_1, t_2) = \frac{c_{xx}(t_1, t_2)}{\sqrt{\sigma_x^2(t_1)\,\sigma_x^2(t_2)}} \tag{11.10}$$

and for the stationary case,

$$R_{xx}(\tau) = \frac{c_{xx}(\tau)}{\sigma_x^2} \tag{11.11}$$

in which σ_x^2 is no longer a function of time due to the stationary condition. We notice that

$$c_{xx}(0) = \sigma_x^2 \tag{11.12}$$

hence

$$R_{xx}(0) = 1 \tag{11.13}$$

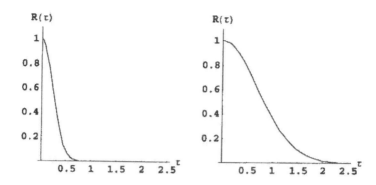

FIGURE 11.4. Correlation coefficient with small (left diagram) and large (right diagram) correlation time scale.

This means that the relative correlation of a quantity with itself is unity, which is an obvious requirement. Also, the following bounds exist

$$-1 \leq R_{xx}(\tau) \leq 1 \tag{11.14}$$

For most natural phenomena, the autocovariance diminishes to zero as the lag time separating the two measurements becomes large. The time (or distance) that $R_{xx}(\tau)$ drops to an insignificant magnitude is referred to as the *correlation time scale* (or *correlation length scale*). More rigorously, it can be defined in terms of an *integral scale*

$$I = \int_0^\infty R_{xx}(\tau) \, d\tau \tag{11.15}$$

Figure 11.4 shows two autocorrelation coefficients, one with small, and the other with large correlation time scale.

Consider hydraulic conductivity as a function of distance. If the formation is largely made of the same material, the hydraulic conductivity can be statistically homogeneous. According to the definition, this implies that the mean and the standard deviation are not functions of location. This does not mean that the hydraulic conductivity does not vary. In fact the standard deviation is a measure of how large the variation is. In addition, there exists an autocorrelation coefficient $R_{KK}(r)$, where r is a lag distance that describes how

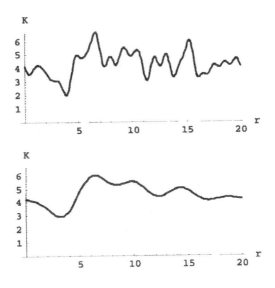

FIGURE 11.5. Two hydraulic conductivity logs with the same stationary mean and standard deviation, but different correlation lengths.

rapidly the fluctuation takes place over distance. If we take a measurement of hydraulic conductivity at a location, and ask the likelihood that the hydraulic conductivity a small distance away takes a similar value, we get an idea of the correlation coefficient at that distance. That correlation is likely to be strong and the value near unity. On the other hand, if we ask the value of hydraulic conductivity at a large distance, the correlation is likely to be weak and close to zero. Hence $R_{KK}(r) = 1$ at $r = 0$ and approaches zero as r gets large. The correlation length scale gives a measure of how far away that the hydraulic conductivity becomes largely uncorrelated. To illustrate the above idea, in Figure 11.5 we present two hydraulic conductivity profiles as taken in well logs. We notice that the two profiles have the same stationary mean and standard deviation, but different correlation length. Hence the effect of correlation length is clearly illustrated.

11.3   Temporal Statistics and Ergodicity

For a single, but very long time series, we can perform statistical analysis in the temporal, i.e. time, direction (see Figure 11.1). The *temporal mean* of a random variable $x(t)$ is defined as

$$\langle x \rangle (t) = \frac{1}{T} \int_{t-T/2}^{t+T/2} x(t)\, dt \tag{11.16}$$

where the brackets $\langle\, \rangle$ are used to denote *temporal average*, and T is a time span large enough such that the average is independent of the selection of T, but small enough such that the average can exhibit a time variation for the nonstationary case similar to that obtained in ensemble average (see the lower diagram of Figure 11.3). A similar concept can be applied to the second order statistics. A *temporal autocorrelation* is defined as

$$\begin{aligned} \Psi_{xx}(t,\tau) &= \langle x(t)x(t+\tau) \rangle \\ &= \frac{1}{T} \int_{t-T/2}^{t+T/2} x(t)\, x(t+\tau)\, dt \end{aligned} \tag{11.17}$$

A *temporal autocovariance* is given by

$$\psi_{xx}(t,\tau) = \langle [x(t) - \langle x \rangle (t)]\, [x(t+\tau) - \langle x \rangle (t+\tau)] \rangle \tag{11.18}$$

A variance is

$$\psi_{xx}(t,0) = \left\langle [x(t) - \langle x \rangle (t)]^2 \right\rangle \tag{11.19}$$

If the process is stationary, then

$$\langle x \rangle = \lim_{t \to \infty} \frac{1}{T} \int_{t-T/2}^{t+T/2} x(t)\, dt \tag{11.20}$$

which is not a function of time. Similar operation applies to the autocorrelation

$$\psi_{xx}(\tau) = \langle [x(t) - \langle x \rangle]\, [x(t+\tau) - \langle x \rangle] \rangle \tag{11.21}$$

and so forth.

In a stationary process, if the statistics based on ensemble and temporal average are the same,

$$\bar{x} = \langle x \rangle \tag{11.22}$$

$$c_{xx}(\tau) = \psi_{xx}(\tau) \tag{11.23}$$

it is known as *ergodic*. An ergodic process must be stationary, but a stationary process is not necessarily ergodic.

Ergodicity is an essential assumption in most engineering applications. We notice that all the theories and applications presented in the sections to follow are based on ensemble statistics. To utilize these results, data input based on ensemble average is needed. In reality, however, sufficient samples needed for performing a meaningful ensemble average are rarely available. Often there exist only a few time series. If the series is long enough, and the process is stationary, then temporal average can be performed. In the absence of ensemble statistics, temporal statistics is used instead. This requires the ergodicity assumption. Although often employed in engineering applications, typically this assumption is neither proven physically, nor demonstrated by observed data. It is merely a pragmatic approach taken by engineers.

11.4   Two Stream Problem—Steady State Solution

Consider the simple problem of an unconfined aquifer flanked by two rivers as shown in Figure 11.6. The piezometric head at the two sides of aquifer coincides with the river stages, H_1 and H_2. On top of the aquifer, there is a spatially varying recharge of rate w. First consider the steady state problem, i.e., the problem is independent of time. The unsteady problem will be investigated in Sec. 11.6. The governing equation for this one-dimensional, steady state problem is

$$\frac{d}{dx}\left(Kh\frac{dh}{dx}\right) = -w(x) \tag{11.24}$$

We shall make the further assumption that the aquifer is deterministic and homogeneous, i.e., K is a known constant. The effect of random aquifer properties will be discussed in Sec. 11.8.

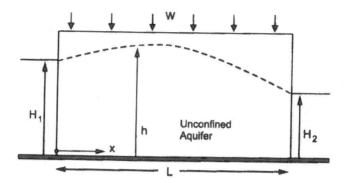

FIGURE 11.6. Definition sketch of the two stream problem.

To have a linear governing equation, we define a new variable ϕ and rewrite the above equation as

$$\frac{d^2\phi}{dx^2} = -f(x) \tag{11.25}$$

where

$$\phi = h^2 \tag{11.26}$$

$$f = \frac{2w}{K} \tag{11.27}$$

For convenience, we also express

$$H_1^2 = \phi_1$$
$$H_2^2 = \phi_2 \tag{11.28}$$

The boundary conditions are

$$\phi = \phi_1; \quad \text{at} \quad x = 0$$
$$\phi = \phi_2; \quad \text{at} \quad x = L \tag{11.29}$$

The deterministic solution of this problem is obtained by integrating Eq. (11.25) twice and using the boundary conditions. We find

$$\phi(x) = -\int_0^x \int_0^{x''} f(x')\,dx'\,dx''$$
$$+ \left[\phi_2 - \phi_1 + \int_0^L \int_0^{x''} f(x')\,dx'\,dx''\right]\frac{x}{L} + \phi_1 \tag{11.30}$$

We can consider the above solution as either a deterministic solution, or a single realization of the infinite number of solutions in the probability space.

Let us assume that the river stages H_1 and H_2 are random variables. Their ensemble statistical characteristics can be obtained by processing historically observed river stages in time, by invoking the ergodicity assumption. The data are processed into mean values $\bar{\phi}_1$ and $\bar{\phi}_2$, and variances $\sigma_{\phi_1}^2$ and $\sigma_{\phi_2}^2$.

The river stages may not be independent of each other. For example, the watersheds may be near enough to be under the influence of the same weather pattern. If the stage of river 1 is higher than average, then there is a good chance that the stage of river 2 is higher than average too. Hence the observed stages should be processed for their covariance $\sigma_{\phi_1 \phi_2}$.

The precipitation recharge has a spatial variation, and is treated as nonstationary (statistically inhomogeneous). Data can be processed to yield $\bar{f}(x)$, $\sigma_f^2(x)$, and $c_{ff}(x_1, x_2)$. There also exists the possibility that river stages and recharge are correlated. Hence the covariances $\sigma_{f\phi_1}(x)$ and $\sigma_{f\phi_2}(x)$ need to be introduced. To simplify the analysis, different levels of statistical homogeneity can be assumed. For example, we can have mean $\bar{f}$, variance σ_f^2, and covariances $\sigma_{f\phi_1}$ and $\sigma_{f\phi_2}$ as constants, and autocovariance $c_{ff}(\chi)$ as a function of the separating distance. These are introduced whenever appropriate.

Given these statistical measures as boundary conditions and forcing terms, we are ready to predict the piezometric head in statistical terms. The first solution to be found is the expectation. This is obtained by performing ensemble average on all terms in Eq. (11.30):

$$\bar{\phi}(x) = -\int_0^x \int_0^{x''} \bar{f}(x') \, dx' \, dx''$$
$$+ \left[\bar{\phi}_2 - \bar{\phi}_1 + \int_0^L \int_0^{x''} \bar{f}(x') \, dx' \, dx'' \right] \frac{x}{L} + \bar{\phi}_1 \quad (11.31)$$

where we recall that the overbar denotes the ensemble average operation as defined in Eq. (11.1). In the above we have assumed that the random variable is sufficiently smooth in space such that the expectation and the integration are commutable.[38] Equation (11.31) implies that, to find the expected value of ϕ, we need only to use the expected values of the input data in the deterministic solution, Eq. (11.30). It should be cautioned that although this is correct for the present case, it is generally *not* true. It works for the present case only because the solution in Eq. (11.30) does not involve any product of random variables. Otherwise the expectation cannot be obtained by using the mean input data in the deterministic solution.

If we make the further assumption of statistical homogeneity such that $\bar{f}$ is no longer a function of spatial variable (although f can vary in space in a single realization), Eq. (11.31) reduces to

$$\bar{\phi}(x) = -\frac{\bar{f}}{2}x^2 + \left(\frac{\bar{\phi}_2 - \bar{\phi}_1}{L} + \frac{\bar{f}L}{2}\right)x + \bar{\phi}_1 \qquad (11.32)$$

This solution is the same as the homogeneous aquifer case in the deterministic solution.

Next we shall investigate the second order statistics. We can decompose all uncertain quantities into an expectation, and a fluctuation denoted by a prime superscript,

$$\begin{aligned}
\phi &= \bar{\phi} + \phi' \\
\phi_1 &= \bar{\phi}_1 + \phi_1' \\
\phi_2 &= \bar{\phi}_2 + \phi_2' \\
f &= \bar{f} + f'
\end{aligned} \qquad (11.33)$$

We can substitute the above into Eq. (11.31), and subtract Eq. (11.32) from it. This gives an identical equation, but in the fluctuation quantities:

$$\begin{aligned}
\phi'(x) &= -\int_0^x \int_0^{x''} f'(x')\, dx'\, dx'' \\
&\quad + \left[\phi_2' - \phi_1' + \int_0^L \int_0^{x''} f'(x')\, dx'\, dx''\right]\frac{x}{L} + \phi_1' \quad (11.34)
\end{aligned}$$

To reduce the lengthy algebraic work, yet not to lose sight of the important physics, we shall present two simplified cases.

In the first case, the recharge is assumed to be deterministic, i.e., $f' = 0$. But the recharge itself need not be zero, and can be a variable in space. Equation (11.34) simplifies to

$$\phi'(x_1) = \frac{\phi'_2 - \phi'_1}{L} x_1 + \phi'_1 \qquad (11.35)$$

In the above, we have written the equation at a location x_1. Next, we write the same equation, but at a location x_2. We multiply these two equations, left side to left side, and right side to right side, and then take the ensemble mean. In doing so, we find terms such as

$$
\begin{aligned}
\overline{\phi'(x_1)\phi'(x_2)} &= c_{\phi\phi}(x_1, x_2) \\
\overline{\phi'_1 \phi'_1} &= \sigma^2_{\phi_1}
\end{aligned}
\qquad (11.36)
$$

etc. The final result in terms of autocovariance of ϕ is

$$
\begin{aligned}
c_{\phi\phi}(x_1, x_2) &= \frac{1}{L^2}(L - x_1)(L - x_2)\sigma^2_{\phi_1} + \frac{x_1 x_2}{L^2}\sigma^2_{\phi_2} \\
&\quad + \frac{1}{L^2}(Lx_1 + Lx_2 - 2x_1 x_2)\sigma_{\phi_1 \phi_2} \qquad (11.37)
\end{aligned}
$$

When $x_1 = x_2$, we obtain variance

$$\sigma^2_\phi(x) = \frac{1}{L^2}(L - x)^2 \sigma^2_{\phi_1} + \frac{x^2}{L^2}\sigma^2_{\phi_2} + 2\frac{x}{L^2}(L - x)\sigma_{\phi_1 \phi_2} \qquad (11.38)$$

Following similar procedure, we can obtain the expected value of flux as

$$\overline{Q} = -\frac{K}{2L}(\bar{\phi}_2 - \bar{\phi}_1) \qquad (11.39)$$

where

$$Q = qh = -Kh\frac{\partial h}{\partial x} = -\frac{1}{2}K\frac{\partial \phi}{\partial x} \qquad (11.40)$$

is the discharge per unit length (in the direction parallel to the length of the river) of aquifer. The autocovariance and variance are

$$c_{QQ}(x_1, x_2) = \sigma^2_Q(x) = \frac{K^2}{4L^2}\left(\sigma^2_{\phi_1} + \sigma^2_{\phi_2} - 2\sigma_{\phi_1 \phi_2}\right) \qquad (11.41)$$

We shall examine a specific case to understand the implication of these solutions.

Example: *Assume that the variance of ϕ at $x = L$ is twice that at $x = 0$, i.e., $\sigma_{\phi_2}^2 = 2\sigma_{\phi_1}^2$. Examine the statistics of head and discharge in the aquifer.*

As indicated in Eqs. (11.31) and (11.32), the mean piezometric head is the same as the deterministic solution, provided that the mean river stages and recharge are used as input. Hence only the second order statistics need to be examined.

It is obvious that the variance must take the value of $\sigma_{\phi_1}^2$ at $x = 0$ and $\sigma_{\phi_2}^2$ at $x = L$. The only question is how does it vary in between? According to Eq. (11.38), this relation is dependent on the covariance of the two river stages $\sigma_{\phi_1 \phi_2}$. We can write Eq. (11.38) in a normalized form as

$$\frac{\sigma_\phi^2(x)}{\sigma_{\phi_1}^2} = \left(1 - \frac{x}{L}\right)^2 + \frac{x^2}{L^2}\frac{\sigma_{\phi_2}^2}{\sigma_{\phi_1}^2} + 2\frac{x}{L}\left(1 - \frac{x}{L}\right)\frac{\sigma_{\phi_2}}{\sigma_{\phi_1}}R_{\phi\phi}(0, L)$$

where

$$R_{\phi\phi}(0, L) = \frac{\sigma_{\phi_1 \phi_2}}{\sigma_{\phi_1}\sigma_{\phi_2}} = \frac{c_{\phi\phi}(0, L)}{\sigma_{\phi_1}\sigma_{\phi_2}}$$

is the autocorrelation coefficient between the left and right river stages.

In Figure 11.7 we plot the normalized head variance versus x/L for a few $R_{\phi\phi}(0, L)$ values, with the given condition $\sigma_{\phi_2}^2 / \sigma_{\phi_1}^2 = 2$. The top curve corresponds to the case that the two river stages are fully correlated, $R_{\phi\phi}(0, L) = 1$. We notice that $\sigma_\phi^2(x)$ changes nearly linearly from $\sigma_{\phi_1}^2$ to $\sigma_{\phi_2}^2$. If the two stages are totally uncorrelated, $R_{\phi\phi}(0, L) = 0$, $\sigma_\phi^2(x)$ is shown as the middle curve in Figure 11.7. We observe that there exists a region in the aquifer where the head variance, hence the uncertainty in head prediction, is smaller than both of the boundary variances. The lowest curve corresponds to the fully, but negatively, correlated case, $R_{\phi\phi}(0, L) = -1$. This means that if stream 1 rises, then stream two drops. This could happen if the

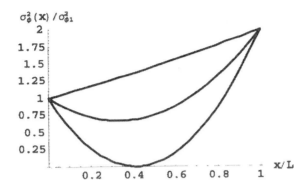

FIGURE 11.7. Head variance for the two stream problem. From top curve down: $R_{\phi\phi}(0, L) = 1, 0, -1$.

two streams are connected by a diversion channel. The uncertainty of head prediction is even smaller in this case. Generally speaking, we expect that large and positive correlations among boundary values tend to increases the magnitude of head variance inside the aquifer. Uncorrelated boundary values reduce the magnitude of uncertainty inside the aquifer.

We next examine the head autocorrelation coefficient at a distance x and the left side of aquifer $(x = 0)$,

$$R_{\phi\phi}(0, x) = \frac{c_{\phi\phi}(0, x)}{\sigma_\phi(0)\,\sigma_\phi(x)} = \frac{c_{\phi\phi}(0, x)}{\sigma_{\phi_1}\,\sigma_\phi(x)}$$

$$= \frac{\left(1 - \dfrac{x}{L}\right)\dfrac{\sigma_{\phi_1}}{\sigma_{\phi_2}} + \dfrac{x}{L}R_{\phi\phi}(0, L)}{\sqrt{\left(1 - \dfrac{x}{L}\right)^2\dfrac{\sigma_{\phi_1}^2}{\sigma_{\phi_2}^2} + \dfrac{x^2}{L^2} + 2\dfrac{x}{L}\left(1 - \dfrac{x}{L}\right)\dfrac{\sigma_{\phi_1}}{\sigma_{\phi_2}}R_{\phi\phi}(0, L)}}$$

We may consider $R_{\phi\phi}(0, x)$ as a measure of the likelihood that the river stage at location x will be higher or lower than normal, given the observation that the river stage on the left side is higher or lower than normal. In Figure 11.8 we plot $R_{\phi\phi}(0, x)$ as a function of x/L for several values of $R_{\phi\phi}(0, L)$. We observe that the autocorrelation coefficient is bound between the value 1 and $R_{\phi\phi}(0, L)$.

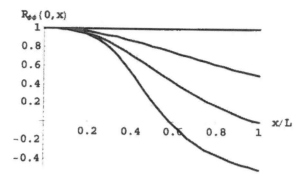

FIGURE 11.8. Head correlation coefficient between location x and the left side of aquifer ($x = 0$). From top curve down: $R_{\phi\phi}(0, L) = 1$, 0.5, 0, and -0.5.

For the discharge, the variance σ_Q^2 is given by Eq. (11.41). It is found to be independent of x. Figure 11.9 plots the normalized σ_Q^2 versus river stage autocorrelation coefficient $R_{\phi\phi}(0, L)$. We observe that the variance of discharge decreases with the increase of river stage autocorrelation coefficient. This trend is opposite to that of head variance. The reason for this effect is clear. When the two stages are positively correlated, they tend to be higher or lower together. The discharge, which is induced by the difference in head, is reduced.

In the second special case, we assume that the recharge is a random function of space, while the river stages are deterministic. Due to the linearity of the system, the solutions obtained for the two cases, random recharge only and deterministic river stages only, can be superimposed. Between these two solutions, the only terms that are missing from the complete solution are those due to the covariances $\sigma_{f\phi_1}$ and $\sigma_{f\phi_2}$. That solution can be obtained following the same procedure.

For the present case of random recharge only, Eq. (11.34) becomes

$$\phi'(x_1) = -\int_0^{x_1} \int_0^{x''} f'(x') \, dx' \, dx'' + \frac{x_1}{L} \int_0^L \int_0^{x''} f'(x') \, dx' \, dx''$$

$$(11.42)$$

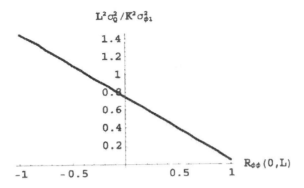

FIGURE 11.9. Normalized discharge variance versus autocorrelation coefficient of river stage.

We rewrite Eq. (11.42) at location x_2 and multiply the two equations together. The ensemble average of the product yields the autocovariance equation involving the quadruple integrals

$$
\begin{aligned}
c_{\phi\phi}(x_1, x_2) = {} & \frac{x_1 x_2}{L^2} \int_0^L \int_0^L \int_0^{y''} \int_0^{x''} c_{ff}(x', y')\, dx'\, dy'\, dx''\, dy'' \\
& - \frac{x_1}{L} \int_0^{x_2} \int_0^L \int_0^{y''} \int_0^{x''} c_{ff}(x', y')\, dx'\, dy'\, dx''\, dy'' \\
& - \frac{x_2}{L} \int_0^L \int_0^{x_1} \int_0^{y''} \int_0^{x''} c_{ff}(x', y')\, dx'\, dy'\, dx''\, dy'' \\
& + \int_0^{x_2} \int_0^{x_1} \int_0^{y''} \int_0^{x''} c_{ff}(x', y')\, dx'\, dy'\, dx''\, dy'' \qquad (11.43)
\end{aligned}
$$

The above is formally the solution of head autocovariance.

Due to the lack of good quality, high-resolution data, the random recharge is typically assumed to be statistically homogeneous. This implies that $\bar{f}$ and σ_f^2 are constants, and $c_{ff}(x_1, x_2) = c_{ff}(|x_1 - x_2|)$ is a function of lag distance only. In this case, the mean is given by

$$
\bar{\phi}(x) = \frac{1}{2} \bar{f} x (L - x) \qquad (11.44)
$$

The autocovariance $c_{\phi\phi}$ requires the recharge autocovariance be given as input. A typical autocovariance function that is used to fit the ob-

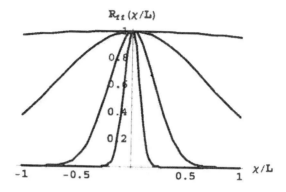

FIGURE 11.10. Autocorrelation coefficient with different length scale. From top curve down: $\ell/L = 5, 1, 0.3$ and 0.1.

servation data takes the form

$$c_{ff}(\chi) = \sigma_f^2\, e^{-\chi^2/\ell^2} \tag{11.45}$$

where $\chi = |x_1 - x_2|$ is the lag distance, and ℓ is a correlation length scale that is related to the integral scale I (see Eq. (11.15)) as

$$\ell = \frac{2}{\sqrt{\pi}}\, I \tag{11.46}$$

Figure 11.10 plots the autocorrelation coefficient $R_{ff}(\chi) = c_{ff}(\chi)/\sigma_f^2$ as a function of χ/L for various ℓ/L values. The top curve corresponds to $\ell = 5L$, i.e., the correlation length scale is 5 times the aquifer length. The autocorrelation coefficient is nearly unity within the range of aquifer. In this case, the recharge is considered as practically fully correlated. When ℓ decreases, we observe that the correlation coefficient quickly drops to zero outside the range of ℓ.

Unfortunately, despite the simplicity of the function in Eq. (11.45), the integration in Eq. (11.43) cannot be carried out analytically. We shall examine two limiting cases in the example below.

Example: *For the aquifer problem described above, examine the limiting cases of a fully-correlated, and a fully-uncorrelated (white noise) recharge pattern.*

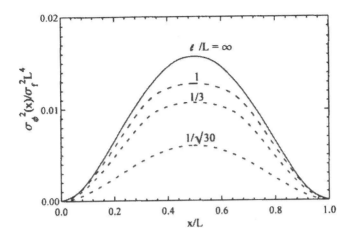

FIGURE 11.11. Head variance for the fully correlated case (solid line) and various recharge correlation length cases (dashed lines).

The fully correlated case is given by taking the correlation length to infinity, $\ell \to \infty$. Equation (11.45) becomes

$$c_{ff}(\chi) = \sigma_f^2$$

which is a constant. Equation (11.43) can now be easily integrated to give

$$c_{\phi\phi}(x_1, x_2) = \frac{\sigma_f^2}{4} x_1 x_2 (L - x_1)(L - x_2)$$

And the variance is

$$\sigma_\phi^2(x) = \frac{\sigma_f^2}{4} x^2 (L - x)^2$$

In Figure 11.11 we plot the head variance normalized by the recharge variance, $\sigma_\phi^2 / \sigma_f^2 L^4$ versus x/L in solid line. We observe that the variance is largest in the middle of aquifer, and drops to zero at the ends. The end values are fixed by the assumed boundary condition of deterministic river stages.

To elaborate further, we present in Figure 11.11 the aquifer head variance corresponding to various correlation lengths ℓ, in dashed

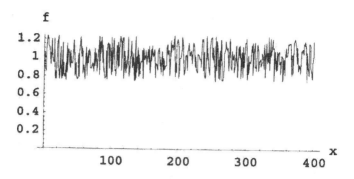

FIGURE 11.12. A spatially distributed recharge with nearly zero correlation length.

lines. These results were obtained elsewhere by numerical method.[23] We observe that as the correlation length decreases, the head variance decreases with it. Hence spatially uncorrelated recharge reduces the uncertainty in head prediction. In the limiting case, as $\ell \to 0$ the autocovariance c_{ff} becomes zero everywhere, except that the variance remains a constant. This means that the recharge distribution over space becomes highly ragged. The recharge intensity at any one point bears no relation to the immediate next. Figure 11.12 gives such an illustration. This situation of course is not possible in reality; hence can be viewed only in an approximate sense. Nevertheless, the mathematical solution indicates that in the extreme case as the spatial variation of recharge gets infinitely ragged, while its amplitude of fluctuation remains constant, the head variation is zero everywhere.

In natural phenomena, however, it has been observed that as the "time series" becomes rougher, its amplitude of fluctuation increases, rather than stays as a constant. This situation can be approximated by a theoretical case known as *white noise*, in which the "signal" is infinitely rough, and its amplitude of fluctuation becomes infinitely large. This is modeled as an autocorrelation function that is zero everywhere, except at $\chi = 0$, where it is infinite. The Dirac delta function (see Sec. 3.5) is used to express the autocorrelation function,

$$c_{ff}(\chi) = S_o \, \delta \left(\frac{\chi}{L} - 0 \right)$$

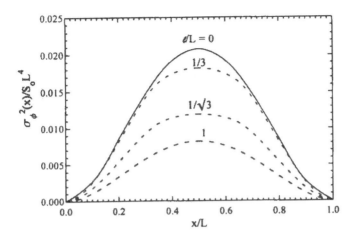

FIGURE 11.13. Head variance for the white noise case (solid line) and various correlation length cases (dashed lines).

where S_o is the magnitude of the white noise *power spectral density function*, whose meaning will be explained in Sec. 11.5 on spectral analysis. In this case, we can again analytically integrate Eq. (11.43) and find

$$c_{\phi\phi}(x_1, x_2) = \frac{S_o}{6} x_2 (x_1 - L)(x_1^2 + x_2^2 - 2Lx_1); \quad \text{for} \quad x_1 > x_2$$

$$= \frac{S_o}{6} x_1 (x_2 - L)(x_1^2 + x_2^2 - 2Lx_2), \quad \text{for} \quad x_2 > x_1$$

and

$$\sigma_\phi^2(x) = \frac{S_o}{3} x^2 (L - x)^2$$

We observe that the head variance is of finite value, despite the input of an infinite recharge variance. The above variance differs from that of the fully correlated case only by a multiplication factor. The normalized variance, $\sigma_\phi^2(x) \big/ S_o L^4$, is presented in Figure 11.13 as the solid line.

A different way to present the white noise case is by way of the following autocovariance:

$$c_{ff}(\chi) = \frac{LS_o}{\ell\sqrt{\pi}} e^{-\chi^2/\ell^2}$$

which should be compared with Eq. (11.45). We observe that the recharge variance

$$\sigma_f^2 = \frac{LS_o}{\ell\sqrt{\pi}}$$

approaches infinity as the correlation length ℓ approaches zero. In addition, the autocorrelation coefficient possesses the following property

$$\lim_{\ell/L \to 0} c_{ff}(\chi) = S_o \delta \left(\frac{\chi}{L} - 0 \right)$$

The white noise case can hence be approximated by using the above autocovariance with a small ℓ. In Figure 11.13 we present the numerical solution[23] using various ℓ/L values in dashed lines.

11.5   Spectral Analysis

In Sec. 11.4 we considered only the spatial dimensions and ignored the time dimension. When the randomness in both time and space come into play, the problem becomes more complicated. To simplify the problem, it is customary to assume that the randomness in time is stationary, or more precisely, weakly stationary.

In the two stream problem analyzed in the preceding section, the stationary (statistically homogeneous) conditions are generally not assumed. In the input data involving random recharge, we have occasionally used the stationary assumption to enable finding a solution in closed form. On the output prediction, no such assumption was imposed. Indeed, the problem was posed as a boundary value problem; hence the solution must have a spatial variation such that the boundary conditions can be satisfied.

With the introduction of time dimension, however, we take a different approach. The randomness in time will not be investigated as an initial value problem. To simplify the analysis, both the input data and the output solution will be assumed to be stationary, i.e.

not a function of time. This however does not mean that time is not present in the governing equation. Indeed, the solution of each single realization is a function of time. The solution of the mean will also be a function of time. But the second order statistics will be stationary. In other words, the autocovariances will be independent of time itself, although they are functions of the lag time.

A powerful tool for analyzing random functions of time is the *Fourier Transform*. We shall assume that a random variable $f(\mathbf{x}, t)$ has a mean $\bar{f}(\mathbf{x}, t)$ and a fluctuation $f'(\mathbf{x}, t)$, such that

$$f'(\mathbf{x}, t) = f(\mathbf{x}, t) - \bar{f}(\mathbf{x}, t) \tag{11.47}$$

Although the focus here is time analysis, we shall keep the variable $\mathbf{x}$ here as a reminder of the spatial dependence that we will encounter. We can apply the *Finite Fourier Transform* to remove the time variable

$$F(\mathbf{x}, \omega, T) = \int_0^T f'(\mathbf{x}, t)\, e^{-i\omega t}\, dt \tag{11.48}$$

where ω is the Fourier transform parameter (also known as *frequency*), $i = \sqrt{-1}$, and T the finite transform period. Capital letters are used to denote Fourier transform quantities. We apply the finite Fourier transform because the infinite Fourier transform is unbounded for non-vanishing f' as $t \to \infty$. The transform is however bounded in the following sense

$$\lim_{T \to \infty} \frac{1}{T} \int_0^T f'(\mathbf{x}, t)\, e^{-i\omega t}\, dt \tag{11.49}$$

We recall the definition of autocovariance in Eq. (11.38). It is written in the following form

$$c_{ff}(\mathbf{x}, \mathbf{y}, \tau) = \overline{f'(\mathbf{x}, t)\, f'(\mathbf{y}, t + \tau)} \tag{11.50}$$

In the above we have automatically assumed that c_{ff} is stationary in time. If the covariance involves two different quantities, then a *cross covariance* is

$$c_{fg}(\mathbf{x}, \mathbf{y}, \tau) = \overline{f'(\mathbf{x}, t)\, g'(\mathbf{y}, t + \tau)} \tag{11.51}$$

Given the weakly stationary assumption, the *cross power spectral density function* is defined as the Fourier transform of crosscorrelation function

$$S_{FG}(\mathbf{x}, \mathbf{y}, \omega) = \int_{-\infty}^{\infty} c_{fg}(\mathbf{x}, \mathbf{y}, \tau) \, e^{-i\omega\tau} \, d\tau \qquad (11.52)$$

In the above we actually used covariance rather than correlation because only the fluctuation quantities will be dealt with. (See Eqs. (11.3) and (11.4) for the difference between correlation and covariance.) Naturally, c_{fg} is the inverse transform of S_{FG}

$$c_{fg}(\mathbf{x}, \mathbf{y}, \tau) = \frac{1}{2\pi} \int_{-\infty}^{\infty} S_{FG}(\mathbf{x}, \mathbf{y}, \omega) \, e^{i\omega\tau} \, d\omega \qquad (11.53)$$

Similarly, corresponding to the autocovariance, $c_{ff}(\mathbf{x}, \mathbf{y}, \tau)$, we can define its Fourier transform pair as the auto power spectral density $S_{FF}(\mathbf{x}, \mathbf{y}, \omega)$.

A special case is worth mentioning. If the autocorrelation function is given as a Dirac delta function, as encountered in the previous example in Sec. 11.4,

$$c_{ff}(\tau) = 2\pi S_o \delta(\tau - 0) \qquad (11.54)$$

its power spectral density is simply a constant,

$$S_{FF}(\omega) = S_o \qquad (11.55)$$

Since S_{FF} is a constant for all frequencies, it is called "white noise," taking the analogy with white light which has a broad frequency spectrum.

For a weakly stationary process, a very important property exists[15]

$$S_{FG}(\mathbf{x}, \mathbf{y}, \omega) = \lim_{T \to \infty} \overline{\frac{1}{T} F^*(\mathbf{x}, \omega, T) G(\mathbf{y}, \omega, T)} \qquad (11.56)$$

in which the asterisk stands for a complex conjugate, and F and G are respectively the finite Fourier transforms of f and g, defined in Eq. (11.48). We observe that the power spectral density function is not reciprocal about its indices. In fact, the following relation exists

$$S_{FG}(\mathbf{x}, \mathbf{y}, \omega) = S_{GF}^*(\mathbf{y}, \mathbf{x}, \omega) \qquad (11.57)$$

This property in Eq. (11.56) is indeed important as it allows us to solve for f and g in the Fourier transform space in terms of F and G, which greatly simplifies the mathematics due to the elimination of time. Once these Fourier transform quantities are obtained, we can find their statistics in terms of power spectral density following Eq. (11.56). Finally, we apply the inverse transform, Eq. (11.53), to find the covariance as a function of lag time, thus accomplishing the solution. We note that the above Fourier transform and inverse procedure is possible because the stationary assumption has reduced the variables from functions of two different times to functions of a single lag time, τ.

11.6   Two Stream Problem—Unsteady Solution

We shall re-examine the two stream problem described in Sec. 11.4, but this time as a transient problem. The transient governing equation for unconfined aquifer is nonlinear, as shown in Sec. 3.2. Based on the discussion in Sec. 3.2, a linearized version is used

$$\frac{\partial}{\partial x}\left(Kh\frac{\partial h}{\partial x}\right) - \frac{S_y h}{H_o}\frac{\partial h}{\partial t} = -w \qquad (11.58)$$

where we note the approximation $h/H_o \approx 1$, with H_o denoting the average water table height. Expressed in terms of the variable ϕ, Eq. (11.58) is written as

$$\frac{\partial^2 \phi(x,t)}{\partial x^2} - \frac{1}{c}\frac{\partial \phi(x,t)}{\partial t} = -f(x,t) \qquad (11.59)$$

where ϕ and f are defined in Eqs. (11.26) and (11.27), and $c = H_o K/S_y$. This is the basic governing equation to use.

Taking the ensemble mean of Eq. (11.59), we find the same governing equation for the mean

$$\frac{\partial^2 \bar{\phi}(x,t)}{\partial x^2} - \frac{1}{c}\frac{\partial \bar{\phi}(x,t)}{\partial t} = -\bar{f}(x,t) \qquad (11.60)$$

in which we assume that the random variables are sufficiently smooth in space such that the differentiation and summation operators are commutable. Equation (11.60) is subject to the boundary conditions,

$$\overline{\phi}(0,t) = \overline{H_1^2}(t) = \bar{\phi}_1(t)$$
$$\overline{\phi}(L,t) = \overline{H_2^2}(t) = \bar{\phi}_2(t) \tag{11.61}$$

where $H_1(t)$ and $H_2(t)$ are river stages, and the initial condition

$$\overline{\phi}(x,0) = \overline{\phi}_o(x) \tag{11.62}$$

The solution of this system, Eqs. (11.60)–(11.62) is formally identical with the deterministic solution, except that all quantities are referred to as statistical mean. With explicitly given boundary and initial conditions, and the recharge function, the above system can be solved using standard methods of partial differential equations. This is, however, not the focus of the present chapter. We shall skip over the solution of the specific cases for the mean, and concentrate on the second order statistical measures.

Subtracting Eq. (11.60) from Eq. (11.59), we find identical governing equation for the fluctuation quantity

$$\frac{\partial^2 \phi'(x,t)}{\partial x^2} - \frac{1}{c}\frac{\partial \phi'(x,t)}{\partial t} = -f'(x,t) \tag{11.63}$$

To simplify the analysis, we shall examine only the case of random recharge, with deterministic river stages. Other cases, including numerical solution, can be found in the literature.[27] With deterministic river stages, the boundary conditions are

$$\phi'(0,t) = \phi'(L,t) = 0 \tag{11.64}$$

Since we are seeking a stationary solution, initial condition is not needed. The spectral technique as described in Sec. 11.25 will be used for the solution of Eqs. (11.63) and (11.64).

Applying Fourier transform to Eq. (11.63), we obtain

$$\frac{d^2\Phi(x,\omega)}{dx^2} - \frac{i\omega}{c}\Phi(x,\omega) = -F(x,\omega) \tag{11.65}$$

where the capital letters denote Fourier transform quantities. (Strictly speaking, it should be finite Fourier transform. It becomes Fourier transform only in the process of taking limit as in Eq. (11.56).) The solution of the above ordinary differential equation, subject to null boundary conditions at $x = 0$ and L, is

$$\Phi(x,\omega) = \frac{2}{L} \sum_{n=1}^{\infty} \frac{\sin \frac{n\pi x}{L}}{\left(\frac{n^2\pi^2}{L^2} + \frac{i\omega}{c}\right)} \int_0^L F(x',\omega) \sin \frac{n\pi x'}{L} \, dx' \quad (11.66)$$

For stationary process, the power spectral density function is given as (see Eq. (11.56))

$$S_{\Phi\Phi}(x_1, x_2, \omega) = \overline{\Phi^*(x_1,\omega)\Phi(x_2,\omega)}$$

$$= \frac{4}{L^2} \sum_{n=1}^{\infty} \sum_{m=1}^{\infty} \frac{\sin \frac{n\pi x_1}{L} \sin \frac{m\pi x_2}{L}}{\left(\frac{n^2\pi^2}{L^2} - \frac{i\omega}{c}\right)\left(\frac{m^2\pi^2}{L^2} + \frac{i\omega}{c}\right)}$$

$$\int_0^L \int_0^L S_{FF}(x_1', x_2', \omega) \, \sin \frac{n\pi x_1'}{L} \sin \frac{m\pi x_2'}{L} \, dx_1' \, dx_2' \quad (11.67)$$

The solution in terms of autocovariance $c_{\phi\phi}(x_1, x_2, \tau)$ is found by performing the Fourier inversion, as indicated in Eq. (11.53). To find the solution, the recharge autocovariance function $c_{ff}(x, y, \tau)$ needs to be explicitly given as input. For a few of the recharge autocovariance functions it is possible to obtain analytical solutions. They are presented in the examples below.

Example: *Recharge is purely random in both time and space.*

By purely random we mean totally uncorrelated data, or the so called white noise case. This case is only of theoretical interest as white noise cannot exist in nature. However, it can be viewed as a theoretical bound. It is also one of the few cases where analytical solutions can be obtained. The autocorrelation is expressed in terms of Dirac delta functions

$$c_{ff}(x_1, x_2, \tau) = S_o \delta\left(\frac{x_1}{L} - \frac{x_2}{L}\right) \delta\left(\frac{c\tau}{L^2}\right)$$

where S_o is the magnitude of the white noise power spectrum. Fourier transform of the above yields

$$S_{FF}(x_1, x_2, \omega) = \frac{S_o L^2}{c} \delta\left(\frac{x_1}{L} - \frac{x_2}{L}\right)$$

Substituting the above into Eq. (11.67), we find

$$S_{\Phi\Phi}(x_1, x_2, \omega) = \frac{2S_oL^2}{c} \sum_{n=1}^{\infty} \frac{\sin \frac{n\pi x_1}{L} \sin \frac{n\pi x_2}{L}}{\left(\frac{n^4\pi^4}{L^4} + \frac{\omega^2}{c^2}\right)}$$

Performing the inverse Fourier transform,[93] we obtain the autocovariance function

$$c_{\phi\phi}(x_1, x_2, \tau) = \frac{S_oL^4}{\pi^2} \sum_{n=1}^{\infty} \frac{1}{n^2} \exp\left(\frac{-n^2\pi^2 c|\tau|}{L^2}\right) \sin \frac{n\pi x_1}{L} \sin \frac{n\pi x_2}{L}$$

At $\tau = 0$ we have the following close form expression

$$\begin{aligned}
c_{\phi\phi}(x_1, x_2, 0) &= \frac{S_oL^4}{\pi^2} \sum_{n=1}^{\infty} \frac{1}{n^2} \sin \frac{n\pi x_1}{L} \sin \frac{n\pi x_2}{L} \\
&= \frac{S_oL^2}{2} x_2(L - x_1); \qquad x_1 > x_2 \\
&= \frac{S_oL^2}{2} x_1(L - x_2); \qquad x_1 < x_2
\end{aligned}$$

The variance is

$$\sigma_\phi^2(x) = c_{\phi\phi}(x, x, 0) = \frac{S_oL^4}{\pi^2} \sum_{n=1}^{\infty} \frac{1}{n^2} \sin^2 \frac{n\pi x}{L} = \frac{S_oL^2}{2} x(L - x)$$

Example: *Recharge is purely random in time and fully correlated in space.*

This case is characterized by the autocovariance function

$$c_{ff}(x_1, x_2, \tau) = S_o \delta\left(\frac{c\tau}{L^2}\right)$$

The power spectral density function is

$$\begin{aligned}
S_{\Phi\Phi}(x_1, x_2, \omega) &= \frac{16S_oL^2}{c\pi^2} \cdot \\
&\sum_{n=1,3,\cdots}^{\infty} \sum_{m=1,3,\cdots}^{\infty} \frac{1}{mn} \frac{\sin \frac{n\pi x_1}{L} \sin \frac{m\pi x_2}{L}}{\left(\frac{n^2\pi^2}{L^2} - \frac{i\omega}{c}\right)\left(\frac{m^2\pi^2}{L^2} + \frac{i\omega}{c}\right)}
\end{aligned}$$

The head autocovariance becomes

$$
\begin{aligned}
c_{\phi\phi}(x_1, x_2, \tau) &= \frac{16 S_o L^4}{\pi^4} \sum_{n=1,3,\cdots}^{\infty} \sum_{m=1,3,\cdots}^{\infty} \frac{1}{mn(m^2 + n^2)} \cdot \\
&\quad \exp\left(\frac{-n^2 \pi^2 c \tau}{L^2}\right) \sin\frac{n\pi x_1}{L} \sin\frac{m\pi x_2}{L}; \quad \tau > 0 \\
&= \frac{16 S_o L^4}{\pi^4} \sum_{n=1,3,\cdots}^{\infty} \sum_{m=1,3,\cdots}^{\infty} \frac{1}{mn(m^2 + n^2)} \cdot \\
&\quad \exp\left(\frac{m^2 \pi^2 c \tau}{L^2}\right) \sin\frac{n\pi x_1}{L} \sin\frac{m\pi x_2}{L}; \quad \tau < 0
\end{aligned}
$$

The variance is

$$
\sigma_\phi^2(x) = \frac{16 S_o L^4}{\pi^4} \sum_{n=1,3,\cdots}^{\infty} \sum_{m=1,3,\cdots}^{\infty} \frac{1}{mn(m^2 + n^2)} \sin\frac{n\pi x}{L} \sin\frac{m\pi x}{L}
$$

Example: *Recharge is fully correlated in time and purely random in space.*

The recharge autocovariance function is

$$
c_{ff}(x_1, x_2, \tau) = S_o \delta\left(\frac{x_1}{L} - \frac{x_2}{L}\right)
$$

The solution of head autocovariance is

$$
\begin{aligned}
c_{\phi\phi}(x_1, x_2) &= \frac{2 S_o L^4}{\pi^4} \sum_{n=1}^{\infty} \frac{1}{n^4} \sin\frac{n\pi x_1}{L} \sin\frac{n\pi x_2}{L} \\
&= \frac{S_o}{6} x_2 (x_1 - L)(x_1^2 + x_2^2 - 2L x_1); \quad x_1 > x_2 \\
&= \frac{S_o}{6} x_1 (x_2 - L)(x_1^2 + x_2^2 - 2L x_2); \quad x_2 > x_1
\end{aligned}
$$

We notice that the above solution is independent of lag time. This is the consequence of the assumption of fully correlated in time. In fact, this solution is identical with the steady state solution of the purely random in space case found in the second example of Sec. 11.4. Hence, for the second order statistics, the "steady state" assumption is equivalent to the fully correlated in time assumption.

Example: *Recharge is fully correlated in both time and space.*

The recharge correlation function is

$$c_{ff}(x_1, x_2, \tau) = \sigma_f^2$$

where σ_f^2 is a constant. As expected, the solution is identical with the steady state, fully correlated in space case as found in the second example of Sec. 11.4:

$$
\begin{aligned}
c_{\phi\phi}(x_1, x_2, \tau) &= \frac{16\sigma_f^2 L^4}{\pi^6} \sum_{n=1,3,\cdots}^{\infty} \sum_{m=1,3,\cdots}^{\infty} \frac{1}{n^3 m^3} \sin\frac{n\pi x_1}{L} \sin\frac{m\pi x_2}{L} \\
&= \frac{\sigma_f^2}{4} x_1 x_2 (L - x_1)(L - x_2)
\end{aligned}
$$

11.7   Perturbation Technique

The stochastic solutions that we have investigated so far are "exact" in the sense that there was no restriction placed on the magnitude of fluctuation of random variables. Hence extreme cases such as autocovariance represented by white noise signal can be modeled. However, those exact solutions are possible only because the random variables that we choose to model are in linear forms in the single realization solution. Otherwise exact treatment is not possible.

In most problems we do not have the luxury of having linear expressions. These problems can only be solved by simulation, such as the Monte Carlo method, or by approximation. One of the most popular approximation methods for solving stochastic problems is the perturbation method.

Consider a function $h(\zeta)$, which is a function of several random variables, $\zeta = (\zeta_1, \zeta_2, \cdots, \zeta_n)$. For a given realization, we can expand the function around the mean values of ζ, denoted as $\bar{\zeta} = (\bar{\zeta}_1, \bar{\zeta}_2, \cdots, \bar{\zeta}_n)$, according to Taylor's series

$$h(\zeta) = h(\bar{\zeta}) + \sum_{i=1}^{n} \frac{\partial h(\bar{\zeta})}{\partial \bar{\zeta}_i} \zeta_i' + \frac{1}{2} \sum_{i=1}^{n} \sum_{j=1}^{n} \frac{\partial^2 h(\bar{\zeta})}{\partial \bar{\zeta}_i \partial \bar{\zeta}_j} \zeta_i' \zeta_j' + O(\zeta^3)$$

(11.68)

where

$$\zeta' = \zeta - \bar{\zeta}$$

(11.69)

is perturbation from the mean. In Eq. (11.68) we have retained up to the second order perturbation terms, and ignored terms of third order and higher. This approximation requires that the perturbed quantities to be small. This becomes the major restriction of the perturbation approach.

In the next step we take the expectation of Eq. (11.68) by applying the following well known rules: the mean of a fluctuation quantity is zero,

$$\overline{\zeta_i'} = 0$$

(11.70)

the mean of any quantity that does not contain fluctuation remains unchanged, such as

$$\overline{h(\bar{\zeta})} = h(\bar{\zeta})$$

(11.71)

We have by definition the covariance

$$\overline{\zeta_i' \zeta_j'} = \sigma_{\zeta_i \zeta_j}$$

(11.72)

and the variance

$$\overline{\zeta_i' \zeta_i'} = \sigma_{\zeta_i \zeta_i} \equiv \sigma_{\zeta_i}^2$$

(11.73)

By dropping the higher order terms, the resultant approximate formula for the mean is

$$\overline{h(\zeta)} \approx h(\bar{\zeta}) + \frac{1}{2} \sum_{i=1}^{n} \sum_{j=1}^{n} \frac{\partial^2 h(\bar{\zeta})}{\partial \bar{\zeta}_i \partial \bar{\zeta}_j} \sigma_{\zeta_i \zeta_j}$$

(11.74)

Eq. (11.74) indicates that in the present second order analysis, the expected outcome of a random process is more than just using the expected values of input parameters in the deterministic solution, i.e., $\overline{h(\zeta)} \neq h(\overline{\zeta})$, generally. It further involves the covariances, which do not exist in a deterministic solution.

On the other hand, if we perform only the first order analysis by dropping the second order terms in Eq. (11.68), we indeed obtain the simple result

$$\overline{h(\zeta)} \approx h(\overline{\zeta}) \tag{11.75}$$

This provides the justification for the common practice in deterministic solution of using the mean boundary conditions to produce the estimate of the mean solution.

We also notice that if $h(\zeta)$ is a linear function of the random variables ζ, the second and higher order terms automatically vanish in Eq. (11.68). Equation (11.75) is then exact. This is what we have observed in Secs. 11.4 and 11.6 where the single realization solutions are linear functions of random variables.

For variances, we subtract Eq. (11.74) from Eq. (11.68) to obtain the equation for the perturbation of h,

$$h'(\zeta) \approx \sum_{i=1}^{n} \frac{\partial h(\overline{\zeta})}{\partial \overline{\zeta}_i} \zeta_i' + \frac{1}{2} \sum_{i=1}^{n} \sum_{j=1}^{n} \frac{\partial^2 h(\overline{\zeta})}{\partial \overline{\zeta}_i \partial \overline{\zeta}_j} \left(\zeta_i' \zeta_j' - \sigma_{\zeta_i \zeta_j} \right) \tag{11.76}$$

We perform the self product of left hand side and right hand side of Eq. (11.76), take the expectation, drop terms of third order and higher in perturbation quantities, and obtain

$$\sigma_h^2 \approx \sum_{i=1}^{n} \sum_{j=1}^{n} \frac{\partial h(\overline{\zeta})}{\partial \overline{\zeta}_i} \frac{\partial h(\overline{\zeta})}{\partial \overline{\zeta}_j} \sigma_{\zeta_i \zeta_j} \tag{11.77}$$

This is approximate equation for variance that is second order accurate. If the product is conducted with the perturbation of a second function $g(\eta)$, where $\eta = (\eta_1, \eta_2, \cdots, \eta_m)$, a crosscovariance formula can be formed between h and g

$$\sigma_{hg} \approx \sum_{i=1}^{n} \sum_{j=1}^{m} \frac{\partial h(\overline{\zeta})}{\partial \overline{\zeta}_i} \frac{\partial g(\overline{\eta})}{\partial \overline{\eta}_j} \sigma_{\zeta_i \eta_j} \tag{11.78}$$

Equations (11.74), (11.76) and (11.77) are extremely powerful tools for stochastic analysis. These formulae show that in order to obtain the mean and covariance of the output prediction, we need the information of mean and covariance of input data, and also the sensitivity coefficients and Hessian matrix components of the deterministic functions. These quantities are then related by the simple relations. The only deficiency of this approach is that, for the approximate formula to be accurate, the perturbation should remain as small quantities. This restriction however is not as severe as it looks, as we shall demonstrate in the next section.

11.8   Uncertainty Analysis of Theis Solution

Consider the aquifer drawdown given by the Theis solution

$$s = \frac{Q_w}{4\pi T} W \left(\frac{r^2 S}{4Tt} \right) \tag{11.79}$$

The aquifer parameters T and S are uncertain, and are estimated to have mean values $\overline{T}$, $\overline{S}$, variances σ_T^2, σ_S^2, and covariance σ_{ST}. We are interested in predicting aquifer drawdown under these parameter uncertainties.

The perturbation method is employed for the solution. Instead of using the parameters T and S for perturbation, we shall use the logarithmic quantities $\log T$ and $\log S$, where log is taken as base 10. Equation (11.79) hence becomes

$$s \left(\log T, \log S \right) = \frac{Q_w}{4\pi 10^{\log T}} W \left(\frac{r^2}{4t} 10^{\log S - \log T} \right) \tag{11.80}$$

There are two reasons that logarithmic quantities are used:

1. The parameters to be perturbed need to be small. The use of logarithmic quantities allows relatively large variations in actual parameter values.
2. Hydrogeological parameters are log-normally distributed.[49] In other words, the logarithms of parameter are normally distributed. Hence it is more convenient to use logarithmic parameters.

We note that for log-normally distributed data, statistic measures based on regular and logarithmically transformed data can be converted as

$$\bar{T} = 10^{\overline{\log T}+(\sigma^2_{\log T}/2)} \tag{11.81}$$

$$\sigma^2_T = \bar{T}^2 \left(10^{\sigma^2_{\log T}} - 1\right) \tag{11.82}$$

and the same for S.

Following the perturbation formulae Eqs. (11.74) and (11.77), we can find the mean and variance of drawdown as

$$\bar{s} \approx \frac{Q_w}{4\pi 10^{\overline{\log T}}} W\left(\frac{r^2}{4t} 10^{\overline{\log S}-\overline{\log T}}\right)$$
$$+ \frac{1}{2} f_1 \sigma^2_{\log T} + \frac{1}{2} f_2 \sigma^2_{\log S} + f_3 \sigma_{\log T \log S} \tag{11.83}$$

and

$$\sigma^2_s \approx f_4^2 \sigma^2_{\log T} + f_5^2 \sigma^2_{\log S} + 2f_4 f_5 \sigma_{\log T \log S} \tag{11.84}$$

where

$$f_1 = \frac{\partial^2 s\left(\overline{\log T}, \overline{\log S}\right)}{\partial \overline{\log T}^2}$$
$$= \frac{Q_w (\ln 10)^2}{4\pi} 10^{-\overline{\log T}} \exp\left(-u^*\right) \cdot$$
$$[u^* + \exp\left(u^*\right) W(u^*) - 2] \tag{11.85}$$

$$f_2 = \frac{\partial^2 s\left(\overline{\log T}, \overline{\log S}\right)}{\partial \overline{\log S}^2}$$
$$= \frac{Q_w (\ln 10)^2}{4\pi} 10^{-\overline{\log T}} u^* \exp\left(-u^*\right) \tag{11.86}$$

$$f_3 = \frac{\partial^2 s\left(\overline{\log T}, \overline{\log S}\right)}{\partial \overline{\log S}\, \partial \overline{\log T}}$$
$$= \frac{Q_w (\ln 10)^2}{4\pi} 10^{-\overline{\log T}} \exp\left(-u^*\right) [1 - u^*] \tag{11.87}$$

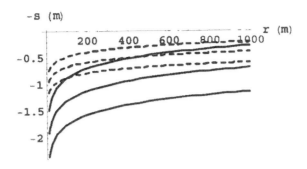

FIGURE 11.14. Mean drawdown versus distance r. Solid curves: stochastic solution; dashed curves: deterministic Theis solution; from top curve down: at $t = 1$ hr, 1 day, and 1 month.

$$f_4 = \frac{\partial s \left(\overline{\log T}, \overline{\log S}\right)}{\partial \log T}$$

$$= \frac{Q_w \ln 10}{4\pi} 10^{-\overline{\log T}} \left[\exp\left(-u^*\right) - W\left(u^*\right)\right] \qquad (11.88)$$

$$f_5 = \frac{\partial s \left(\overline{\log T}, \overline{\log S}\right)}{\partial \log S}$$

$$= \frac{Q_w \ln 10}{4\pi} 10^{-\overline{\log T}} \exp\left(-u^*\right) \qquad (11.89)$$

and

$$u^* = \frac{r^2}{4t} 10^{\overline{\log S} - \overline{\log T}} \qquad (11.90)$$

Hence the mean and variance of drawdown can be evaluated from the mean, variance, and covariance of logarithmic transmissivity and storativity, $\overline{\log T}$, $\overline{\log S}$, $\sigma^2_{\log T}$, $\sigma^2_{\log S}$, and $\sigma_{\log T \log S}$.

Example: *Consider a pumping well located in a confined aquifer with constant discharge of 100 m^3/hr. With the knowledge of aquifer geology, the mean values of logarithmic transmissivity and storativity are estimated to be* $\overline{\log T} = 2$ *and* $\overline{\log S} = -5$, *in which the unit* m^2/hr *is used for* T. *Predict the mean and variance of drawdown.*

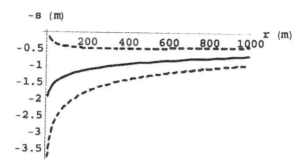

FIGURE 11.15. Mean drawdown (solid line) plus/minus one standard deviation (dashed line), versus distance r, at $t = 1$ day.

Before we proceed, the variance and covariance of transmissivity and storativity need to be estimated. Based on compiled data,[49,51,105] standard deviation of logarithmic hydraulic conductivity, $\sigma_{\log K}$, for various aquifer materials ranges from 0.2 to 2.3. Since the uncertainty for aquifer thickness is not expected to be large, we can pretty much assume that $\sigma_{\log T}$ has the same range as $\sigma_{\log K}$. The higher values of $\sigma_{\log T}$ may raise concern about the validity of perturbation approximation. However, prior experience[37,55] indicates that the perturbation solution can provide reasonable head variance prediction for $\sigma_{\log T}$ value as large as 4. Hence the perturbation solution is expected to be valid for all practical data ranges.

For the present example, a value of $\sigma_{\log T} = 0.5$ is chosen. This means that $\overline{\log T} \pm \sigma_{\log T}$ spans one order of magnitude of T. The standard deviation of logarithmic storativity $\sigma_{\log S}$ is typically smaller than $\sigma_{\log T}$. Dagan[36] suggested

$$\sigma_{\log S} \approx \frac{1}{3}\sigma_{\log K}$$

Without further knowledge about the material , we arbitrary choose $\sigma_{\log S} = 0.2$. Freeze[49] suggested that specific storage is strongly, but negatively, correlated to the hydraulic conductivity. In other words, larger T normally leads to smaller S. We hence expect the cross-

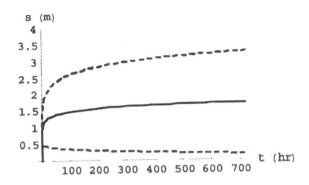

FIGURE 11.16. Mean drawdown (solid line) plus/minus one standard deviation (dashed line), versus time t, at $r = 100$ m.

correlation coefficient between $\log T$ and $\log S$ to be

$$-1 \leq \frac{\sigma_{\log T \log S}}{\sqrt{\sigma_{\log T}^2 \sigma_{\log S}^2}} \leq 0$$

Referring to the variance data given above, we find for fully (but negatively) correlated transmissivity and storativity data, $\sigma_{\log T \log S} = -0.1$, and for uncorrelated data, $\sigma_{\log T \log S} = 0$. We shall choose $\sigma_{\log T \log S} = -0.08$ for the present study. The input data set is now complete.

To assist the solution, a *Mathematica* package *Stochast.m* is programmed and presented in Appendix C. It provides the mean and variance of drawdown, $\bar{s}$ and σ_s^2. Figure 11.14 shows the mean of drawdown at three different times, 1 hr, 1 day and 1 month, for r from 10 to 1000 m. The drawdown is plotted in negative values to show the cone of depression.

In the same diagram, we also present the "deterministic solution" in dashed lines. With the presence of uncertainty, it is not clear what a deterministic solution is. We shall assume that it is the Theis solution using mean transmissivity and storativity. Using Eqs. (11.81) and (11.82) and the statistical data provided above, we can calculate $\bar{T} = 133$ m^2/hr and $\bar{S} = 0.0000105$, which are used in the calculation of the deterministic solution. We observe that this solution significantly underestimates the mean drawdown.

Figure 11.15 demonstrates the mean drawdown plus/minus one standard deviation as function of r, at $t = 1$ day. We observe that the uncertainty of drawdown is larger near the pumping well than away from it. Sensitivity analysis provides similar confirmation. Figures 9.3 and 9.4 show that the drawdown at small distance from the pumping well is more sensitive to transmissivity and storativity variation than at large distance.

Figure 11.16 presents the mean drawdown plus/minus one standard deviation as a function of t, at $r = 100$ m. We observe that the standard deviation increases with time.

Chapter 12
FRACTURED AQUIFER

12.1 Fractured Porous Media

Intact rocks are generally impermeable. However, natural rocks often contain fractures due to temperature and stress changes. When an interconnected fracture network exists, fluid can be transmitted. If fractures are large in size and few in number, they can be modeled as discrete cracks with given size, orientation, aperture width, etc.[123,124] If they are small and numerous, statistical averaging should be applied to model them as a continuum. With the exception of cases in which the fracture location and width can be mapped, fractured aquifers are generally modeled as porous continua.

In rocks, a few types of porosities can be identified:[99] an *intergranular type*, consisting of void spaces between mineral grains of the rock; a *vesicular type*, resulting from leaking due to weathering; and *fractured porosity*, consisting of large-scale openings such as vugs, fissures, joints, etc.

According to Barenblatt,[6,9] and Warren and Root,[120] a fractured porous formation can be modeled by a *double porosity medium*. A distinction is made between two types of porosities: an interconnected fracture network system, and an intergranular pore system (see Figure 12.1). In the principle of continuum mechanics, these two systems occupy the same space. Typically, the *fracture system* accounts for a small volume fraction of the rock, but is responsible for most of the permeability. When a pressure gradient is applied,

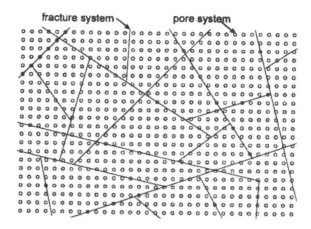

FIGURE 12.1. A fractured porous medium.

water is rapidly transmitted or extracted from the fractures. But due to their small volume fraction, little can be produced from the storage. The *porous block system*, on the other hand, is of negligible permeability, yet can occupy a significant portion of the rock volume to provide fluid storage. When water is being depleted in the fracture system, a large pressure drop is created between the two systems. If the density of the fissures is high, the distance that water needs to travel from the center of a porous block to the nearest fracture is not large. Hence the porous block can readily replenish the fracture, despite its low permeability. Additional fluid will then be delivered by this combined system.

From the above description, we realize that there exists similarity between the double porosity system and the multilayered aquifer system.[54] The two porosities, a fracture system and a porous block system, are equivalent to two aquifers. As one aquifer is being depleted, a pressure difference is created between the two aquifers. The second aquifer replenishes the first just as the porous block system supplies water to the fracture system. A review of some of the equivalent systems can be found in Chen.[21]

In the section below, we shall construct the most general case of a *multiple porosity model*. It is constructed as a full mathematical equivalence to the multilayered aquifer system. In this case we can utilize the mathematical development presented so far for the multilayered aquifer system. We then reduce the solution to the simpler cases such as the double porosity model of Barenblatt.

12.2 Multiple Porosity Model

We assume that the various porosities of rock media can be lumped into groups according to their hydraulic properties. Each of the porosity groups is modeled as a continuum defined over the entire space, and has its own mass conservation and pressure diffusion law. Whenever a pressure difference exists between two groups, flow exchange takes place. The number of porosities modeled should not be large due to our insufficient knowledge about the pore structure. Also, a number of porosities of different physical origins can be lumped into a single type if their mechanical characteristics are not widely distinguishable. Traditionally only two porosities are considered. However, in some cases up to three porosities have been modeled.[34]

In the derivation below, we shall preserve the general form of a multiple porosity model as it can be easily reduced to the special cases. In each of the porosity system, a full diffusion process with storage effect is modeled. The interaction law among the porosity systems is assumed to be time-dependent. In other words, the flow exchange depends on not only the pressure difference between two porosities, but also a storage capacity of an intervening medium. This intervening medium is envisioned as a *skin layer* that exists between the fracture surface and the porous block, as proposed by Moench.[84] Due to the large areal to thickness aspect ratio of the skin, a one-dimensional diffusion law is sufficient for the flow process. We recognize that not all the above-mentioned mechanisms are important for all porosities modeled, due to their different physical origins. However, the simpler models can always be achieved as a reduction of the more general models.

Adopting the petroleum engineering convention, we shall use pore pressure p, instead of piezometric head h, as the primary modeling parameter. The pore pressure referred to is the *dynamic pore pressure*, namely the perturbation of pore pressure from the pre-existing, background pressure, which includes those of hydrostatic and *in situ* stress origins.

Darcy's law, Eq. (2.13) (see also Eq. (2.9)), can be written for each of the i-th porosity in an n-porosity model as

$$\mathbf{q}_i = -\kappa_i \nabla p_i; \qquad i = 1, \cdots, n \tag{12.1}$$

where $\mathbf{q}_i$ is the specific flux, p_i is the dynamic pore pressure, and $\kappa_i = K_i/\gamma = k_i/\mu$ is the mobility coefficient (see Eq. (2.10)). We note that the tensor convention of summation over repeated indices is not enacted here.

Next, consider the continuity condition. Referring to an infinitesimal spatial frame as shown in Figure 2.4, we note that these n porous media *coexist* in the same space. Mass conservation is enforced for fluid residing in each of the n porous media. For each medium, we must account for three types of contributions:

1. the fluid flux entering or leaving the i-th medium across the surfaces of the frame,

2. the fluid exchanged with other media occupying the same frame due to pressure differences, and

3. the gain or loss of fluid inside the medium, which is possible due to the compressibility of fluid and pore matrix.

The continuity equation is written for medium i as

$$-\nabla \mathbf{q}_i = \sum_{j=1}^{n} C_{ij} \frac{\partial p_j}{\partial t} + \sum_{j=1}^{n} \Gamma_{ij} \tag{12.2}$$

The left hand side of Eq. (12.2) accounts for the fluid flux across the enclosing surfaces. The first term on the right hand side corresponds to the rate of fluid mass gained due to the compressibility of the fluid and solid matrix. It is analogous to the $S_s \partial h/\partial t$ term in Eq. (2.28). Recalling that the specific storage S_s is the volume of fluid released from a unit frame due to a unit decline in piezometric head, we recognize that C in Eq. (12.2) is the volume of fluid released from the unit frame due to a unit decline in pressure. Hence C is the bulk compressibility of the combined fluid/solid matrix system. For a single porosity medium, the bulk compressibility is related to the specific storage by a factor of specific weight of fluid,

$$C = \gamma S_s \qquad (12.3)$$

For an n-porosity medium, however, we have not only C_i for each of the pore systems, but also the cross terms C_{ij} due to the coupling of matrix deformation[76] between medium i and j. In other words, the pressure change in one medium can induce compression of matrix in not only the same medium, but also all other media occupying the same space. Hence we need to sum over the compression effect related to p_i for the n pore systems. We note that there should exist the symmetric condition

$$C_{ij} = C_{ji} \qquad (12.4)$$

The cross terms can be neglected if one considers that the solid compressibility is much smaller than that of fluid.[76]

Next, we need to consider the fluid exchanged between medium i and j due to the pressure differential between them. This is denoted as Γ_{ij} and given as the last term in Eq. (12.2). We note that $\Gamma_{ii} = 0$.

Combining Eqs. (12.1) and (12.2), we obtain the governing equations

$$\nabla (\kappa_i \nabla p_i) - \sum_{j=1}^{n} C_{ij} \frac{\partial p_j}{\partial t} - \sum_{j=1}^{n} \Gamma_{ij} = 0; \qquad i,j = 1, \cdots, n \quad (12.5)$$

for an n-porosity model.

So far we have not explicitly provided a model for the *interporosity flow* Γ_{ij}, other than stating that it should be dependent on the pressure differential between two media. Taking the most general case, we assume that between a fracture and porous block there exists a *skin layer* lining the fracture surface due to the deposit of minerals[84] (see Figure 12.2). Because the skin is very thin as compared to its areal extent, the flow through the skin is basically one-dimensional. A one-dimensional diffusion equation can be introduced

$$\kappa'_{ij} \frac{\partial^2 p'_{ij}}{\partial \xi^2} = C'_{ij} \frac{\partial p'_{ij}}{\partial t} \qquad (12.6)$$

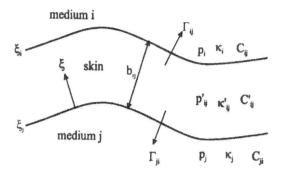

FIGURE 12.2. Flow exchange through skin layer.

where the symbols κ'_{ij}, p'_{ij}, C'_{ij} take their former meaning, but are associated with the skin layer interlining the two media i and j, and ξ is a measure of distance in the thickness direction of the skin layer. By definition, there exists a symmetry between these quantities: $\kappa'_{ij} = \kappa'_{ji}$, $C'_{ij} = C'_{ji}$, and $p'_{ij} = p'_{ji}$. Equation (12.6) is subject to the boundary conditions

$$\begin{aligned}
p'_{ij}(x,y,z,\xi) &= p_i(x,y,z) && \text{at} \quad \xi = \xi_i \\
p'_{ij}(x,y,z,\xi) &= p_j(x,y,z) && \text{at} \quad \xi = \xi_j
\end{aligned} \qquad (12.7)$$

Once the pressure in the skin is known, the flow exchange is obtained from Darcy's law

$$\Gamma_{ij} = -\beta_{ij}\kappa'_{ij}\frac{\partial p'_{ij}}{\partial \xi}\bigg|_{\xi=\xi_i} \qquad (12.8)$$

where β_{ij} represents skin area per unit volume which has the dimension of $[L^{-1}]$, because the amount of fluid exchanged is proportional to the skin area in the unit volume. In reality, the combined coefficient $\beta_{ij}\kappa'_{ij}$ is an empirical coefficient that can only be determined through data fitting. From the definition in Eq. (12.15), we notice that symmetry does not exist for the interporosity flow term, $\Gamma_{ij} \neq \Gamma_{ji}$.

Substitution of (12.8) into (12.5) produces

$$\nabla\left(\kappa_i\nabla p_i\right) - \sum_{j=1}^{n} C_{ij}\frac{\partial p_j}{\partial t} + \sum_{j=1}^{n}\beta_{ij}\kappa'_{ij}\frac{\partial p'_{ij}}{\partial \xi}\bigg|_{\xi=\xi_i} = 0 \qquad (12.9)$$

Equations (12.6) and (12.9) form the governing equations of the multiple porosity system.[24] These equations can be compared to the multilayered aquifer equations (6.3) and (6.5). Among the similarities, we also notice the differences:

1. Equation (6.3) is defined in two spatial dimensions, whereas Eq. (12.9) is generally in three dimensions.

2. Unlike the multiple porosity model, the multilayered aquifer system does not contain cross terms for aquifer storativity (compressibility), because each aquifer occupies a different space.

3. Each aquifer can exchange flow only with the two adjacent layers. For the multiple porosity model, flow exchange can take place between any two porosities, because they occupy the same space.

4. The solution of aquitard head defined in Eq. (6.5) is important for practical purposes. But the pressure distribution in the skin layer is unimportant, as it is largely a conceptual model simulating the delayed pressure equilibrium response.

Since the multilayered aquifer theory and the multiple porosity theory are mathematically equivalent, we can follow the derivation in Sec. 6.3 to eliminates the skin layer pressure p'_{ij} from the solution system. In doing so, the governing equations in the preceding section are transformed into an integro-differential equation system (see Eq. (6.24)):

$$
\nabla \left(\kappa_i \nabla p_i \right) - \sum_{j=1}^{n} C_{ij} \frac{\partial p_j}{\partial t}
$$

$$
- \sum_{j=1}^{n} \frac{\beta_{ij} \kappa'_{ij}}{b'_{ij}} \left[\int_0^t \frac{\partial p_i(x,y,z,t-\tau)}{\partial t} f_{ij}(\tau) \, d\tau \right.
$$

$$
\left. - \int_0^t \frac{\partial p_j(x,y,z,t-\tau)}{\partial t} h_{ij}(\tau) \, d\tau \right] = 0 \qquad (12.10)
$$

In the above, f and h are the memory and influence functions as introduced in Eqs. (6.21) and (6.22),

$$f_{ij}(t) = 1 + 2 \sum_{m=1}^{\infty} \exp\left(-\frac{m^2 \pi^2 \kappa'_{ij} t}{C'_{ij} b'^2_{ij}}\right); \quad i \neq j \qquad (12.11)$$

$$h_{ij}(t) = 1 + 2 \sum_{m=1}^{\infty} (-1)^m \exp\left(-\frac{m^2 \pi^2 \kappa'_{ij} t}{C'_{ij} b'^2_{ij}}\right); \quad i \neq j \quad (12.12)$$

$$f_{ii}(t) = h_{ii}(t) = 0 \qquad (12.13)$$

Continuing on with the simplification, we apply the Laplace transform to the above system, and obtain the system equivalent to Eq. (6.33),

$$\nabla\left(\kappa_i \nabla \tilde{p}_i\right) - \sum_{j=1}^{n} C_{ij} s \tilde{p}_j - \sum_{j=1}^{n} \frac{\beta_{ij} \kappa'_{ij}}{b'_{ij}} \left[s \tilde{f}_{ij} \tilde{p}_i - s \tilde{h}_{ij} \tilde{p}_j\right] = 0 \quad (12.14)$$

where s is the Laplace transform parameter. Here we note a change in notation. Up to the preceding chapter we have been using the symbol p for the Laplace transform parameter to avoid the conflict with the drawdown notation, s. In this chapter, however, the notation p is reserved for pressure, and we return to the conventional symbol s for the Laplace transform.

The memory and influence functions used in the above are

$$\tilde{f}_{ij}(s) = \sqrt{\frac{C'_{ij} b'^2_{ij}}{\kappa'_{ij} s}} \coth\left(\sqrt{\frac{C'_{ij} b'^2_{ij} s}{\kappa'_{ij}}}\right) \qquad (12.15)$$

$$\tilde{h}_{ij}(s) = \sqrt{\frac{C'_{ij} b'^2_{ij}}{\kappa'_{ij} s}} \operatorname{csch}\left(\sqrt{\frac{C'_{ij} b'^2_{ij} s}{\kappa'_{ij}}}\right) \qquad (12.16)$$

This completes the construction of a solution system for a multiple porosity system.

12.3 Pumping Well in Double Porosity Medium

The pumping well solution for multiple porosity medium in two-dimensional geometry is completely analogous to the multilayered

aquifer system, except that the coefficient matrix needs to be redefined. The three-dimensional solution of a point sink in multiple
porosity medium can be derived following the same mathematical
procedure for the two-dimensional solution. Both the two- and three-
dimensional cases are presented below. However, for the briefness of
presentation, only the double porosity model is treated here.

Consider two cases, a point sink located at the origin, $\mathbf{x} = (0,0,0)$,
and a line sink at $\mathbf{x} = (0,0)$, respectively corresponding to a three-
dimensional and a two-dimensional problem. With the usual assumption of homogeneity, isotropy, and infinite aquifer extent, we first
consider an constant extraction from medium 1 at the rate of Q_1,
starting at $t = 0$. The governing equation (12.14) is written, for the
three-dimensional case, as

$$\kappa_1 \nabla^2 \tilde{p}_{11} - sC_{11}\tilde{p}_{11} - sC_{12}\tilde{p}_{21}$$

$$-\frac{s\beta\kappa'}{b'}\tilde{f}\tilde{p}_{11} + \frac{s\beta\kappa'}{b'}\tilde{h}\tilde{p}_{21} = \frac{Q_1}{s}\,\delta(\mathbf{x}-0) \qquad (12.17)$$

$$\kappa_2 \nabla^2 \tilde{p}_{21} - sC_{21}\tilde{p}_{11} - sC_{22}\tilde{p}_{21}$$

$$-\frac{s\beta\kappa'}{b'}\tilde{f}\tilde{p}_{21} + \frac{s\beta\kappa'}{b'}\tilde{h}\tilde{p}_{11} = 0 \qquad (12.18)$$

For the two-dimensional case, Q_1 should be replaced by Q_1/b, where
b is the aquifer thickness. In the above we have added a second index
to the pressure notation to denote the medium in which the extraction takes place. We also have dropped all the subscripts associated
with the skin layer quantities, such as κ'_{12}, $\tilde{f}_{12}$, etc. since there is
only one skin layer involved.

Equations (12.17) and (12.18) may be organized into a shorthand
form:

$$\nabla^2 \tilde{p}_{11} - a_{11}\tilde{p}_{11} - a_{12}\tilde{p}_{21} \quad = \quad \frac{Q_1}{s\kappa_1}\delta(\mathbf{x}-0) \qquad (12.19)$$

$$\nabla^2 \tilde{p}_{21} - a_{21}\tilde{p}_{11} - a_{22}\tilde{p}_{21} \quad = \quad 0 \qquad (12.20)$$

in which

$$a_{11} \quad = \quad \frac{s\beta\kappa'\tilde{f}}{b'\kappa_1} + \frac{sC_{11}}{\kappa_1}$$

$$a_{22} \quad = \quad \frac{s\beta\kappa'\tilde{f}}{b'\kappa_2} + \frac{sC_{22}}{\kappa_2}$$

$$a_{12} = -\frac{s\beta\kappa'\tilde{h}}{b'\kappa_1} + \frac{sC_{12}}{\kappa_1}$$

$$a_{21} = -\frac{s\beta\kappa'\tilde{h}}{b'\kappa_2} + \frac{sC_{21}}{\kappa_2} \tag{12.21}$$

When fluid is extracted from the second medium, the solution system becomes

$$\nabla^2\tilde{p}_{12} - a_{11}\tilde{p}_{12} - a_{12}\tilde{p}_{22} = 0 \tag{12.22}$$

$$\nabla^2\tilde{p}_{22} - a_{21}\tilde{p}_{12} - a_{22}\tilde{p}_{22} = \frac{Q_2}{s\kappa_2}\delta(\mathbf{x} - 0) \tag{12.23}$$

where the second index for pressure denotes medium in which the extraction takes place.

We recognize that Eqs. (12.19), (12.20), (12.22) and (12.23) are the same solution system as Eqs. (7.12) and (7.13). The solution in two-dimensional geometry (line sink) is already available (cf. Eqs. (7.33)–(7.36)):

$$\tilde{p}_{11} = -\frac{Q_1}{2\pi sb\kappa_1}\left[\frac{(k_1^2 - a_{22})K_0(k_1r)}{k_1^2 - k_2^2}\right.$$
$$\left. + \frac{(k_2^2 - a_{22})K_0(k_2r)}{k_2^2 - k_1^2}\right] \tag{12.24}$$

$$\tilde{p}_{21} = -\frac{Q_1}{2\pi sb\kappa_1}a_{21}\left[\frac{K_0(k_1r)}{k_1^2 - k_2^2} + \frac{K_0(k_2r)}{k_2^2 - k_1^2}\right] \tag{12.25}$$

$$\tilde{p}_{12} = -\frac{Q_2}{2\pi sb\kappa_2}a_{12}\left[\frac{K_0(k_1r)}{k_1^2 - k_2^2} + \frac{K_0(k_2r)}{k_2^2 - k_1^2}\right] \tag{12.26}$$

$$\tilde{p}_{22} = -\frac{Q_2}{2\pi sb\kappa_2}\left[\frac{(k_1^2 - a_{11})K_0(k_1r)}{k_1^2 - k_2^2}\right.$$
$$\left. + \frac{(k_2^2 - a_{11})K_0(k_2r)}{k_2^2 - k_1^2}\right] \tag{12.27}$$

The eigenvalues in the above are the same as those defined in Eqs. (7.21) and (7.22),

$$k_1^2 = \frac{a_{11} + a_{22} - \sqrt{(a_{11} - a_{22})^2 + 4a_{12}a_{21}}}{2} \tag{12.28}$$

$$k_2^2 = \frac{a_{11} + a_{22} + \sqrt{(a_{11} - a_{22})^2 + 4a_{12}a_{21}}}{2} \tag{12.29}$$

The point sink (three-dimensional) solution can be similarly obtained. As demonstrated in Sec. 7.2, the knowledge about Green's function of the system

$$(\nabla^2 - k_1^2)(\nabla^2 - k_2^2)\phi = -4\pi\delta(\mathbf{x} - \mathbf{0}) \tag{12.30}$$

is the key to the solution. The three-dimensional counterpart to the two-dimensional solution Eq. (7.29) is:[28]

$$\phi = \frac{1}{r}\left[\frac{\exp(-k_1 r)}{k_1^2 - k_2^2} + \frac{\exp(-k_2 r)}{k_2^2 - k_1^2}\right] \tag{12.31}$$

The point source solution is then derived following a similar procedure:

$$\tilde{p}_{11} = -\frac{Q_1}{4\pi s \kappa_1}\frac{1}{r}\left[\frac{(k_1^2 - a_{22})\exp(-k_1 r)}{k_1^2 - k_2^2}\right.$$
$$\left.+\frac{(k_2^2 - a_{22})\exp(-k_2 r)}{k_2^2 - k_1^2}\right] \tag{12.32}$$

$$\tilde{p}_{21} = -\frac{Q_1}{4\pi s \kappa_1}\frac{a_{21}}{r}\left[\frac{\exp(-k_1 r)}{k_1^2 - k_2^2} + \frac{\exp(-k_2 r)}{k_2^2 - k_1^2}\right] \tag{12.33}$$

$$\tilde{p}_{12} = -\frac{Q_2}{4\pi s \kappa_2}\frac{a_{12}}{r}\left[\frac{\exp(-k_1 r)}{k_1^2 - k_2^2} + \frac{\exp(-k_2 r)}{k_2^2 - k_1^2}\right] \tag{12.34}$$

$$\tilde{p}_{22} = -\frac{Q_2}{4\pi s \kappa_2}\frac{1}{r}\left[\frac{(k_1^2 - a_{11})\exp(-k_1 r)}{k_1^2 - k_2^2}\right.$$
$$\left.+\frac{(k_2^2 - a_{11})\exp(-k_2 r)}{k_2^2 - k_1^2}\right] \tag{12.35}$$

In the derivation presented so far, we consider Q_1 and Q_2 as given quantities. In fact, for a well pumping at a total rate of

$$Q = Q_1 + Q_2 \tag{12.36}$$

it is not yet known the partition of discharge between the two media. To determine the partition, we assume that both media are in full contact with water in the well, hence are subjected to the same pressure drop at $r \to 0$. This condition is the same as the two-aquifer

problem in which a well is screened in both aquifers. As discussed in
the last example of Sec. 8.2, the partition is

$$Q_1 = \frac{\kappa_1}{\kappa_1 + \kappa_2} Q \qquad (12.37)$$

$$Q_2 = \frac{\kappa_2}{\kappa_1 + \kappa_2} Q \qquad (12.38)$$

These values are substituted into Eqs. (12.24)–(12.27), and (12.32)–
(12.35). Since these extractions take place simultaneously, the pressure in each medium is the summation of the two sets of solutions:

$$\tilde{p}_1 = \tilde{p}_{11} + \tilde{p}_{12} \qquad (12.39)$$

$$\tilde{p}_2 = \tilde{p}_{21} + \tilde{p}_{22} \qquad (12.40)$$

12.4 Barenblatt Model

In the Barenblatt model[6] the two media are identified as a fracture
system and a porous block system. It is assumed that the flow exchange between the two media is at a *pseudosteady* state. In other
words, the interporosity flow reacts instantaneously to the pressure
differential between the two media. The pseudosteady condition may
be viewed as a special case of diffusion in skin layer, by taking the
storativity C' in Eq. (12.6) to zero. It is easily shown that the pressure distribution in the skin should vary linearly between p_m and p_f,
where the subscript f denotes fracture and m denotes porous block.
The flux is proportional to the pressure difference. Hence we have

$$\begin{aligned}
\Gamma_{mf} &= -\Gamma_{fm} \\
&= \frac{\beta}{b'} \kappa_m (p_m - p_f) \\
&= \alpha \kappa_m (p_m - p_f) \qquad (12.41)
\end{aligned}$$

where b' is the skin thickness. We also assume that the skin material
is the same as porous block, thus $\kappa' = \kappa_m$. It is obvious that β and
b' can be combined to give

$$\alpha = \frac{\beta}{b'} \qquad (12.42)$$

in which α is a geometric factor with the dimension of $[L^{-2}]$. In reality, α is an empirical parameter that can only be found by data fitting. We further assume that the cross matrix compressibility terms, C_{fm} and C_{mf}, are zero. With these simplifications, Eq. (12.5) becomes

$$\kappa_m \nabla^2 p_m - C_m \frac{\partial p_m}{\partial t} - \alpha \kappa_m (p_m - p_f) = 0 \qquad (12.43)$$

$$\kappa_f \nabla^2 p_f - C_f \frac{\partial p_f}{\partial t} + \alpha \kappa_m (p_m - p_f) = 0 \qquad (12.44)$$

This is known as the *complete Barenblatt model*.[6]

If is of interest to find the pumping well solution for this model. The solution takes the same form as that developed in the preceding section and only a few coefficients need to be modified. We first assign the correspondence of subscripts as $1 \to m$ and $2 \to f$. For example, Eqs. (12.39) and (12.40) indicate that the pressure in the porous block system and the fracture system is respectively given by

$$\tilde{p}_m = \tilde{p}_{11} + \tilde{p}_{12} \qquad (12.45)$$

$$\tilde{p}_f = \tilde{p}_{21} + \tilde{p}_{22} \qquad (12.46)$$

where $\tilde{p}_{11}$, $\tilde{p}_{12}$, $\tilde{p}_{21}$, and $\tilde{p}_{22}$ are defined in Eqs. (12.24)–(12.27) and (12.32)–(12.35). Other equations, (12.28), (12.29), (12.37) and (12.38) are interpreted the same way. The coefficients a_{ij} in Eq. (12.21), however, take the new form

$$a_{11} = \frac{sC_m}{\kappa_m} + \alpha$$

$$a_{22} = \frac{sC_f}{\kappa_f} + \frac{\alpha \kappa_m}{\kappa_f}$$

$$a_{12} = -\alpha$$

$$a_{21} = -\frac{\alpha \kappa_m}{\kappa_f} \qquad (12.47)$$

This concludes the two- and three-dimensional pumping well solution for the complete Barenblatt model.

The complete Barenblatt model, Eqs. (12.43) and (12.44), is often simplified. One of the popular models in use in the petroleum industry is the *Warren-Root model*.[120] If the permeability of the porous block is so small such that the flow in the block can be ignored, the first term in Eq. (12.43) is dropped. The pair of equations

$$C_m \frac{\partial p_m}{\partial t} + \alpha \kappa_m (p_m - p_f) = 0 \qquad (12.48)$$

$$\kappa_f \nabla^2 p_f - C_f \frac{\partial p_f}{\partial t} + \alpha \kappa_m (p_m - p_f) = 0 \qquad (12.49)$$

is known as the Warren-Root model. To find the solution of a pumping well with constant discharge, we solve the following Laplace transform equations in three dimensions:

$$C_m s \tilde{p}_m + \alpha \kappa_m (\tilde{p}_m - \tilde{p}_f) = 0 \qquad (12.50)$$

$$\nabla^2 \tilde{p}_f - \frac{sC_f}{\kappa_f} \tilde{p}_f + \frac{\alpha \kappa_m}{\kappa_f} (\tilde{p}_m - \tilde{p}_f) = \frac{Q}{s\kappa_f} \delta(\mathbf{x} - 0) \qquad (12.51)$$

For two-dimensional solution, Q in the above should be replaced by Q/b. We note in the above that the discharge is entirely produced from the fracture system, because the permeability of the porous block has been set to zero. Equation (12.50) can be used to eliminate $\tilde{p}_m$ in Eq. (12.51) to yield

$$\nabla^2 \tilde{p}_f - \left[\frac{\alpha \kappa_m}{\kappa_f} + \frac{sC_f}{\kappa_f} - \frac{\alpha^2 \kappa_m^2}{\kappa_f (C_m s + \alpha \kappa_m)} \right] \tilde{p}_f = \frac{Q}{s\kappa_f} \delta(\mathbf{x} - 0)$$

$$(12.52)$$

Solution of the above is

$$\tilde{p}_f = -\frac{Q}{2\pi sb\kappa_f} K_0(kr); \qquad \text{2-D} \qquad (12.53)$$

$$\tilde{p}_f = -\frac{Q}{4\pi s\kappa_f} \frac{\exp(-kr)}{r}; \qquad \text{3-D} \qquad (12.54)$$

where

$$k = \sqrt{\frac{\alpha \kappa_m}{\kappa_f} + \frac{sC_f}{\kappa_f} - \frac{\alpha^2 \kappa_m^2}{\kappa_f (C_m s + \alpha \kappa_m)}} \qquad (12.55)$$

From Eqs. (12.50), (12.53) and (12.54) we also obtain

$$\tilde{p}_m = -\frac{Q}{2\pi sb\kappa_f} \frac{\alpha \kappa_m}{C_m s + \alpha \kappa_m} K_0(kr); \qquad \text{2-D} \qquad (12.56)$$

$$\tilde{p}_m = -\frac{Q}{4\pi s\kappa_f} \frac{\alpha \kappa_m}{C_m s + \alpha \kappa_m} \frac{\exp(-kr)}{r}; \qquad \text{3-D} \qquad (12.57)$$

Equations (12.53), (12.54), (12.56) and (12.57) are the constant rate pumping well solution for the Warren-Root model.

For the two-dimensional case, Warren and Root[120] presented an approximate solution in the time domain, which is widely used in the petroleum industry for well test analysis.[46] In normalized form, it is expressed as

$$p_f^* = \ln t^* - 0.5772157 - W\left[\frac{\lambda t^*}{4(\omega - \omega^2)}\right] + W\left[\frac{\lambda t^*}{4(1-\omega)}\right] \quad (12.58)$$

where W is the Theis well function, and

$$p_f^* = -\frac{4\pi b\kappa_f}{Q} p_f \quad (12.59)$$

$$t^* = \frac{4\kappa_f}{(C_m + C_f)r^2} t \quad (12.60)$$

$$\lambda = \frac{\alpha\kappa_m r^2}{\kappa_f} \quad (12.61)$$

$$\omega = \frac{C_f}{C_m + C_f} \quad (12.62)$$

In the above, ω is the ratio of fracture system storativity to total storativity, and λ is the *interporosity flow coefficient*. We notice from the above combination of parameters that in a well test analysis, given the pumping rate and drawdown data, only three parameters, λ, ω, and $b\kappa_f$ (known as the *permeability-thickness*), can be independently determined.

Example: *Compare the pressure drawdown based on the Warren-Root approximate solution Eq. (12.58) and the full solution Eq. (12.53).*

Both of the two-dimensional solutions, Eqs. (12.53) and (12.58), are respectively programmed into the *Mathematica* package *DblPoro.m* listed in Appendix C as *PfWarrenRoot1* and *PfWarrenRoot2*. The pressure in the porous block system, Eq. (12.56), is programmed as *PmWarrenRoot1*.

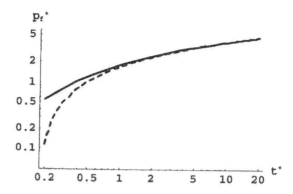

FIGURE 12.3. Comparison of Warren-Root solution as p_f^* vs. t^* in log-log scale, for $\lambda = 10^{-5}$ and $\omega = 0.1$. (Solid line: full solution; dashed line: approximate solution)

The comparison of p_f between the full and the approximate solution is made in dimensionless parameters shown in Eqs. (12.59)–(12.62). The two parameters λ and ω are chosen as $\lambda = 10^{-5}$ and $\omega = 0.1$. The result is plotted in Figure 12.3 as pressure versus time (dimensionless). It is presented in log-log scale and in small time range to enhance the difference between the two solutions. We observe that the Warren-Root solution, Eq. (12.58), is a good approximation of the full solution, except at very small dimensionless times.

Example: *Plot the Warren-Root solution for a range of parameters.*

The *Mathematica* function *PfWarrenRoot1*, corresponding to the full solution Eq. (12.53), is used to evaluate the type curves in Figure 12.4. Two families of curves, $\lambda = 10^{-5}$ and 10^{-7}, are shown, respectively in solid and dashed lines. In each group, three ω values, 0.1, 0.01, and 0.001, are presented.

Soon after the introduction of the general model, Eqs. (12.43) and (12.44), Barenblatt[7,8] proposed a simplified model by assuming that the flow in the porous block system and the storage in the

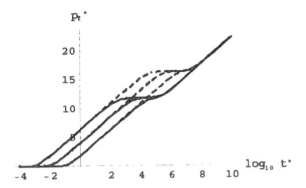

FIGURE 12.4. Warren-Root type curves for various λ and ω values, in semi-log scale. (Solid lines: $\lambda = 10^{-5}$, from the right most curve to the left, $\omega = 0.1$, 0.01, and 0.001; Dashed lines: $\lambda = 10^{-7}$, from the right most curve to the left, $\omega = 0.1$, 0.01, and 0.001.)

fracture system, are negligible. This model is also equivalent to a reduction of the Warren-Root model by ignoring the fracture storage C_f. Setting C_f in Eqs. (12.48) and (12.49) to zero, we obtain the *simplified Barenblatt model*:

$$C_m\frac{\partial p_m}{\partial t} + \alpha\kappa_m(p_m - p_f) = 0 \tag{12.63}$$

$$\kappa_f\nabla^2 p_f + \alpha\kappa_m(p_m - p_f) = 0 \tag{12.64}$$

For solution of a constant rate (step rise) pumping well, we introduce to the right hand side of Eq. (12.64) a Dirac delta function:

$$\kappa_f\nabla^2 p_f + \alpha\kappa_m(p_m - p_f) = Q\delta(\mathbf{x}-0)\mathrm{H}(t-0) \tag{12.65}$$

To emphasize the sudden start-up nature of pumping, a Heaviside unit step function is included in the right hand side. The above equation is for a point sink in a three-dimensional geometry. For the two-dimensional, line sink case, Q needs to be replaced by Q/b.

To solve the two equations, (12.63) and (12.65), we seek to eliminate the porous block pressure p_m between them. First, we differentiate Eq. (12.65) with respect to time and obtain

$$\kappa_f\frac{\partial}{\partial t}\nabla^2 p_f + \alpha\kappa_m\left(\frac{\partial p_m}{\partial t} - \frac{\partial p_f}{\partial t}\right) = Q\delta(\mathbf{x}-0)\delta(t-0) \tag{12.66}$$

In the above we notice that the time derivative of a Heaviside step function yields a Dirac delta function. Equations (12.63) and (12.65) are respectively utilized to eliminate $\partial p_m / \partial t$ and p_m in Eq. (12.66), such that

$$\frac{\partial}{\partial t} \nabla^2 p_f + \frac{\alpha \kappa_m}{C_m} \nabla^2 p_f - \frac{\alpha \kappa_m}{\kappa_f} \frac{\partial p_f}{\partial t} =$$

$$\frac{Q}{\kappa_f} \delta(\mathbf{x}-0)\delta(t-0) + \frac{\alpha \kappa_m Q}{C_m \kappa_f} \delta(\mathbf{x}-0) \mathrm{H}(t-0) \qquad (12.67)$$

To solve the above equation, the Laplace transform is applied

$$s\nabla^2 \tilde{p}_f + \frac{\alpha \kappa_m}{C_m} \nabla^2 \tilde{p}_f - \frac{s\alpha \kappa_m}{\kappa_f} \tilde{p}_f - \nabla^2 p_f(0) + \frac{s\alpha \kappa_m}{\kappa_f} p_f(0) =$$

$$\frac{Q}{\kappa_f} \delta(\mathbf{x}-0) + \frac{\alpha \kappa_m Q}{sC_m \kappa_f} \delta(\mathbf{x}-0) \qquad (12.68)$$

where $p_f(0)$ denotes the initial value of p_f. As pointed out by Barenblatt,[8,9] an effect of dropping the time derivative term in the fracture flow equation (12.64) in the process of simplification is that the initial value of p_f cannot be arbitrarily set. Its value is dependent on the chosen initial value of p_m.

Utilizing Eq. (12.65) at $t = 0^+$, we realize

$$\nabla^2 p_f(0) - \frac{\alpha \kappa_m}{\kappa_f} p_f(0) = -\frac{\alpha \kappa_m}{\kappa_f} p_m(0) + \frac{Q}{\kappa_f} \delta(\mathbf{x}-0) \qquad (12.69)$$

The initial condition for p_m is

$$p_m = 0, \qquad \text{at} \quad t = 0 \qquad (12.70)$$

Using Eqs. (12.69) and (12.70) in Eq. (12.68) and reorganizing, we obtain

$$\nabla^2 \tilde{p}_f - \frac{\alpha \kappa_m C_m s}{\kappa_f(C_m s + \alpha \kappa_m)} \tilde{p}_f = \frac{Q}{sb\kappa_f} \frac{\alpha \kappa_m}{C_m s + \alpha \kappa_m} \delta(\mathbf{x}-0) \qquad (12.71)$$

Utilizing Green's function presented in Eq. (3.84) and its three-dimensional equivalent, we can solve the above equation as

$$\tilde{p}_f = -\frac{Q}{2\pi sb\kappa_f} \frac{\alpha \kappa_m}{C_m s + \alpha \kappa_m} \mathrm{K}_0(kr); \qquad \text{2-D} \qquad (12.72)$$

$$\tilde{p}_f = -\frac{Q}{4\pi s\kappa_f} \frac{\alpha \kappa_m}{C_m s + \alpha \kappa_m} \frac{\exp(-kr)}{r}; \qquad \text{3-D} \qquad (12.73)$$

where

$$k = \sqrt{\frac{\alpha \kappa_m C_m s}{\kappa_f (C_m s + \alpha \kappa_m)}} \qquad (12.74)$$

The pressure in the porous block is then

$$\tilde{p}_m = -\frac{Q}{2\pi s b \kappa_f} \left(\frac{\alpha \kappa_m}{C_m s + \alpha \kappa_m} \right)^2 K_0(kr); \qquad \text{2-D} \quad (12.75)$$

$$\tilde{p}_m = -\frac{Q}{4\pi s \kappa_f} \left(\frac{\alpha \kappa_m}{C_m s + \alpha \kappa_m} \right)^2 \frac{\exp(-kr)}{r}; \qquad \text{3-D} \quad (12.76)$$

The Laplace transform solution Eq. (12.72) can be compared with the two-dimensional line sink solution in the time domain obtained by Barenblatt:[7,8]

$$p_f = -\frac{Q}{4\pi b \kappa_f} \int_0^\infty \frac{1}{\eta} J_0(\sqrt{\frac{\alpha \kappa_m r^2}{\kappa_f}} \eta)$$
$$\left[1 - \exp\left(-\frac{\eta^2}{1+\eta^2} \frac{\alpha \kappa_m t}{C_m} \right) \right] d\eta \quad (12.77)$$

or

$$p_f^* = \int_0^\infty \frac{1}{\eta} J_0(\sqrt{\lambda} \eta) \left[1 - \exp\left(-\frac{\eta^2}{1+\eta^2} \frac{\lambda t^*}{4} \right) \right] d\eta \qquad (12.78)$$

where t^* is redefined from Eq. (12.60) as

$$t^* = \frac{4\kappa_f}{C_m r^2} t \qquad (12.79)$$

Example: *Plot type curves for the Barenblatt solution for a range of parameters.*

Pressure solutions based on Eqs. (12.72) and (12.75) have been programmed in the *Mathematica* macro *DblPoro.m*, listed in Appendix C, respectively as *PfBarenblatt* and *PmBarenblatt*. In Figure 12.5, we plot the normalized pressure p_f^* and p_m^* versus the dimensionless time t^* in log-log scale. These relations should be functions

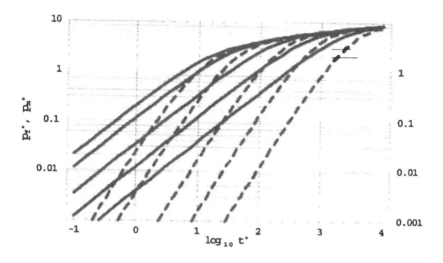

FIGURE 12.5. Type curves for Barenblatt simplified model. (Solid lines: p_f^*; dashed lines: p_m^*; from the rightmost curve to the left: $\lambda = 0.0025$, 0.01, 0.04, 0.25, and 1.)

of the interporosity coefficient λ only. The family of curves are presented for a number of λ values, which can be compared to the published result.[111]

Appendix A
LAPLACE TRANSFORM

The Laplace transform is a powerful tool that can reduce a partial differential equation to an ordinary differential equation, and an ordinary differential equation to an algebraic equation. It is normally applied to the time variable to resolve time derivatives. Given a function of time, $f(t)$, its Laplace transform $\tilde{f}(p)$ is defined as

$$\mathcal{L}\{f(t)\} = \tilde{f}(p) = \int_0^\infty f(t)\, e^{-pt}\, dt \qquad (A.1)$$

where p is the *Laplace transform parameter**. The inverse transformation is symbolized as

$$\mathcal{L}^{-1}\left\{\tilde{f}(p)\right\} = f(t) \qquad (A.2)$$

It is defined by a contour integral in the complex variable plane. In this book, the inverse transform is carried out either by a table of known results (see Table A-1), or by a numerical algorithm. The contour integral formula is not used.

A few fundamental properties of the Laplace transform are summarized below. The reader is referred to Churchill[33] or any other advanced engineering mathematics books for more detail.

---

*It is customary to use the letter s to denote the Laplace transform parameter. However, to avoid confusion with the drawdown variable s used throughout the book, we shall use the notation p in place of s.

The Laplace transform can resolve a time derivative into an algebraic form as

$$\mathcal{L}\left\{\frac{\partial f(t)}{\partial t}\right\} = p\tilde{f}(p) - f(0) \qquad (A.3)$$

where $f(0)$ is the initial condition. The order of the Laplace transform and a differentiation is interchangeable, such that

$$\mathcal{L}\left\{\frac{\partial f(x,t)}{\partial x}\right\} = \frac{\partial \mathcal{L}\{f(x,t)\}}{\partial x} = \frac{\partial \tilde{f}(x,p)}{\partial x} \qquad (A.4)$$

We can obtain from the definition Eq. (A.1) that

$$\mathcal{L}^{-1}\left\{a\tilde{f}(p)\right\} = af(t) \qquad (A.5)$$

and

$$\mathcal{L}^{-1}\left\{\tilde{f}(ap)\right\} = \frac{1}{a}f\left(\frac{t}{a}\right) \qquad (A.6)$$

Other useful formulae are

$$\mathcal{L}^{-1}\left\{e^{-ap}\tilde{f}(p)\right\} = f(t-a)\,\mathrm{H}(t-a) \qquad (A.7)$$

where H is the Heaviside unit step function, and

$$\mathcal{L}^{-1}\left\{\int_p^\infty \tilde{f}(p)\,dp\right\} = \frac{1}{t}f(t) \qquad (A.8)$$

Finally, there is a very powerful property of the Laplace transform, known as the *convolutional theorem*, which is a key to the solution of a number of problems in the book. The theorem takes the form

$$\mathcal{L}^{-1}\left\{\tilde{f}_1(p)\,\tilde{f}_2(p)\right\} = \int_0^t f_1(t-\tau)\,f_2(t)\,d\tau \qquad (A.9)$$

The integral on the right-hand-side is called a *convolutional integral*. Equation (A.9) can be used to find the Laplace inverse of the product of $\tilde{f}_1(p)$ and $\tilde{f}_2(p)$ if the individual inverse transforms of them, $f_1(t)$ and $f_2(t)$, are known. In this book we also use it for a different purpose. When a convolutional integral exists by the application of the Duhamel principle of superposition, we can apply the Laplace transform and the convolutional theorem to reduce such an integral to an algebraic expression:

$$\mathcal{L}\left\{\int_0^t f_1(t - \tau)\, f_2(t)\, d\tau\right\} = \tilde{f}_1(p)\, \tilde{f}_2(p) \tag{A.10}$$

Such technique has been applied to the integro-differential equations found in several of the aquifer theories.

Table A-1: A short table of the Laplace transform

$f(t)$	$\tilde{f}(p)$
1	p^{-1}
t	p^{-2}
$H(t - t_o)$	$p^{-1}\exp\left(-pt_o\right)$
$H(t - 0)$	p^{-1}
$\delta(t - t_o)$	$\exp\left(-pt_o\right)$
$\delta(t - 0)$	1
$\exp(-at)$	$1/(p - a)$
$t\exp(-t)$	$1/(1 + p)^2$
$E_1\left(t^{-1}\right)$	$2p^{-1}K_0\left(2\sqrt{p}\right)$
$t^{-1}\exp(-t^{-1})$	$2K_0\left(2\sqrt{p}\right)$
$-\gamma - \ln t$	$(\ln p)/p$

† $\gamma = 0.5772\ldots$ is the Euler number.

Appendix B

FORTRAN PROGRAMS

B.1 Theis Well Function $W(u)$

B.1.1 Direct Solution in Time

```
      Program TheisW
c Evaluate Theis well function W(u)
10    write (*,*) 'Enter u => '
      read (*,*) u
      if (u.le.0.) stop
      write (*,*) 'W(u) =', w(u)
      go to 10
      end
c----------------------------------------------------------------
      function w(u)
      if (u.le.1.) then
      w = -0.5772157 - log(u)
      ser = -u
      do 10 n=1,10
      w = w - ser
      if (abs(ser/w).lt.1.e-5) return
      ser = ser*(-u)*n/(n+1)**2
10    continue
      else
      w = exp(-u)/u*(u**2+2.334733*u+0.250621)/
     1    (u**2+3.330657*u+1.681534)
      endif
      return
      end
```

B.1.2 Solution by Laplace Inversion

```
      Program TheisWL
c Laplace inverse of W(u) by Stehfest algorithm
      implicit real*8(a-h,o-z)
      common ci(20), n
10    write (*,*) 'Enter u = '
      read (*,*) u
      if (u.le.0) stop
      n = 10
      if (u.ge.1.0) n = 18
      call stehcoef
      write (*,*) 'W(u) =', ft(1/u,0.d0,0.d0)
      go to 10
      end
c------------------------------------------------------------
      include 'Bessel.fi'
      include 'StehCoef.fi'
      include 'StehFt.fi'
      include 'Fctorial.fi'
c------------------------------------------------------------
      function fp(p,arg1,arg2)
c User defined function of Laplace transform expression
      implicit real*8(a-h,o-z)
      fp = 2/p*besselk0(2*dsqrt(p))
      return
      end
```

B.2 Jacob-Lohman Well Functions

B.2.1 $F(u_w, \rho)$

```
      Program JaLoF
c Laplace inverse of Jacob-Lohman well function F(uw,rho)
      implicit real*8(a-h,o-z)
      common ci(20),n
10    write (*,*) 'Enter uw, rho = '
      read (*,*) u, rho
      if (u.le.0) stop
      n = 10
      call stehcoef
```

```
      write (*,*) 'F(uw,rho) =',ft(1/u,rho,0.d0)
      go to 10
      end
c-----------------------------------------------------------------
      include 'Bessel.fi'
      include 'StehCoef.fi'
      include 'StehFt.fi'
      include 'Fctorial.fi'
c-----------------------------------------------------------------
      function fp(p,rho,arg2)
      implicit real*8(a-h,o-z)
      fp = (1/p)*besselk0(2*dsqrt(p)*rho)/
     1     besselk0(2*dsqrt(p))
      return
      end
```

B.2.2 $G(u_w)$

```
      Program JaLoG
c Laplace inverse of Jacob-Lohman well function G(uw)
      implicit real*8(a-h,o-z)
      common ci(20),n
10    write (*,*) 'Enter uw = '
      read (*,*) u
      if (u.le.0) stop
      n = 10
      if (u.ge.1.0) n = 18
      call stehcoef
      write(*,*) 'G(uw) = ', 1.0/ft(1/u,0.d0,0.d0)
      go to 10
      end
c-----------------------------------------------------------------
      include 'Bessel.fi'
      include 'StehCoef.fi'
      include 'StehFt.fi'
      include 'Fctorial.fi'
c-----------------------------------------------------------------
      function fp(p,arg1,arg2)
      implicit real*8(a-h,o-z)
      fp = 0.5/p/besselk0(2*dsqrt(p))
      return
```

```
      end
```

B.3   Papadopulos-Cooper Well Functions

B.3.1 $W(u, \alpha, \rho)$

```
      Program PaCoW
c Laplace inverse of Papadopulos-Cooper well function
c  W(u,alpha,rho)
      implicit real*8(a-h,o-z)
      common ci(20),n
10    write (*,*) 'Enter u, alpha, rho = '
      read (*,*) u, alpha, rho
      if (u.le.0) stop
      n = 10
      if (u.ge.1.0) n = 18
      call stehcoef
      write (*,*) 'W(u,alpha,rho) = ', ft(1/u,alpha,rho)
      go to 10
      end
c----------------------------------------------------------------
      include 'Bessel.fi'
      include 'StehCoef.fi'
      include 'StehFt.fi'
      include 'Fctorial.fi'
c----------------------------------------------------------------
      function fp(p,alpha,rho)
      implicit real*8(a-h,o-z)
      fp = besselk0(2*dsqrt(p))/p/(sqrt(p)/rho*
1         besselk1(2*dsqrt(p)/rho)
2      + (p/alpha/rho**2)*besselk0(2*dsqrt(p)/rho))
      return
      end
```

B.3.2 $F(u_w, \alpha)$

```
      Program PaCoF
c Laplace inverse of Papadopulos-Cooper well function
c  F(uw,alpha)
      implicit real*8(a-h,o-z)
      common ci(20),n
```

```
10      write (*,*) 'Enter uw, alpha = '
        read (*,*) u, alpha
        if (u.le.0) stop
        n = 10
        if (u.ge.1.0) n = 18
        call stehcoef
        write (*,*) 'F(uw,alpha) = ',ft(1/u,alpha,0.d0)
        go to 10
        end
c------------------------------------------------------------
        include 'Bessel.fi'
        include 'StehCoef.fi'
        include 'StehFt.fi'
        include 'Fctorial.fi'
c------------------------------------------------------------
        function fp(p,alpha,arg2)
        implicit real*8(a-h,o-z)
        fp = besselk0(2*dsqrt(p))/p/(sqrt(p)*besselk1
     1       (2*dsqrt(p)) + (p/alpha)*besselk0(2*dsqrt(p)))
        return
        end
```

B.4   Hantush-Jacob Well Function $W(u, \beta)$

```
        Program HantJcbW
c Laplace inverse of W(u,beta)
        implicit real*8(a-h,o-z)
        common ci(20),n
10      write (*,*) 'Enter u, beta = '
        read (*,*) u, beta
        if (u.le.0) stop
        n = 10
        if (u.ge.1.0) n = 18
        call stehcoef
        write(*,*) 'W(u,beta) = ', ft(1/u,beta,0.d0)
        go to 10
        end
c------------------------------------------------------------
        include 'Bessel.fi'
        include 'StehCoef.fi'
```

```
      include 'StehFt.fi'
      include 'Fctorial.fi'
```
```
c-------------------------------------------------------------------
      function fp(p,beta,arg2)
      implicit real*8(a-h,o-z)
      fp = 2/p*besselk0(dsqrt(beta**2+4*p))
      return
      end
```

B.5 Neuman Well Functions

B.5.1 $W(u, u_y, \Gamma)$

```
      Program NeumanW1
c Neuman's unconfined aquifer well function W(u,uy,Gamma)
      implicit real*8(a-h,o-z)
      dimension r(0:200)
c Define integration ranges using even roots of J0
      nr = 100
      r(0) = 0.0d00
      r(1) = 5.520078110080565
      do 5 i=2, nr
      r(i) = bj0root(r(i-1)+2*3.14159)
5     continue
c Evaluate and print the well function
10    write (*,*) 'Enter u, uy and Gamma = '
      read (*,*) u, uy, gamma
      if (u.le.0.0) stop
      sigma=u/uy
      write(*,*)'W(u,uy,gamma)= ',wf(u,sigma,gamma,nr,r)
      goto 10
      end
```
```
c-------------------------------------------------------------------
      include 'Bessel.fi'
      include 'BJ0Root.fi'
      include 'Gauss.fi'
      include 'epsilon.fi'
```
```
c-------------------------------------------------------------------
      function wf(u,sigma,gamma,nr,r)
c Neuman's well function
      implicit real*8(a-h,o-z)
```

```
      parameter (ng=30)
      dimension x(ng), w(ng), wsum(0:200), r(0:200)
      gammasq=dsqrt(gamma)
      wsum(0)=0.0d0
c Integrate by subintervals
      do 20 j=1,nr
      call gauleg(r(j-1)/gammasq,r(j)/gammasq,x,w,ng)
      dwf=0.0
c Perform Gaussian quadrature
      do 12 i=1,ng
      dwf=dwf+w(i)*func(u,sigma,gamma,x(i))
12    continue
      wsum(j) = wsum(j-1) + dwf
20    continue
c Use epsilon algorithm to extrapolate
      wf = epsilonn(nr,wsum)
      return
      end
c------------------------------------------------------------
      function func(u,sigma,gamma,x)
      implicit real*8(a-h,o-z)
      func=4.d0*x*besselj0(x*gamma**0.5)*summ(x,u,sigma,gamma)
      return
      end
c------------------------------------------------------------
      function summ(x,u,sigma,gamma)
      implicit real*8(a-h,o-z)
      parameter (nsum=12,pi=3.141592654)
      dimension psum(0:200)
      psum(0)= an(u,sigma,gamma,x,0)
      do 20 m=1,nsum
      dsum=an(u,sigma,gamma,x,m)
      psum(m)=psum(m-1) + dsum
20    continue
      summ = epsilonn(nsum,psum)
      return
      end
c------------------------------------------------------------
      function an(u,sigma,gamma,x,n)
      implicit real*8(a-h,o-z)
      if(n.eq.0) then
```

```fortran
      g0=groot(sigma,x,0)
      if (g0.lt.50) an=(1.d0-dexp(-gamma*(x**2-g0**2)/
     1   4.d0/u))*dtanh(g0)/(x**2+(1.0+sigma)*g0**2-
     2   (x**2-g0**2)**2/sigma)/g0
      if (g0.ge.50) an=(1.d0-dexp(-gamma*(x**2-g0**2)/4.d0/u))
     1   /(x**2+(1.0+sigma)*g0**2-(x**2-g0**2)**2/sigma)/g0
      else
      gn=groot(sigma,x,n)
      an=(1.d0-dexp(-gamma*(x**2+gn**2)/4.d0/u))*dtan(gn)/
     1   (x**2-(1.0+sigma)*gn**2-(x**2+gn**2)**2/sigma)/gn
      endif
      return
      end
c----------------------------------------------------------------
      function groot(sigma,x,k)
c Find root of gamma_n by Newton-Raphson method
      implicit real*8(a-h,o-z)
      parameter (xacc=1.0d-8)
      if(k.eq.0) then
      groot=x
      do 11 j=1,100
      if(groot.lt.50) dg=(sigma*groot*dtanh(groot)-(x**2
     1   -groot**2))/(2*groot+groot*sigma/dcosh(groot)**2
     2   +sigma*dtanh(groot))
      if (groot.ge.50) dg=(sigma*groot-(x**2-groot**2))/
     1   (2*groot+sigma)
      groot=groot-dg
      if(dabs(dg/groot).lt.xacc) return
11    continue
      else
      groot=(k-0.5)*3.1415926
      do 12 j=1,100
      dg=(sigma*groot*dsin(groot)+(x**2+groot**2)*
     1   dcos(groot))/((2+sigma)*groot*dcos(groot)-
     2   (x**2+groot**2-sigma)*dsin(groot))
      groot=groot-dg
      if(dabs(dg/groot).lt.xacc) return
12    continue
      endif
      return
      end
```

B.5.2 $W(u, \Gamma)$ and $W(u_y, \Gamma)$

```
      Program NeumanW2
c Neuman's type A and B unconfined aquifer well function
c  W(uy,Gamma)
      implicit real*8(a-h,o-z)
      dimension r(0:200)
      write(*,*)'Enter 1 for W(u,Gamma);2 for W(uy,Gamma)'
      read(*,*) ncase
c Define integration ranges using even roots of J0
      nr = 100
      r(0) = 0.0d00
      r(1) = 5.520078110080565
      do 5 i=2, nr
      r(i) = bj0root(r(i-1)+2*3.14159)
5     continue
c Evaluate and print the well function
10    write (*,*) 'Enter u (or uy) and Gamma = '
      read (*,*) u, gamma
      if (u.le.0.0) stop
      write(*,*)'W(u(uy),gamma)=',wf(u,gamma,nr,r,ncase)
      goto 10
      end
c-----------------------------------------------------------
      include 'Bessel.fi'
      include 'BJ0Root.fi'
      include 'Gauss.fi'
      include 'epsilon.fi'
c-----------------------------------------------------------
      function wf(u,gamma,nr,r,ncase)
c Neuman's well function
      implicit real*8(a-h,o-z)
      parameter (ng=30)
      dimension x(ng), w(ng), wsum(0:200), r(0:200)
      gammasq=dsqrt(gamma)
      wsum(0)=0.0d0
c Integrate by subintervals
      do 20 j=1,nr
      call gauleg(r(j-1)/gammasq,r(j)/gammasq,x,w,ng)
      dwf=0.0
c Perform Gaussian quadrature
      do 12 i=1,ng
```

```
      dwf=dwf+w(i)*func(u,gamma,x(i),ncase)
12    continue
      wsum(j) = wsum(j-1) + dwf
20    continue
c Use epsilon algorithm to extrapolate
      wf = epsilonn(nr,wsum)
      return
      end
c--------------------------------------------------------------
      function func(u,gamma,x,ncase)
      implicit real*8(a-h,o-z)
      if (ncase.eq.1)func=64.d0*x*besselj0(x*gamma**0.5)
1         *summ(x,u,gamma,ncase)
      if (ncase.eq.2)func=4.0d0*x*besselj0(x*gamma**0.5)
1         *((1.0d0-dexp(-gamma*x*dtanh(x)/(4.0d0*u)))*
2         dtanh(x)/(2.0d0*x**3)+summ(x,u,gamma,ncase))
      return
      end
c--------------------------------------------------------------
      function summ(x,u,gamma,ncase)
      implicit real*8(a-h,o-z)
      parameter (nsum=12,pi=3.141592654)
      dimension psum(0:200)
      psum(0)= 0.0d0
      do 20 m=1,nsum
      a = 4.0d0*x**2+(2.0d0*m-1.0d0)**2*pi**2
      denom = (2.0d0*m-1.0d0)**2*pi**2*a
      if(ncase.eq.1)dsum=(1-dexp(-gamma/16.d0/u*a))/denom
      if(ncase.eq.2)dsum=16.d0/denom
      psum(m)=psum(m-1) + dsum
20    continue
      summ = epsilonn(nsum,psum)
      return
      end
```

B.6 Hantush-Neuman Well Functions $W_{1,2}(u, \beta, \eta)$

```
      Program HantNeuW
c Laplace inverse of W1(u,beta,eta) and W2(u,beta,eta)
      implicit real*8(a-h,o-z)
```

```
      common ci(20),n
      common/case/ncase
      write (*,*) 'Enter 1 for W1, 2 for W2 '
      read (*,*) ncase
10    write (*,*) 'Enter u, beta, eta = '
      read (*,*) u, beta, eta
      if (u.le.0) stop
      n = 10
      if (u.ge.1.0) n = 18
      call stehcoef
      write(*,*) 'W(u,beta,eta) = ',ft(1/u,beta,eta)
      go to 10
      end
c-------------------------------------------------------------
      include 'Bessel.fi'
      include 'StehCoef.fi'
      include 'StehFt.fi'
      include 'Fctorial.fi'
c-------------------------------------------------------------
      function fp(p,beta,eta)
      implicit real*8(a-h,o-z)
      common/case/ncase
      goto (10,20), ncase
10    fp = 2/p*besselk0(dsqrt(4*p+2*beta*dsqrt(eta*p)/
     1     dtanh(2*dsqrt(eta*p)/beta)))
      return
20    fp = 2/p*besselk0(dsqrt(4*p+2*beta*dsqrt(eta*p)*
     1     dtanh(2*dsqrt(eta*p)/beta)))
      return
      end
```

B.7   Hantush Small Time Leaky Aquifer Well Function $H(u, \beta')$

```
      Program HantushH
c Laplace inverse of H(u,beta')
      implicit real*8(a-h,o-z)
      common ci(20),n
10    write (*,*) "Enter u, beta' = "
      read (*,*) u, beta
```

```
      if (u.le.0) stop
      n = 10
      if (u.ge.1.0) n = 18
      call stehcoef
      write(*,*) "W(u,beta') =",ft(1/u,beta,0.d0)
      go to 10
      end
c----------------------------------------------------------------
      include 'Bessel.fi'
      include 'StehCoef.fi'
      include 'StehFt.fi'
      include 'Fctorial.fi'
c----------------------------------------------------------------
      function fp(p,beta,arg2)
      implicit real*8(a-h,o-z)
      fp = 2/p*besselk0(dsqrt(4*p+8*beta*dsqrt(p)))
      return
      end
```

B.8   Variable Pumping Rate

B.8.1   By Convolutional Integral

```
      Program VarPump
c Evaluating drawdown from a variable pumping schedule
      implicit real*8(a-h,o-z)
      dimension x(200), w(200)
      write(*,*) 'Transmissivity (T) = '
      read(*,*) Tr
      write(*,*) 'Storativity (S) = '
      read(*,*) St
      write(*,*) 'Distance from pumping well (r) ='
      read(*,*) r
      write(*,*) 'Pumping rate parameters (Q0, k) ='
      read(*,*) q0, ak
      write(*,*) 'Number of Gauss integration points ='
      read(*,*) ng
10    write(*,*) 'Enter time (t) (enter 0 to stop) ='
      read(*,*) t
      if (t.le.0) stop
      call gauleg(0,t,x,w,ng)
```

```
      s=0.
      do 20 i=1,ng
      s=s+w(i)*qw(q0,ak,x(i))*dw(St,Tr,r,t,x(i))
20 continue
      write(*,*) 's = ', s/(4.*3.14159*Tr)
      go to 10
      end
c--------------------------------------------------------------
      include 'gauss.fi'
c--------------------------------------------------------------
      function qw(q0,ak,t)
c Pumping rate as function of time (replace by your own)
      implicit real*8(a-h,o-z)
      qw = q0*(1.-exp(-ak*t))
      return
      end
c--------------------------------------------------------------
      function dw(St,Tr,r,t,tau)
c Derivative of the well function
c For Theis well function
      implicit real*8(a-h,o-z)
      dw=exp(-r**2*St/4./Tr/(t-tau))/(t-tau)
      return
      end
```

B.8.2   By Laplace Transform

```
      Program VarPumpL
c Evaluating drawdown from a variable pumping schedule
c using Laplace Transform technique
      implicit real*8(a-h,o-z)
      common ci(20),n
      common/para/St,Tr,r,q0,ak
      write(*,*) 'Transmissivity (T) ='
      read(*,*) Tr
      write(*,*) 'Storativity (S) ='
      read(*,*) St
      write(*,*) 'Distance from pumping well (r) ='
      read(*,*) r
      write(*,*) 'Pumping rate parameters (Q0, k) ='
      read(*,*) q0, ak
```

```
         n = 15
         call stehcoef
10       write(*,*) 'Enter time (t) ='
         read(*,*) t
         if (t.le.0) stop
         write(*,*)'s = ', ft(t,0.d0,0.d0)
         go to 10
         end
c----------------------------------------------------------------------
         include 'Bessel.fi'
         include 'StehCoef.fi'
         include 'StehFt.fi'
         include 'Fctorial.fi'
c----------------------------------------------------------------------
         function fp(p,arg1,arg2)
         implicit real*8(a-h,o-z)
         common/para/St,Tr,r,q0,ak
c Laplace transform of pumping rate (replace by your own)
         qw = q0*(1./p-1./(p+ak))
c Laplace transform of Theis solution
         sp = 1./(4*3.14159*Tr)*(2/p)*besselk0(r*dsqrt(p*St/Tr))
         fp = qw*sp*p
         return
         end
```

B.9   Three-Aquifer System

```
         Program Multi
c For three-aquifer-four-aquitard system
c p:    Laplace transform parameter
c Q:    Pumping rate
c r:    Distance from pumping well
c t:    time from start of pumping
c Tr:   transmissivity of aquifer
c St:   storativity of aquifer
c Stp:  storativity of aquitard
c bp:   thickness of aquitard
c aKp:  hydraulic conductivity of aquitard
         implicit real*8(a-h,o-z)
         common/pr/r,p,Tr(3),St(3),Stp(0:3),bp(0:3),aKp(0:3)
```

```
      common/coef/eksq(3),phi(3),a(3,3),sp(3,3),s(3,3)
      common ci(20), n
      open (1,file='multi.dat')
      read (1,*) aKp(0), Stp(0), bp(0)
      do 10 layer=1,3
      read (1,*) Tr(layer), St(layer)
      read (1,*) aKp(layer), Stp(layer), bp(layer)
10    continue
      write(*,*)"Pumping rate for aquifer 1, 2, 3 => "
      read (*,*) Q1, Q2, Q3
      n = 10
      call stehcoef
20    write(*,*)"Enter r and t (enter 0, 0 to stop) =>"
      read (*,*) r, t
      if (r.eq.0) stop
      call ft(t)
      write(*,*)"s(1) = ", Q1*s(1,1)+Q2*s(1,2)+Q3*s(1,3)
      write(*,*)"s(2) = ", Q1*s(2,1)+Q2*s(2,2)+Q3*s(2,3)
      write(*,*)"s(3) = ", Q1*s(3,1)+Q2*s(3,2)+Q3*s(3,3)
      go to 20
      end
c-------------------------------------------------------
      include 'Bessel.fi'
      include 'StehCoef.fi'
      include 'Fctorial.fi'
c-------------------------------------------------------
      subroutine ft(t)
      implicit real*8(a-h,o-z)
      common ci(20), n
      common/pr/r,p,Tr(3),St(3),Stp(0:3),bp(0:3),aKp(0:3)
      common/coef/eksq(3),phi(3),a(3,3),sp(3,3),s(3,3)
      do 20 i=1,3
      do 20 j=1,3
      s(i,j) = 0.0d0
20    continue
      do 10 k=1,n
      p = k*dlog(2.0d0)/t
      call fp
      do 10 i=1,3
      do 10 j=1,3
      s(i,j) = s(i,j)+(dlog(2.0d0)/t)*ci(k)*sp(i,j)
```

```
10    continue
      return
      end
c-----------------------------------------------------------------
      subroutine fp
      implicit real*8(a-h,o-z)
      common/pr/r,p,Tr(3),St(3),Stp(0:3),bp(0:3),aKp(0:3)
      common/coef/eksq(3),phi(3),a(3,3),sp(3,3),s(3,3)
      call coefa
      call eigenval
      call green
      sp(1,1)=0.159155/Tr(1)/p*(phi(1)+(eksq(2)+eksq(3)-
     1    a(2,2)-a(3,3))*phi(2)+((eksq(3)-a(2,2))*
     2    (eksq(3)-a(3,3))-a(2,3)*a(3,2))*phi(3))
      sp(2,2)=0.159155/Tr(2)/p*(phi(1)+(eksq(2)+eksq(3)-
     1    a(1,1)-a(3,3))*phi(2)+(eksq(3)-a(1,1))*
     2    (eksq(3)-a(3,3))*phi(3))
      sp(3,3)=0.159155/Tr(3)/p*(phi(1)+(eksq(2)+eksq(3)-
     1    a(1,1)-a(2,2))*phi(2)+((eksq(3)-a(1,1))*
     2    (eksq(3)-a(2,2))-a(1,2)*a(2,1))*phi(3))
      sp(1,2)=0.159155/Tr(2)/p*a(1,2)*(phi(2)+(eksq(3)-
     1    a(3,3))*phi(3))
      sp(1,3)=0.159155/Tr(3)/p*a(1,2)*a(2,3)*phi(3)
      sp(2,1)=0.159155/Tr(1)/p*a(2,1)*(phi(2)+(eksq(3)-
     1    a(3,3))*phi(3))
      sp(2,3)=0.159155/Tr(3)/p*a(2,3)*(phi(2)+(eksq(3)-
     1    a(1,1))*phi(3))
      sp(3,1)=0.159155/Tr(1)/p*a(2,1)*a(3,2)*phi(3)
      sp(3,2)=0.159155/Tr(2)/p*a(3,2)*(phi(2)+(eksq(3)-
     1    a(1,1))*phi(3))
      return
      end
c-----------------------------------------------------------------
      subroutine coefa
      implicit real*8(a-h,o-z)
      common/pr/r,p,Tr(3),St(3),Stp(0:3),bp(0:3),aKp(0:3)
      common/coef/eksq(3),phi(3),a(3,3),sp(3,3),s(3,3)
      a(1,1)=(p/Tr(1))*(aKp(0)*g(0)/bp(0)+aKp(1)*f(1)/
     1    bp(1)+St(1))
      a(2,2)=(p/Tr(2))*(aKp(1)*f(1)/bp(1)+aKp(2)*f(2)/
     1    bp(2)+St(2))
```

```fortran
      a(3,3)=(p/Tr(3))*(aKp(2)*f(2)/bp(2)+aKp(3)*g(3)/
     1    bp(3)+St(3))
      a(1,2)=-(p/Tr(1))*(aKp(1)*h(1)/bp(1))
      a(2,1)=-(p/Tr(2))*(aKp(1)*h(1)/bp(1))
      a(2,3)=-(p/Tr(2))*(aKp(2)*h(2)/bp(2))
      a(3,2)=-(p/Tr(3))*(aKp(2)*h(2)/bp(2))
      a(1,3)=0.d0
      a(3,1)=0.d0
      return
      end
c------------------------------------------------------------------
      subroutine eigenval
      implicit real*8(a-h,o-z)
      complex*16 c0,c1,c2,d
      common/pr/r,p,Tr(3),St(3),Stp(0:3),bp(0:3),aKp(0:3)
      common/coef/eksq(3),phi(3),a(3,3),sp(3,3),s(3,3)
      c0=a(1,1)*a(2,3)*a(3,2)+a(1,2)*a(2,1)*a(3,3)-
     1    a(1,1)*a(2,2)*a(3,3)
      c1=-a(1,2)*a(2,1)+a(1,1)*a(2,2)-a(2,3)*a(3,2)+
     1    a(1,1)*a(3,3)+a(2,2)*a(3,3)
      c2=-a(1,1)-a(2,2)-a(3,3)
      d=(-27*c0+9*c1*c2-2*c2**3+cdsqrt(4*(3*c1-c2**2)**3
     1 +(-27*c0+9*c1*c2-2*c2**3)**2))**(1/3.)
      eksq(1)=dreal(-c2/3.-(2**(1/3.)*(3*c1-c2**2))/
     1 (3.*d)+d/(3.*2**(1/3.)))
      eksq(2)=dreal(-c2/3.+((1-(0,1)*dsqrt(3.d0))*
     1    (3*c1-c2**2))/(3.*2**(2/3.)*d)-((1+(0,1)*
     2    dsqrt(3.d0))*d)/(6.*2**(1/3.)))
      eksq(3)=dreal(-c2/3.+((1+(0,1.d0)*dsqrt(3.d0))*
     1    (3*c1-c2**2))/(3.*2**(2/3.)*d)-((1-(0,1)*
     2    dsqrt(3.d0))*d)/(6.*2**(1/3.)))
      return
      end
c------------------------------------------------------------------
      subroutine green
      implicit real*8(a-h,o-z)
      common/pr/r,p,Tr(3),St(3),Stp(0:3),bp(0:3),aKp(0:3)
      common/coef/eksq(3),phi(3),a(3,3),sp(3,3),s(3,3)
      phi(1)=besselk0(dsqrt(eksq(1))*r)
      phi(2)=(besselk0(dsqrt(eksq(1))*r)-
     1 besselk0(dsqrt(eksq(2))*r))/(eksq(1)-eksq(2))
```

```fortran
      phi(3)=besselk0(dsqrt(eksq(1))*r)/(eksq(1)-eksq(2))/
     1    (eksq(1)-eksq(3))+besselk0(dsqrt(eksq(2))*r)/
     2    (eksq(2)-eksq(1))/(eksq(2)-eksq(3))+besselk0(dsqrt
     3    (eksq(3))*r)/(eksq(3)-eksq(1))/(eksq(3)-eksq(2))
      return
      end
c--------------------------------------------------------------
      function f(i)
      implicit real*8(a-h,o-z)
      common/pr/r,p,Tr(3),St(3),Stp(0:3),bp(0:3),aKp(0:3)
      if (aKp(i).le.0) then
      f=0
      else
      f=dsqrt(Stp(i)*bp(i)/aKp(i)/p)/
     1  dtanh(dsqrt(Stp(i)*bp(i)*p/aKp(i)))
      endif
      return
      end
c--------------------------------------------------------------
      function g(i)
      implicit real*8(a-h,o-z)
      common/pr/r,p,Tr(3),St(3),Stp(0:3),bp(0:3),aKp(0:3)
      if (aKp(i).le.0) then
      g=0
      else
      g=dsqrt(Stp(i)*bp(i)/aKp(i)/p)*
     1  dtanh(dsqrt(Stp(i)*bp(i)*p/aKp(i)))
      endif
      return
      end
c--------------------------------------------------------------
      function h(i)
      implicit real*8(a-h,o-z)
      common/pr/r,p,Tr(3),St(3),Stp(0:3),bp(0:3),aKp(0:3)
      if (aKp(i).le.0) then
      h=0
      else
      h=dsqrt(Stp(i)*bp(i)/aKp(i)/p)/
     1  dsinh(dsqrt(Stp(i)*bp(i)*p/aKp(i)))
      endif
      return
```

```
      end
```

Sample Data File: Multi.dat

```
0.005     0.0002     50.    K'(0), S'(0), b'(0)
3000.     0.0001             T(1),  S(1)
0.02      0.0008     30.    K'(1), S'(1), b'(1)
500.      0.00005            T(2),  S(2)
0.05      0.0005     20.    K'(2), S'(2), b'(2)
300.      0.00002            T(3),  S(3)
0.0008    0.00008    45.    K'(3), S'(3), b'(3)
Data for a three-aquifer-four-aquitard system
```

B.10   Parameter Determination

InvHJ.for

```
      Program InvHJ
c Parameter determination for Hantush-Jacob solution
      implicit real*8(a-h,o-z)
      PARAMETER(GTOL=1.0d-4)
      dimension para(3)
      EXTERNAL func,dfunc
      common ci(20),n
      common/dd/t(100),s(100),Q,r,Pi,ndat
      open (1,file=''drawdown.dat'')
      read(1,*) Q, r, ndat
      read(1,*) (t(i),i=1,ndat)
      read(1,*) (s(i),i=1,ndat)
      write(*,*) ''Enter initial T, S, beta ''
      read(*,*) para(1), para(2), para(3)
      do 10 i=1,3
10    para(i)=dlog10(para(i))
      n=10
      Pi=3.141592654
      call stehcoef
      call dfpmin(para,3,gtol,iter,fret,func,dfunc)
      write(*,*) 'Iterations:',iter
      write(*,*) 'T = ',10**para(1),' S = ',10**para(2),
     1    '' beta = '',10**para(3)
      write(*,*) 'Average Error',dsqrt(fret/ndat)
```

```
      stop
      end
c----------------------------------------------------------------
      include 'Bessel.fi'
      include 'StehCoef.fi'
      include 'Fctorial.fi'
      include 'BFGS.fi'
c----------------------------------------------------------------
      FUNCTION func(x)
c Define objective function
c Log10(T) = x(1), Log10(S) = x(2), Log10(beta) = x(3)
      implicit real*8(a-h,o-z)
      dimension x(3)
      common/dd/t(100),s(100),Q,r,Pi,ndat
      func=0.d0
      Tr=10**x(1)
      St=10**x(2)
      beta=10**x(3)
      do 10 i=1,ndat
      u=r**2*St/4/Tr/t(i)
      call wellfunc(1/u,beta,w,wd)
      func=func+(dlog10(Q/4/Pi/Tr*w)-dlog10(s(i)))**2
10    continue
      return
      end
c----------------------------------------------------------------
      SUBROUTINE dfunc(x,df)
c Define derivatives of objective function
      implicit real*8(a-h,o-z)
      dimension x(3),df(3)
      common/dd/t(100),s(100),Q,r,Pi,ndat
      do 20 i=1,3
20 df(i)=0.d0
      Tr=10**x(1)
      St=10**x(2)
      beta=10**x(3)
      do 10 i=1,ndat
      u=r**2*St/4/Tr/t(i)
      call wellfunc(1/u,beta,w,wd)
      err=2*(dlog10(Q/4/Pi/Tr*w)-dlog10(s(i)))
      df(1)=df(1)-err
```

```
      df(2)=df(2)-dexp(-u-beta**2/4/u)*err/w
      df(3)=df(3)-beta**2/2*wd*err/w
10    continue
      df(1)=df(1)-df(2)
      return
      end
c----------------------------------------------------------------
      subroutine wellfunc(tstar,beta,w,wd)
c Evaluate Hantush-Jacob well function and derivative
      implicit real*8(a-h,o-z)
      common ci(20), n
      w = 0.0d0
      wd = 0.0d0
      a = dlog(2.0d0)/tstar
      do 10 i=1,n
      p = i*a
      call fp(p,beta,wp,wpd)
      w = w + ci(i)*wp
      wd = wd + ci(i)*wpd
10    continue
      w = a*w
      wd = a*wd
      return
      end
c----------------------------------------------------------------
      subroutine fp(p,beta,wp,wpd)
      implicit real*8(a-h,o-z)
      arg=dsqrt(beta**2+4*p)
      wp=2/p*besselk0(arg)
      wpd=4/p*besselk1(arg)/arg
      return
      end
```

Sample Data File: Drawdown.dat

```
3.34  96.0 11                        (Q, r, ndat)
5.  28.  41.  60.  75.  244.  493.  669.  958.  1129.
1185.                                      (time)
0.76  3.3  3.59  4.08  4.39  5.47  5.96  6.11  6.27  6.4
6.42                                    (drawdown)
(* Units: ft, min *)
(* Data from Walton [1987] p. 91 *)
```

B.11   Common Subroutines

B.11.1   Stehfest Inverse Algorithm

StehCoef.fi

```
      subroutine stehcoef
c Coefficients of Stehfest inversion
      implicit real*8(a-h,o-z)
      common ci(20), n
      nhlf = n/2
      do 10 i=1,n
      ci(i) = 0.0d0
      k1 = (i+1)/2
      k2 = min(i,nhlf)
      do 20 k=k1,k2
      ci(i) = ci(i) + k**nhlf*fac(2*k)/fac(nhlf-k)/
     1 fac(k)/fac(k-1)/fac(i-k)/fac(2*k-i)
20    continue
      ci(i) = (-1)**(i+nhlf)*ci(i)
10    continue
      return
      end
```

StehFt.fi

```
      function ft(t,arg1,arg2)
c Perform Stehfest Laplace inversion
      implicit real*8(a-h,o-z)
      common ci(20), n
      ft = 0.0d0
      a = dlog(2.0d0)/t
      do 10 i=1,n
      p = i*a
      ft = ft + ci(i)*fp(p,arg1,arg2)
10    continue
      ft = a*ft
      return
      end
```

Fctorial.fi

```
      function fac(i)
c Factorial i!
```

```
      implicit real*8(a-h,o-z)
      fac = 1.0d0
      if (i.eq.1) return
      do 10 j=2,i
      fac = fac*j
10    continue
      return
      end
```

B.11.2 Bessel Functions

Bessel.fi

```
      function besselk0(x)
c Modified Bessel function K_0
      implicit real*8(a-h,o-z)
      if (x.le.2.0) then
      y=x*x/4.0
      besselk0=(-dlog(x/2.)*besseli0(x))+(-0.57721566+
     1  y*(0.4227842+y*(0.23069756+y*(0.348859d-1+y*
     2  (0.262698d-2+y*(0.1075d-3+y*0.74d-5))))))
      else
      y=(2.0/x)
      besselk0=(dexp(-x)/dsqrt(x))*(1.25331414+y*
     1  (-0.7832358d-1+y*(0.2189568d-1+y*(-0.1062446d-1+
     2  y*(0.587872d-2+y*(-0.25154d-2+y*0.53208d-3))))))
      endif
      return
      end
c--------------------------------------------------------------
      function besselk1(x)
c Modified Bessel function K_1
      implicit real*8(a-h,o-z)
      if (x.le.2.0) then
      y=x*x/4.0
      besselk1=(dlog(x/2.)*besseli1(x))+(1./x)*(1.+y*
     1  (0.15443144+y*(-0.67278579+y*(-0.18156897+y*
     2  (-0.1919402d-1+y*(-0.110404d-2+y*(-0.4686d-4)))))))
      else
      y=2.0/x
      besselk1=(dexp(-x)/dsqrt(x))*(1.25331414+y*
     1  (0.23498619+y*(-0.3655620d-1+y*(0.1504268d-1+y*
```

```fortran
      2 (-0.780353d-2+y*(0.325614d-2+y*(-0.68245d-3))))))))
        endif
        return
        end
c--------------------------------------------------------------------
        function besseli0(x)
c Modified Bessel function I_0
        implicit real*8(a-h,o-z)
        if (x.lt.3.75) then
        y=(x/3.75)**2
        besseli0=1.+y*(3.5156229+y*(3.0899424+y*(1.2067492
      1  +y*(0.2659732+y*(0.360768d-1+y*0.45813d-2)))))
        else
        y=3.75/x
        besseli0=(dexp(x)/dsqrt(x))*(0.39894228+y*
      1 (0.1328592d-1+y*(0.225319d-2+y*(-0.157565d-2+y*
      2 (0.916281d-2+y*(-0.2057706d-1+y*(0.2635537d-1+y*
      3 (-0.1647633d-1+y*0.392377d-2)))))))))
        endif
        return
        end
c--------------------------------------------------------------------
        function besseli1(x)
c Modified Bessel function I_1
        implicit real*8(a-h,o-z)
        if (x.lt.3.75) then
        y=(x/3.75)**2
        besseli1=x*(0.5d0+y*(0.87890594+y*(0.51498869+y*
      1 (0.15084934+y*(0.2658733d-1+y*(0.301532d-2+y*
      2 0.32411d-3))))))
        else
        y = 3.75/x
        besseli1=(dexp(x)/dsqrt(x))*(0.39894228+y*
      1 (-0.3988024d-1+y*(-0.362018d-2+y*(0.163801d-2+y*
      2 (-0.1031555d-1+y*(0.2282967d-1+y*(-0.2895312d-1+
      3 y*(0.1787654d-1+y*(-0.420059d-2)))))))))
        endif
        return
        end
c--------------------------------------------------------------------
        function besselj0(x)
```

```fortran
c Bessel function J_0
      implicit real*8(a-h,o-z)
      if(dabs(x).lt.8.)then
      y=x**2
      besselj0=(57568490574.+y*(-13362590354.+y*(651619640.7+
     1  y*(-11214424.18+y*(77392.33017+y*(-184.9052456))))))/
     2  (57568490411.+y*(1029532985.+y*(9494680.718+y*
     3  (59272.64853+y*(267.8532712+y)))))
      else
      ax=dabs(x)
      z=8.d0/ax
      y=z**2
      xx=ax-.785398164
      besselj0=dsqrt(.636619772/ax)*(dcos(xx)*(1.+y*
     1  (-.1098628627d-2+y*(.2734510407d-4+y*(-.2073370639d-5+
     2  y*.2093887211d-6))))-z*dsin(xx)*(-.1562499995d-1+y*
     3  (.1430488765d-3+y*(-.6911147651d-5+y*(.7621095161d-6+y
     4  *(-.934945152d-7))))))
      endif
      return
      end
c-------------------------------------------------------------
      function besselj1(x)
c Bessel function J_1
      implicit real*8(a-h,o-z)
      if(dabs(x).lt.8.)then
      y=x**2
      besselj1=x*(72362614232.+y*(-7895059235.+y*(242396853.1+
     1  y*(-2972611.439+y*(15704.48260+y*(-30.16036606))))))/
     2  (144725228442.+y*(2300535178.+y*(18583304.74+y*
     3  (99447.43394+y*(376.9991397+y)))))
      else
      ax=dabs(x)
      z=8.0d0/ax
      y=z**2
      xx=ax-2.356194491
      besselj1=dsqrt(.636619772/ax)*(dcos(xx)*(1.+y*
     1  (.183105d-2+y*(-.3516396496d-4+y*(.2457520174d-5
     2  +y*(-.240337019d-6)))))-z*dsin(xx)*(.04687499995+y
     3  *(-.2002690873d-3+y*(.8449199096d-5+y*
     4  (-.88228987d-6+y*.105787412d-6)))))*dsign(1.d0,x)
```

```
      endif
      return
      end
```

B.11.3   Root of Bessel Function

BJ0Root.fi

```
      function bj0root(rstart)
c Find a root of J_0 near rstart by Newton-Raphson method
      implicit real*8(a-h,o-z)
      parameter (xacc=1.0d-8)
      bj0root=rstart
      do 11 j=1,100
      dx=-besselj0(bj0root)/besselj1(bj0root)
      bj0root=bj0root-dx
      if(abs(dx).lt.xacc) return
11    continue
      return
      end
```

B.11.4   Gaussian Quadrature Coefficients

Gauss.fi

```
      subroutine gauleg(x1,x2,x,w,n)
c Calculate Gaussian quadrature nodes and weights
c Source: Numerical Recipes, Press, et al. 1992
c Reprinted with permission
      implicit real*8(a-h,o-z)
      dimension x(n), w(n)
      parameter (eps=3.d-14)
      m=(n+1)/2
      xm=0.5d0*(x2+x1)
      xl=0.5d0*(x2-x1)
      do 12 i=1,m
      z=dcos(3.141592654d0*(i-.25d0)/(n+.5d0))
1     continue
      p1=1.d0
      p2=0.d0
      do 11 j=1,n
      p3=p2
```

```
        p2=p1
        p1=((2.d0*j-1.d0)*z*p2-(j-1.d0)*p3)/j
11      continue
        pp=n*(z*p1-p2)/(z*z-1.d0)
        z1=z
        z=z1-p1/pp
        if(abs(z-z1).gt.eps)goto 1
        x(i)=xm-xl*z
        x(n+1-i)=xm+xl*z
        w(i)=2.d0*xl/((1.d0-z*z)*pp*pp)
        w(n+1-i)=w(i)
12      continue
        return
        end
```

B.11.5   Epsilon Series Extrapolation Algorithm

Epsilon.fi

```
        function epsilonn(n,psum)
c Extrapolate a series by epsilon algorithm
        implicit real*8(a-h,o-z)
        dimension eps(-1:200,0:200),psum(0:200)
        do 10 m=0,n
        eps(-1,m) = 0.0d0
        eps(0,m) = psum(m)
10      continue
        do 30 i=1,n
        do 30 m=n-i,0,-1
        eps(i,m)=eps(i-2,m+1)+1./(eps(i-1,m+1)-eps(i-1,m))
30      continue
        epsilonn=eps(n,0)
        return
        end
```

B.11.6   BFGS Optimization Scheme

BFGS.fi

```
        SUBROUTINE dfpmin(p,n,gtol,iter,fret,func,dfunc)
C Source: Numerical Recipes, Press, et al., 1992
c Reprinted with permission
```

```
        INTEGER iter,n,NMAX,ITMAX
        REAL*8 fret,gtol,p(n),func,EPS,STPMX,TOLX
        PARAMETER (NMAX=50,ITMAX=200,STPMX=100.,EPS=3.d-8,
     1   TOLX=4.*EPS)
        EXTERNAL dfunc,func
CU      USES dfunc,func,lnsrch
        INTEGER i,its,j
        LOGICAL check
        REAL*8 den,fac,fad,fae,fp,stpmax,sum,sumdg,sumxi,
     1   temp,test,dg(NMAX),g(NMAX),hdg(NMAX),
     2   hessin(NMAX,NMAX),pnew(NMAX),xi(NMAX)
        fp=func(p)
        call dfunc(p,g)
        sum=0.
        do 12 i=1,n
        do 11 j=1,n
        hessin(i,j)=0.
11      continue
        hessin(i,i)=1.
        xi(i)=-g(i)
        sum=sum+p(i)**2
12      continue
        stpmax=STPMX*max(dsqrt(sum),dfloat(n))
        do 27 its=1,ITMAX
        iter=its
        call lnsrch(n,p,fp,g,xi,pnew,fret,stpmax,check,func)
        fp=fret
        do 13 i=1,n
        xi(i)=pnew(i)-p(i)
        p(i)=pnew(i)
13      continue
        test=0.
        do 14 i=1,n
        temp=dabs(xi(i))/max(dabs(p(i)),1.)
        if(temp.gt.test)test=temp
14      continue
        if(test.lt.TOLX)return
        do 15 i=1,n
        dg(i)=g(i)
15      continue
        call dfunc(p,g)
```

```fortran
      test=0.
      den=max(fret,1.)
      do 16 i=1,n
      temp=dabs(g(i))*max(dabs(p(i)),1.)/den
      if(temp.gt.test)test=temp
16    continue
      if(test.lt.gtol)return
      do 17 i=1,n
      dg(i)=g(i)-dg(i)
17    continue
      do 19 i=1,n
      hdg(i)=0.
      do 18 j=1,n
      hdg(i)=hdg(i)+hessin(i,j)*dg(j)
18    continue
19    continue
      fac=0.
      fae=0.
      sumdg=0.
      sumxi=0.
      do 21 i=1,n
      fac=fac+dg(i)*xi(i)
      fae=fae+dg(i)*hdg(i)
      sumdg=sumdg+dg(i)**2
      sumxi=sumxi+xi(i)**2
21    continue
      if(fac**2.gt.EPS*sumdg*sumxi)then
      fac=1./fac
      fad=1./fae
      do 22 i=1,n
      dg(i)=fac*xi(i)-fad*hdg(i)
22    continue
      do 24 i=1,n
      do 23 j=1,n
      hessin(i,j)=hessin(i,j)+fac*xi(i)*xi(j)-fad*hdg(i)
     1 *hdg(j)+fae*dg(i)*dg(j)
23    continue
24    continue
      endif
      do 26 i=1,n
      xi(i)=0.
```

```
      do 25 j=1,n
      xi(i)=xi(i)-hessin(i,j)*g(j)
25    continue
26    continue
27    continue
      pause 'too many iterations in dfpmin'
      return
      END
c---------------------------------------------------------------
      SUBROUTINE lnsrch(n,xold,fold,g,p,x,f,stpmax,check,func)
C Source: Numerical Recipes, Press, et al., 1992
c Reprinted with permission
      INTEGER n
      LOGICAL check
      REAL*8 f,fold,stpmax,g(n),p(n),x(n),xold(n),func,
     1  ALF,TOLX
      PARAMETER (ALF=1.d-4,TOLX=1.d-7)
      EXTERNAL func
CU    USES func
      INTEGER i
      REAL*8 a,alam,alam2,alamin,b,disc,f2,fold2,rhs1,
     1  rhs2,slope,sum,temp,test,tmplam
      check=.false.
      sum=0.
      do 11 i=1,n
      sum=sum+p(i)*p(i)
11    continue
      sum=dsqrt(sum)
      if(sum.gt.stpmax)then
      do 12 i=1,n
      p(i)=p(i)*stpmax/sum
12    continue
      endif
      slope=0.
      do 13 i=1,n
      slope=slope+g(i)*p(i)
13    continue
      test=0.
      do 14 i=1,n
      temp=dabs(p(i))/max(dabs(xold(i)),1.)
      if(temp.gt.test)test=temp
```

```
14      continue
        alamin=TOLX/test
        alam=1.
1       continue
        do 15 i=1,n
        x(i)=xold(i)+alam*p(i)
15      continue
        f=func(x)
        if(alam.lt.alamin)then
        do 16 i=1,n
        x(i)=xold(i)
16      continue
        check=.true.
        return
        else if(f.le.fold+ALF*alam*slope)then
        return
        else
        if(alam.eq.1.)then
        tmplam=-slope/(2.*(f-fold-slope))
        else
        rhs1=f-fold-alam*slope
        rhs2=f2-fold2-alam2*slope
        a=(rhs1/alam**2-rhs2/alam2**2)/(alam-alam2)
        b=(-alam2*rhs1/alam**2+alam*rhs2/alam2**2)/(alam-alam2)
        if(a.eq.0.)then
        tmplam=-slope/(2.*b)
        else
        disc=b*b-3.*a*slope
        tmplam=(-b+dsqrt(disc))/(3.*a)
        endif
        if(tmplam.gt..5*alam)tmplam=.5*alam
        endif
        endif
        alam2=alam
        f2=f
        fold2=fold
        alam=max(tmplam,.1*alam)
        goto 1
        END
```

B.12   Summary

A summary of *Fortran* programs developed in this book and their page references are presented Table B.1.

TABLE B.1. Summary of Fortran programs.

Program/Subroutine	Page
TheisW	78
TheisWL	94
JaLoF	95
JaLoG	95
PaCoW	101
PaCoF	103
NeumanW1	112
NeumanW2	112
HantJcbW	132
HantNeuW	139
HantushH	143
VarPump	150
VarPumpL	152
Multi	217
InvHJ	258

Appendix C

MATHEMATICA MACROS

C.1 Preparatory Work

Once a *Mathematica* "Notebook" is open, the following commands should be entered:

```
In[1]:= SetDirectory["\mydir"];
In[2]:= <<"nlapinv.m"
In[3]:= <<"wfield.m"
In[4]:= << Graphics'Graphics'
```

In the above, the first input line changes the directory to where the macro files (nlapinv.m and wfield.m) are stored. The entry \mydir needs to be replaced by the actual directory name. The second and third lines load the two user supplied macros. The fourth line reads in a graphic package (part of the *Mathematica* software) for plotting use.

C.2 Numerical Inversion of Laplace Transform

NlapInv.m

```
BeginPackage["nlapinv'"]
(* Approximate inversion of Laplace transform by
   Stehfest algorithm. *)
```

```
NLapInv::usage = "NLapInv[expr, s, t, n] \n
expr =Laplace transform expression to be inverted\n
s =Laplace transform parameter\n
t = time\n
n = number of terms in series"

Begin["'Private'"]
csteh[n_, i_] := (-1)^(i + n/2) Sum[ k^(n/2) (2 k)! /
    ( (n/2 - k)! k! (k - 1)! (i - k)! (2 k - i)! ),
    { k, Floor[ (i+1)/2 ], Min[ i, n/2] } ] //N
NLapInv[F_, s_, t_, n_] := Log[2]/t Sum[csteh[2 Floor[n/2],i]*
    F /. s -> i Log[2]/t, {i, 1, 2 Floor[n/2]} ] //N
End[]
EndPackage[ ]
```

C.3   Well Functions and Drawdown for One-Aquifer Systems

WField.m

```
BeginPackage["wfield'", {"nlapinv'"} ]
(* Groundwater well field simulation package *)

(* Summary: This package provides the following well
   functions and their associated drawdowns:
Theis well function -> W(u)
Jacob-Lohman well function -> G(uw), F(uw,rho)
Papadopulos-Cooper well function -> W(u,alpha,rho),F(uw,alpha)
Hantush-Jacob well function -> W(u,beta)
Hantush well function -> H(u,betap)
Hantush-Neuman well functions->W1(u,beta,eta),W2(u,beta,eta)*)

WField::usage = " Well functions: \n
  WTheis[u]\n
  WJacobLohmanG[uw]\n
  WJacobLohmanF[uw,rho]\n
  WPapaCooper[u,alpha,rho]\n
  WPapaCooperF[uw,alpha]\n
  WHantushJacob[u,beta]\n
```

```
  WHantush[u,betap]\n
  WHantushNeuman1[u,beta,eta]\n
  WHantushNeuman2[u,beta,eta]\n
Drawdown functions:\n
  STheis[Q,T,S,x,y,t,xo(opt),yo(opt),to(opt)]\n
  SJacobLohman[sw,T,S,rw,x,y,t,xo(opt),yo(opt),to(opt)]\n
  SPapaCooper[Q,T,S,rw,rc,x,y,t,xo(opt),yo(opt),to(opt)]\n
  SHantushJacob[Q,T,S,lambda,x,y,t,xo(opt),yo(opt),to(opt)]\n
  SHantushP[Q,S,Kh,Kv,b,ell,d, r,z,t]\n
  SHantush[Q,T,S,lambdap,x,y,t,xo(opt),yo(opt),to(opt)]\n
  SHantushNeuman1[Q,T,S,lambda,eta,x,y,t,xo(opt),yo(opt),\n
     to(opt)]\n
  SHantushNeuman2[Q,T,S,lambda,eta,x,y,t,xo(opt),yo(opt),\n
     to(opt)]\n
Velocity functions:\n
  VTheis[Q,T,S,b,phi,x,y,t,xo(opt),yo(opt),to(opt)]\n
Arguments:\n
 u = (r^2 S)/(4 T t)\n
 uw = (rw^2 S)/(4 T t)\n
 alpha = rw^2 S/rc^2\n
 beta = r/lambda\n
 beta' = r/lambda' = r/(4 lambda)*(S'/S)^(1/2)\n
 rho = r/rw\n
 lambda = leakage factor = (b'T/K')^(1/2)\n
 lambda'=modified leakage factor=4 lambda (S/S')^(1/2)\n
 eta = S'/S\n
 Q = pumping rate\n
 T = transmissivity\n
 S = storativity\n
 x, y = location at which the drawdown is evaluated\n
 t = time at which the drawdown is evaluated\n
 xo, yo = well location (default = 0, 0)\n
 to = pumping startup time (default = 0)\n
 r=distance from the well=((x - xo)^2+(y - yo)^2)^(1/2)\n
 rw = well radius\n
 rc = large diameter well radius\n
 b'= thickness of aquitard\n
 K'= conductivity of aquitard\n
 S'= storativity of aquitard"

WTheis::usage = "Type ?WField"
```

```
WJacobLohmanG::usage = "Type ?WFfield"
WJacobLohmanF::usage = "Type ?WFfield"
WPapaCooper::usage = "Type ?WFfield"
WPapaCooperF::usage = "Type ?WFfield"
WHantushJacob::usage = "Type ?WFfield"
WHantush::usage = "Type ?WFfield"
WHantushNeuman1::usage = "Type ?WFfield"
WHantushNeuman2::usage = "Type ?WFfield"
STheis::usage = "Type ?WFfield"
SJacobLohman::usage = "Type ?WFfield"
SPapaCooper::usage = "Type ?WFfield"
SHantushJacob::usage = "Type ?WFfield"
SHantushP::usage = "Type ?WFfield"
SHantush::usage = "Type ?WFfield"
SHantushNeuman1::usage = "Type ?WFfield"
SHantushNeuman2::usage = "Type ?WFfield"
VTheis::usage = "Type ?WFfield"
Begin["'Private'"]

(* Theis *)
WTheis[u_] := - ExpIntegralEi[-u] //N
STheis[Q_,T_,S_,x_,y_,t_,xo_:0,yo_:0,to_:0]:=If[t<=to,0,
  Q/(4 Pi T) WTheis[((x-xo)^2+(y-yo)^2)S/(4 T(t-to))]]//N
VTheis[Q_,T_,S_,b_,phi_,x_,y_,t_,xo_:0,yo_:0,to_:0]:=If[
  t <= to,{0,0},-Q/(2 Pi b phi) / ((x-xo)^2+(y-yo)^2)*
  Exp[-((x-xo)^2+(y-yo)^2) S/(4 T(t-to))]{x-xo,y-yo}]//N

(* Jacob-Lohman *)
WJacobLohmanG[uw_]:= 1/NLapInv[1/(2 p) /
  BesselK[0,2 Sqrt[p]],p,1/uw,20] //N
WJacobLohmanF[uw_,rho_]:= NLapInv[ BesselK[0, 2 Sqrt[p]*
  rho] /p/BesselK[0, 2 Sqrt[p]],p,1/uw, 20] //N
SJacobLohman[sw_,T_,S_,rw_,x_,y_,t_,xo_:0,yo_:0,to_:0]:=
  If[t<=to,0,sw NLapInv[1/p BesselK[0,Sqrt[p S((x-xo)^2+
  (y-yo)^2)/T]]/BesselK[0,Sqrt[p S rw^2/T]],p,t-to,20]]//N

(* Papadopulos-Cooper *)
WPapaCooper[u_,alpha_,rho_]:= NLapInv[1/p BesselK[0,2*
  Sqrt[p]]/(Sqrt[p]/rho BesselK[1,2 Sqrt[p]/rho] +
  (p/alpha/rho^2) BesselK[0,2 Sqrt[p]/rho]),p,1/u,20]//N
WPapaCooperF[uw_,alpha_]:= NLapInv[1/p BesselK[0,2*
```

```
  Sqrt[p]]/(Sqrt[p] BesselK[1,2 Sqrt[p]] + (p/alpha)*
  BesselK[0,2 Sqrt[p]]),p,1/uw,20]//N
SPapaCooper[Q_,T_,S_,rw_,rc_,x_,y_,t_,xo_:0,yo_:0,
  to_:0]:=If[ t <= to, 0, Q/(4 Pi T) NLapInv[1/p 4*
  BesselK[0,Sqrt[p S((x-xo)^2+(y-yo)^2)/T]] / (2 *
  Sqrt[p S rw^2/T] BesselK[1,Sqrt[p S rw^2/T]] + p rc^2/
  T BesselK[0,Sqrt[p S rw^2/T]]),p,t-to,20] ]//N


(* Hantush-Jacob *)
WHantushJacob[u_,beta_]:= NLapInv[2/s*
  BesselK[0,Sqrt[beta^2+4 s]],s,1/u,20]
SHantushJacob[Q_,T_,S_,lambda_, x_,y_,t_,xo_:0,yo_:0,to_:0]:=
  If[t <= to, 0, Q/(4 Pi T) WHantushJacob[((x-xo)^2+(y-yo)^2)*
  S/(4 T(t-to)), Sqrt[(x-xo)^2+(y-yo)^2]/lambda] ]//N


(* Hantush Partially Penetrating Well*)
SHantushP[Q_,S_,Kh_,Kv_,b_,ell_,d_,r_,z_,t_]:= Q/(4 Pi Kh b)*
  (WTheis[ r^2 S/(4 Kh b t)] + 2 b/Pi/(ell-d) Sum[1/n (
  Sin[n Pi ell/b]-Sin[n Pi d/b]) Cos[n Pi (b-z)/b]*
  WHantushJacob[r^2 S/(4 Kh b t), n Pi r/b Sqrt[Kv/Kh]],
  {n,1,5}]) //N


(* Hantush Leaky Aquifer Small Time H*)
WHantush[u_,beta_]:=NLapInv[2/s *
  BesselK[0, Sqrt[8 beta Sqrt[s] + 4 s]], s, 1/u, 20]
SHantush[Q_,T_,S_,Bp_,x_,y_,t_,xo_:0,yo_:0,to_:0]:=
  If[t <= to, 0, Q/(4 Pi T) WHantush[((x-xo)^2+(y-yo)^2)S/
  (4 T(t-to)), Sqrt[(x-xo)^2+(y-yo)^2]/Bp] ]//N


(* Hantush-Neuman Case 1 *)
WHantushNeuman1[u_,beta_,eta_]:= NLapInv[2/s BesselK[0,
  Sqrt[4 s + 2 beta Sqrt[eta s]*
  Coth[2 Sqrt[eta s]/beta]]],s,1/u,20]
SHantushNeuman1[Q_,T_,S_,lambda_,eta_,x_,y_,t_,xo_:0,yo_:0,
  to_:0]:= If[t <= to, 0, Q/(4 Pi T)*
  WHantushNeuman1[((x-xo)^2+(y-yo)^2) S/(4 T(t-to)),
  Sqrt[(x-xo)^2+(y-yo)^2]/lambda, eta] ]//N


(* Hantush-Neuman Case 2 *)
WHantushNeuman2[u_,beta_,eta_]:= NLapInv[2/s BesselK[0,
  Sqrt[4 s + 2 beta Sqrt[eta s]*
```

```
    Tanh[2 Sqrt[eta s]/beta]]],s,1/u,20]
SHantushNeuman2[Q_,T_,S_,lambda_,eta_,x_,y_,t_,xo_:0,
  yo_:0,to_:0]:=  If[t <= to, 0, Q/(4 Pi T)*
  WHantushNeuman2[((x-xo)^2+(y-yo)^2) S/(4 T(t-to)),
  Sqrt[(x-xo)^2+(y-yo)^2]/lambda, eta] ]//N


End[]
EndPackage[ ]
```

C.4   Drawdown for Multilayered Aquifer Systems

MWField.m

```
BeginPackage["mwfield'", {"nlapinv'"} ]

(* Groundwater well field simulation package *)
(*    for multilayered aquifer system *)

(* Naming convention:
One-aquifer system:
S1: drawdown for aquifer 1
Sp0, Sp1: drawdown for aquitard 0 and 1
Two-aquifer system:
Sij: drawdown in aquifer i due to pumping in aquifer j
Q: pumping rate
Kp,Sp,bp: hydraulic conducivity, storativity, and thickness of
    aquitard (The first and last aquitard can be eliminated by
    assigning Kp = 0, and by keeping bp at an arbitrary nonzero
    constant.)
T, S: transmissivity and storativity of aquifer
x,y,z,t: x, y, z, t
xo,yo: pump location (can be omitted if zero)
zo: bottom elevation of aquitard
to: pump starting time (can be omitted if zero) *)

MWField::usage = " Drawdown functions:\n
S1[Q,Kp0,Sp0,bp0,T1,S1,Kp1,Sp1,bp1,x,y,t,xo(opt),yo(opt),\n
  to(opt)]\n
Sp0[Q,Kp0,Sp0,bp0,T1,S1,Kp1,Sp1,bp1,x,y,z,t,xo(opt),yo(opt),\n
  zo(opt),to(opt)]\n
Sp1[...]\n
```

```
S11[Q,Kp0,Sp0,bp0,T1,S1,Kp1,Sp1,bp1,T2,S2,Kp2,Sp2,bp2,x,y,t,\n
   xo(opt),yo(opt),to(opt)]\n
S21[...], S12[...], S22[...]"

S1::usage = "Type ?MWField"
Sp0::usage = "Type ?MWField"
Sp1::usage = "Type ?MWField"
S11::usage = "Type ?MWField"
S21::usage = "Type ?MWField"
S12::usage = "Type ?MWField"
S22::usage = "Type ?MWField"

Begin["'Private'"]

(* One-aquifer-two-aquitard system *)

S1[Q_,Kp0_,Sp0_,bp0_,T1_,S1_,Kp1_,Sp1_,bp1_,x_,y_,t_,xo_:0,
   yo_:0,to_:0]:= If[ t <= to, 0, Q/(2 Pi T1) NLapInv[(1/p)*
   BesselK[0,Sqrt[(p/T1)(Kp0 g[Kp0,Sp0,bp0,p] / bp0 + Kp1*
   g[Kp1,Sp1,bp1,p] / bp1 + S1)] r[x,y,xo,yo]], p, t-to, 10] ]
Sp0[Q_,Kp0_,Sp0_,bp0_,T1_,S1_,Kp1_,Sp1_,bp1_,x_,y_,z_,t_,
   xo_:0,yo_:0,zo_:0,to_:0]:= If[t <= to,0,Q/(2 Pi T1)*
   NLapInv[BesselK[0,Sqrt[(p/T1)(Kp0 g[Kp0,Sp0,bp0,p] /
   bp0 + Kp1 g[Kp1,Sp1,bp1,p] / bp1 + S1)] r[x,y,xo,yo]]*
   G2[Kp1,Sp1,bp1,z,p,zo], p, t-to, 10] ]
Sp1[Q_,Kp0_,Sp0_,bp0_,T1_,S1_,Kp1_,Sp1_,bp1_,x_,y_,z_,t_,
   xo_:0,yo_:0,zo_:0,to_:0]:=If[t <= to,0,Q/(2 Pi T1)*
   NLapInv[BesselK[0,Sqrt[(p/T1)(Kp0 g[Kp0,Sp0,bp0,p] /
   bp0 + Kp1 g[Kp1,Sp1,bp1,p] / bp1 + S1)] r[x,y,xo,yo]]*
   G1[Kp1,Sp1,bp1,z,p,zo], p, t-to, 10] ]

(* Two-aquifer-three-aquitard system *)

S11[Q1_,Kp0_,Sp0_,bp0_,T1_,S1_,Kp1_,Sp1_,bp1_,T2_,S2_,Kp2_,
   Sp2_,bp2_, x_,y_,t_,xo_:0,yo_:0,to_:0] := If[t<=to, 0, Q1/
   (2 Pi T1) NLapInv[(1/p)(-(k1[Kp0,Sp0,bp0,T1,S1,Kp1,Sp1,bp1,
   T2,S2,Kp2,Sp2,bp2,p]^2 - a22[Kp1,Sp1,bp1,T2,S2,Kp2,Sp2,bp2,
   p]) BesselK[0,k1[Kp0,Sp0,bp0,T1,S1,Kp1,Sp1,bp1,T2,S2,Kp2,
   Sp2,bp2,p] r[x,y,xo,yo]] + (k2[Kp0,Sp0,bp0,T1,S1,Kp1,Sp1,
   bp1,T2,S2,Kp2,Sp2,bp2,p]^2 - a22[Kp1,Sp1,bp1,T2,S2,Kp2,Sp2,
   bp2,p]) BesselK[0,k2[Kp0,Sp0,bp0,T1,S1,Kp1,Sp1,bp1,T2,S2,
```

```
      Kp2,Sp2,bp2,p] r[x,y,xo,yo]]) / (Sqrt[(a11[Kp0,Sp0,bp0,T1,
      S1,Kp1,Sp1,bp1,p] - a22[Kp1,Sp1,bp1,T2,S2,Kp2,Sp2,bp2,p])^2+
      4 a12[Kp1,Sp1,bp1,T1,p] a21[Kp1,Sp1,bp1,T2,p]]),p,t-to,10]]
S22[Q2_,Kp0_,Sp0_,bp0_,T1_,S1_,Kp1_,Sp1_,bp1_,T2_,S2_,Kp2_,
      Sp2_,bp2_,x_,y_,t_,xo_:0,yo_:0,to_:0] := If[t<=to, 0, Q2/
      (2 Pi T2) NLapInv[(1/p)(-(k1[Kp0,Sp0,bp0,T1,S1,Kp1,Sp1,bp1,
      T2,S2,Kp2,Sp2,bp2,p]^2 - a11[Kp0,Sp0,bp0,T1,S1,Kp1,Sp1,bp1,
      p]) BesselK[0,k1[Kp0,Sp0,bp0,T1,S1,Kp1,Sp1,bp1,T2,S2,Kp2,
      Sp2,bp2,p] r[x,y,xo,yo]] + (k2[Kp0,Sp0,bp0,T1,S1,Kp1,Sp1,
      bp1,T2,S2, Kp2,Sp2,bp2,p]^2 - a11[Kp0,Sp0,bp0,T1,S1,Kp1,Sp1,
      bp1,p]) BesselK[0,k2[Kp0,Sp0,bp0,T1,S1,Kp1,Sp1,bp1,T2,S2,
      Kp2,Sp2,bp2,p] r[x,y,xo,yo]]) / (Sqrt[(a11[Kp0,Sp0,bp0,T1,
      S1,Kp1,Sp1,bp1,p] - a22[Kp1,Sp1,bp1,T2,S2,Kp2,Sp2,bp2,p])^2+
      4 a12[Kp1,Sp1,bp1,T1,p] a21[Kp1,Sp1,bp1,T2,p]]),p,t-to,10]]
S12[Q2_,Kp0_,Sp0_,bp0_,T1_,S1_,Kp1_,Sp1_,bp1_,T2_,S2_,Kp2_,
      Sp2_,bp2_, x_,y_,t_,xo_:0,yo_:0,to_:0] := If[t<=to, 0, Q2/
      (2 Pi T2) NLapInv[(1/p) a12[Kp1,Sp1,bp1,T1,p](-BesselK[0,
      k1[Kp0,Sp0,bp0, T1,S1,Kp1,Sp1,bp1,T2,S2,Kp2,Sp2,bp2,p]*
      r[x,y,xo,yo]] + BesselK[0,k2[Kp0,Sp0,bp0,T1,S1,Kp1,Sp1,bp1,
      T2,S2,Kp2,Sp2,bp2,p] r[x,y,xo,yo]]) / (Sqrt[(a11[Kp0,Sp0,
      bp0,T1,S1, Kp1,Sp1,bp1,p] - a22[Kp1,Sp1,bp1,T2,S2,Kp2,Sp2,
      bp2,p])^2 + 4 a12[Kp1,Sp1,bp1,T1,p] a21[Kp1,Sp1,bp1,T2,p]]),
      p, t-to, 10] ]
S21[Q1_,Kp0_,Sp0_,bp0_,T1_,S1_,Kp1_,Sp1_,bp1_,T2_,S2_,Kp2_,
      Sp2_,bp2_,x_,y_,t_, xo_:0,yo_:0,to_:0] := If[t<=to, 0, Q1/
      (2 Pi T1) NLapInv[(1/p) a21[Kp1,Sp1,bp1,T2,p](-BesselK[0,
      k1[Kp0,Sp0,bp0,T1,S1,Kp1,Sp1,bp1,T2,S2,Kp2,Sp2,bp2,p]*
      r[x,y,xo,yo]] + BesselK[0,k2[Kp0,Sp0,bp0,T1,S1,Kp1,Sp1,bp1,
      T2,S2,Kp2,Sp2,bp2,p] r[x,y,xo,yo]]) / (Sqrt[(a11[Kp0,Sp0,
      bp0,T1,S1, Kp1,Sp1,bp1,p] - a22[Kp1,Sp1,bp1,T2,S2,Kp2,Sp2,
      bp2,p])^2 + 4 a12[Kp1,Sp1,bp1,T1,p] a21[Kp1,Sp1,bp1,T2,p]]),
      p, t-to, 10] ]
(* Matrix coefficients *)
a11[Kp0_,Sp0_,bp0_,T1_,S1_,Kp1_,Sp1_,bp1_,p_]:= (p/T1)*
   (Kp0 g[Kp0,Sp0,bp0,p]/bp0+Kp1 f[Kp1,Sp1,bp1,p]/bp1+S1)
a22[Kp1_,Sp1_,bp1_,T2_,S2_,Kp2_,Sp2_,bp2_,p_]:= (p/T2)*
   (Kp1 f[Kp1,Sp1,bp1,p]/bp1+Kp2 g[Kp2,Sp2,bp2,p]/bp2+S2)
a12[Kp1_,Sp1_,bp1_,T1_,p_]:= -(p/T1) Kp1 h[Kp1,Sp1,bp1,p]/bp1
a21[Kp1_,Sp1_,bp1_,T2_,p_]:= -(p/T2) Kp1 h[Kp1,Sp1,bp1,p]/bp1
(* Eigen values *)
k1[Kp0_,Sp0_,bp0_,T1_,S1_,Kp1_,Sp1_,bp1_,T2_,S2_,Kp2_,Sp2_,
```

```
bp2_,p_] := Sqrt[(a11[Kp0,Sp0,bp0,T1,S1,Kp1,Sp1,bp1,p] +
  a22[Kp1,Sp1,bp1,T2,S2,Kp2,Sp2,bp2,p] - Sqrt[
  (a11[Kp0,Sp0,bp0,T1,S1,Kp1,Sp1,bp1,p] - a22[Kp1,Sp1,
  bp1,T2,S2,Kp2,Sp2,bp2,p])^2 + 4 a12[Kp1,Sp1,bp1,T1,p]*
  a21[Kp1,Sp1,bp1,T2,p]]) / 2]
k2[Kp0_,Sp0_,bp0_,T1_,S1_,Kp1_,Sp1_,bp1_,T2_,S2_,Kp2_,Sp2_,
  bp2_,p_] := Sqrt[(a11[Kp0,Sp0,bp0,T1,S1,Kp1,Sp1,bp1,p] +
  a22[Kp1,Sp1,bp1,T2,S2,Kp2,Sp2,bp2,p] + Sqrt[
  (a11[Kp0,Sp0,bp0,T1,S1,Kp1,Sp1,bp1,p] - a22[Kp1,Sp1,
  bp1,T2,S2,Kp2,Sp2,bp2,p])^2 + 4 a12[Kp1,Sp1,bp1,T1,p]*
  a21[Kp1,Sp1,bp1,T2,p]]) / 2 ]


(* General Auxiliary functions *)
f[Kp_,Sp_,bp_,p_]:= Sqrt[Sp bp/(Kp p)] Coth[Sqrt[Sp bp p/Kp]]
h[Kp_,Sp_,bp_,p_]:= Sqrt[Sp bp/(Kp p)] Csch[Sqrt[Sp bp p/Kp]]
g[Kp_,Sp_,bp_,p_]:= If[Kp <= 0, 0, Sqrt[Sp bp/(Kp p)]*
  Tanh[ Sqrt[Sp bp p/Kp]]]
G1[Kp_,Sp_,bp_,z_,p_,zo_]:= (1/p)Sech[Sqrt[Sp bp p/Kp]]*
  Cosh[Sqrt[Sp bp p/Kp] (zo+bp-z)/bp ]
G2[Kp_,Sp_,bp_,z_,p_,zo_]:= (1/p)Sech[Sqrt[Sp bp p/Kp]]*
  Cosh[Sqrt[Sp bp p/Kp] (z-zo)/bp ]
r[x_,y_,xo_,yo_] := Sqrt[(x-xo)^2+(y-yo)^2]


End[]
EndPackage[ ]
```

C.5   Aquifer Parameter Identification

WInv.m

```
BeginPackage["WInv'", {"wfield'"} ]

(* Aquifer parameter determination by pumping test *)
(* Summary: This package performs parameter determination of
storativity (S) and transmissivity (T) for Theis aquifer by
visual or nonlinear least square method. For visual method,
ContourPlotLSqError[...] and Plot3DLSqError[...] provide
contour and surface plot, respectively. For nonlinear least
square, ParInv[...] performs the inverse and plot the match.*)

WInv::usage="
```

```
!!! Requires a data file named PumpData.m !!! \n
ContourPlotLSqError[Tmin_,Tmax_,Smin_,Smax_], \n
Plot3DLSqError[Tmin_,Tmax_,Smin_,Smax_], \n
where Tmin,Tmax = plotting range of transmissivity T\n
Smin, Smax = plotting range of storativity S \n
ParInv[IniT_,IniS_], \n
where IniT, IniS = initial guess of T and S "
ContourPlotLSqError::usage="Type ?WInv"
Plot3DLSqError::usage="Type ?WInv"
ParInv::usage="Type ?WInv"


Begin["'Private'"]

(* Input drawdown data *)
Get[ "PumpData'" ];

(* number of data *)
ndat = Length[drawdown];

(* Define square error *)
LSqError[T_,S_] :=Sum[ ( STheis[Q,T,S,r,0,time[[i]]] -
   drawdown[[i]] )^2, {i,1,ndat} ]

(* Contour plot and surface plot functions *)
ContourPlotLSqError[Tmin_,Tmax_,Smin_,Smax_] := ContourPlot[
  Log[10, LSqError[10^T, 10^S]], {T, Log[10, Tmin],
  Log[10, Tmax]}, {S, Log[10, Smin], Log[10, Smax]},
  PlotPoints -> 20, FrameLabel->
  {"\!\(log\_10\ T\)", "\!\(log\_10\ S\)"}]
Plot3DLSqError[Tmin_,Tmax_,Smin_,Smax_] := Plot3D[Log[10,
  LSqError[10^T, 10^S]], {T, Log[10, Tmin], Log[10, Tmax]},
  {S,Log[10,Smin],Log[10,Smax]}, PlotPoints-> 20, AxesLabel->
  {"\!\(log\_10\ T\)","\!\(log\_10\ S\)","\!\(log\_10\ E\)"}]

(* Function for nonlinear least square inversion *)
ParInv[IniT_,IniS_] := CompoundExpression[
IniST = IniS/IniT;
c1 = Sum[drawdown[[i]] Exp[-r^2 st/(4 time[[i]])],{i,1,ndat}]/
  Sum[ WTheis[r^2 st/(4 time[[i]])] Exp[-r^2 st/
  (4 time[[i]])], {i,1,ndat}];
ST = st /. FindRoot[Sum[( c1 WTheis[r^2 st/(4 time[[i]])] -
```

```
    drawdown[[i]]) WTheis[ r^2 st/(4 time[[i]])], {i,1,ndat}],
    {st,IniST}];
T = 1/(4 Pi/Q c1) /. st -> ST;
S = ST T;
Print[StringForm["Transmissivity = '',Storativity = '''",T,S]];
fieldata=Table[ {time[[i]], drawdown[[i]]}, {i,1,ndat}];
plot1=Plot[STheis[Q,T,S,r,0,t], {t,0.0001,time[[ndat]]},
    DisplayFunction-> Identity];
plot2 = ListPlot[fieldata, DisplayFunction -> Identity];
Show[{plot1,plot2},AxesLabel-> {"time", "drawdown"},Prolog->
    PointSize[.02], DisplayFunction -> $DisplayFunction] ]


End[]
EndPackage[ ]
```

PumpData.m

```
(* Data from Walton [1970] as cited in McElwee [1980] *)
(* The units must be consistent, as no unit conversion
    is performed in the package.  Example: Q = m^3/hr, r = m,
    s = m, t = hr *)

Q = 29.41 (*ft^3/min*)
r = 824 (*ft*)

drawdown = {0.3, 0.7, 1.3, 2.1, 3.2, 3.6, 4.1, 4.7, 5.1, 5.3,
5.7, 6.1, 6.3, 6.7, 7.0, 7.5, 8.3, 8.5, 9.2, 9.7, 10.2, 10.9}
(*ft*)

time = {3, 5, 8, 12, 20, 24, 30, 38, 47, 50, 60, 70, 80,
90, 100, 130, 160, 200, 260, 320, 380, 500} (*min*)
```

C.6   Stochastic Theis Solution

Stochast.m

```
BeginPackage["stochast'"]
(* Stochastic pumping well solution *)
(* Functions:
Ms[]: Mean of logarithmic drawdown
```

Vs[]: Variance of Logarithmic drawdown *)

```
Stochast::usage =
"Ms[MLogT,MLogS,VLogT,VLogS,VLogTLogS,Q,r,t]\n
Vs[MLogT,MLogS,VLogT,VLogS,VLogTLogS,Q,r,t]\n
MLogT: Mean of logarithmic transmissivity\n
MLogS: Mean of logarithmic storativity\n
VLogT: Variance of logarithmic transmissivity\n
VLogS: Variance of logarithmic storativity\n
VLogTLogS: Covariance of logarithmic transmissivity and\n
    storativity\n
Q: discharge\n
r: radial distance\n
t: time"
Ms:usage = "Type ?Stochast"
Vs:usage = "Type ?Stochast"

Begin["'Private'"]

Ms[MLogT_,MLogS_,VLogT_,VLogS_,VLogTLogS_,Q_,r_,t_]:=
    -Q/4/Pi/10^MLogT ExpIntegralEi[r^2/4/t 10^(MLogS-MLogT)] +
    1/2 f1[MLogT,MLogS,Q,r,t] VLogT + 1/2 f2[MLogT,MLogS,Q,r,t]*
    VLogS + f3[MLogT,MLogS,Q,r,t] VLogTLogS;
Vs[MLogT_,MLogS_,VLogT_,VLogS_,VLogTLogS_,Q_,r_,t_]:=
    f4[MLogT,MLogS,Q,r,t]^2 VLogT+f5[MLogT,MLogS,Q,r,t]^2 VLogS+
    2 f4[MLogT,MLogS,Q,r,t] f5[MLogT,MLogS,Q,r,t] VLogTLogS;

f1[MLogT_,MLogS_,Q_,r_,t_]:= Q Log[10]^2/4/Pi*
    10^(-MLogT) Exp[-ustar[MLogT,MLogS,r,t]]*
    (ustar[MLogT,MLogS,r,t] - Exp[ustar[MLogT,MLogS,r,t]]*
    ExpIntegralEi[-ustar[MLogT,MLogS,r,t]] -2)
f2[MLogT_,MLogS_,Q_,r_,t_]:= Q Log[10]^2/4/Pi 10^(-MLogT)*
    ustar[MLogT,MLogS,r,t] Exp[-ustar[MLogT,MLogS,r,t]]
f3[MLogT_,MLogS_,Q_,r_,t_]:= Q Log[10]^2/4/Pi 10^(-MLogT)*
    Exp[-ustar[MLogT,MLogS,r,t]] (1-ustar[MLogT,MLogS,r,t])
f4[MLogT_,MLogS_,Q_,r_,t_]:= Q Log[10]/4/Pi 10^(-MLogT)*
    (Exp[-ustar[MLogT,MLogS,r,t]] +
    ExpIntegralEi[-ustar[MLogT,MLogS,r,t]])
f5[MLogT_,MLogS_,Q_,r_,t_]:= Q Log[10]/4/Pi 10^(-MLogT)*
    Exp[-ustar[MLogT,MLogS,r,t]]
ustar[MLogT_,MLogS_,r_,t_]:= r^2/4/t 10^(MLogS-MLogT)
```

```
End[]
EndPackage[ ]
```

C.7   Double Porosity Medium

DblPoro.m

```
BeginPackage["dblporo'", {"nlapinv'"} ]

(* Groundwater well field simulation package *)
(*    for double porosity medium *)

(* Naming convention:
pf: pressure for fracture system
pm: pressure for porous block system
Q: pumping rate
r: radial distance
t: time
b: thickness of aquifer
km: mobility coefficient of porous block
kf: mobility coefficient of fracture
Cm: bulk compressibility of porous block
Cf: bulk compressibility of fracture
alpha: geometric factor *)

DblPoro::usage = "Pressure Solutions:\n
PfWarrenRoot1[Q,b,km,kf,Cm,Cf,alpha,r,t]\n
PmWarrenRoot1[Q,b,km,kf,Cm,Cf,alpha,r,t]\n
PfWarrenRoot2[Q,b,km,kf,Cm,Cf,alpha,r,t]\n
PfBarenblatt[Q,b,km,kf,Cm,alpha,r,t]\n
PmBarenblatt[Q,b,km,kf,Cm,alpha,r,t]"
PfWarrenRoot1::usage = "Type ?DblPoro"
PmWarrenRoot1::usage = "Type ?DblPoro"
PfWarrenRoot2::usage = "Type ?DblPoro"
PfBarenblatt::usage = "Type ?DblPoro"
PmBarenblatt::usage = "Type ?DblPoro"

Begin["'Private'"]

(* Warren-Root Model *)
```

```
PfWarrenRoot1[Q_,b_,km_,kf_,Cm_,Cf_,alpha_,r_,t_]:=
  -Q/(2Pi b kf) NLapInv[(1/s) BesselK[0,Sqrt[alpha km/kf
  + s Cf/kf - alpha^2 km^2/kf/(Cm s+alpha km)]r],s,t,10]
PmWarrenRoot1[Q_,b_,km_,kf_,Cm_,Cf_,alpha_,r_,t_]:=
  -Q alpha km/(2Pi b kf) NLapInv[1/s/(Cm s + alpha km)*
  BesselK[0,Sqrt[alpha km/kf + s Cf/kf - alpha^2 km^2/
  kf/(Cm s+ alpha km)] r],s,t,10]
PfWarrenRoot2[Q_,b_,km_,kf_,Cm_,Cf_,alpha_,r_,t_]:=
  -Q/(4Pi b kf) (Log[4 kf t/(Cm+Cf)/r^2] - EulerGamma +
  ExpIntegralEi[-alpha km t (Cm+Cf)/Cf/Cm] -
  ExpIntegralEi[-alpha km t/Cm])

(* Barenblatt Model *)
PfBarenblatt[Q_,b_,km_,kf_,Cm_,alpha_,r_,t_]:= -Q alpha km /
  (2Pi b kf) NLapInv[1/s/(Cm s + alpha km) BesselK[0, Sqrt[
  alpha km/kf - alpha^2 km^2/kf/(Cm s + alpha km)] r],s,t,10]
PmBarenblatt[Q_,b_,km_,kf_,Cm_,alpha_,r_,t_]:= -Q alpha^2 *
  km^2/(2Pi b kf) NLapInv[1/s/(Cm s+alpha km)^2 BesselK[0,
  Sqrt[alpha km/kf-alpha^2 km^2/kf/(Cm s+alpha km)]r],s,t,10]

End[]
EndPackage[ ]
```

Bibliography

[1] Abramowitz, M. and Stegun, I.A., *Handbook of Mathematical Functions*, Dover, 1972.

[2] Anderson, M.P. and Woessner, W.W., *Applied Groundwater Modeling, Simulation of Flow and Advective Transport*, Academic Press, 1992.

[3] Aral, M.M., *Ground Water Modeling in Multilayer Aquifers, Steady Flow*, Lewis Publ., 1990.

[4] Aral, M.M., *Ground Water Modeling in Multilayer Aquifers, Unsteady Flow*, Lewis Publ., 1990.

[5] Aris, R., *Vector, Tensors, and the Basic Equations of Fluid Mechanics*, Prentice-Hall, 1962.

[6] Barenblatt, G.I. and Zheltov, Y.P., "On fundamental equations of flow of homogeneous liquids in naturally fractured rocks," (in Russian), *Dokl. Akad. Nauk.*, **132**, 545-548, 1960.

[7] Barenblatt, G.I., Zheltov, Y.P. and Kochina, I.N., "Basic concepts in the theory of seepage of homogeneous liquids in fractured rocks," (in Russian), *Prik. Matem. i Mekh.*, **24**, 852-864, 1960.

[8] Barenblatt, G.I., "On certain boundary value problems for the equations of seepage of a liquid in fractured rocks," (in Russian), *Prik. Matem. i Mekh.*, **27**, 348-350, 1963.

[9] Barenblatt, G.I., Entov, V.M. and Ryzhik, V.M., *Theory of Fluid Flows through Natural Rocks*, Kluwer, 1990.

[10] Batu, V., *Aquifer Hydraulics, A Comprehensive Guide to Hydrogeologic Data Analysis*, Wiley, 1998.

[11] Bear, J., *Dynamics of Fluids in Porous Media*, American Elsevier, New York, 1972.

[12] Bear, J., *Hydraulics of Groundwater*, McGraw-Hill, 1979.

[13] Bear, J. and Bachmat, Y., Introduction to modeling of transport phenomena in porous media, Kluwer Academic Publishers, Dordrecht/Boston/London, 1990.

[14] Bear, J., Cheng, A.H.-D., Sorek, S., Ouazar, D. and Herrera, I., (eds.), *Seawater Intrusion in Coastal Aquifers—Concepts, Methods and Practices*, Kluwer Academic Publishers, Dordrecht/Boston/London, 625 p., 1999.

[15] Bendat, J. and Piersol, G.A., *Random Data Analysis and Measurement Procedures*, 2nd ed., Wiley, 1986.

[16] Biswas, A.K., *History of Hydrology*, Elsevier, New York, 1970.

[17] Bochevr, F.M., *Designing Groundwater Extraction Facilities* (in Russian), Stroiizdzt, Moscow, 1976.

[18] Boulton, N.S., "Analysis of data from non-equilibrium pumping tests allowing for delayed yield from storage," *Proc. Inst. Civil Engrs. (London)*, **26**, 469-482, 1963.

[19] Carslaw, H.S. and Jaeger, J.C., *Conduction of Heat in Solids, 2nd ed.*, Oxford Science, 1959.

[20] Cedergren, H.R., *Seepage, Drainage and Flow Nets*, 2nd ed., Wiley, New York, 1977.

[21] Chen, Z.X., "Transient flow of slightly compressible fluids through double-porosity, double permeability systems—A state-of-the-art review," *Transport in Porous Media*, **4**, 147-184, 1989.

[22] Cheng, A.H.-D. and Ou, K., "An efficient Laplace transform solution for multiaquifer systems," *Water Resources Research*, **25**, 742–748, 1989.

[23] Cheng, A.H.-D. and Lafe, O.E., "Boundary element solution for stochastic groundwater flow: Random boundary condition and recharge," *Water Resour. Res.*, **27**, 231–242, 1991.

[24] Cheng, A.H.-D., Abousleiman, Y., Detournay, C. and Roegiers, J-C., "Source solution for a generalized dual porosity model," *Structure et Comportement Mécanique des Géomatériaux-Colloque René Houpert*, Nancy, France, eds. F. Homand, F. Masrouri and J-P. Tisot, 109-116, 1992.

[25] Cheng, A.H.-D. and Morohunfola, O.K., "Multilayered leaky aquifer systems: I. Pumping well solution," *Water Resources Research*, **29**, 2787-2800, 1993.

[26] Cheng, A.H.-D. and Morohunfola, O.K., "Multilayered leaky aquifer systems: II. Boundary element solution," *Water Resources Research*, **29**, 2801-2811, 1993.

[27] Cheng, A.H.-D., Abousleiman, Y., Ruan, F. and Lafe, O.E., "Boundary element solution for stochastic groundwater flow: Temporal weakly stationary problems," *Water Resour. Res.*, **29**, 2893–2908, 1993.

[28] Cheng, A.H.-D., Antes, H. and Ortner, N., "Fundamental solutions of products of Helmholtz and polyharmonic operators," *Eng. Analy. Boundary Elements*, **14**, 187–191, 1994.

[29] Cheng, A.H.-D., Sidauruk, P. and Abousleiman, Y., "Approximate inversion of the Laplace transform," *Mathematica Journal*, **4**, 76-82, 1994.

[30] Cheng, A.H.-D. and Ouazar, D., "Theis solution under aquifer parameter uncertainty," *Ground Water*, **33**, 11–15, 1995.

[31] Cheng, A.H.-D. and Sidauruk, P., "A groundwater flow Mathematica package," *Ground Water*, **34**, 41–48, 1996.

[32] Chorley, D.W. and Frind, E.O., "An iterative quasi three-dimensional finite element model for heterogeneous multiaquifer systems," *Water Resour. Res.*, **14**, 943-952, 1978.

[33] Churchill, R.V., *Operational Mathematics*, 2nd ed., McGraw-Hill, 1966.

[34] Closmann, P.J., "An aquifer model for fissured reservoirs," *Soc. Pet. Eng. J., Trans. AIME*, **259**, 385-398, 1975.

[35] Cooper, H.H. and Jacob, C.E., "A generalized graphical method for evaluating formation constants and summarizing well-field history," *Trans. Am. Geophys. Union*, **27**, 526-534, 1946.

[36] Dagan, G., "Models of groundwater flow in statistically homogeneous porous formation," *Water Resour. Res.*, **15**, 47-63, 1979.

[37] Dagan, G., "A note on higher-order corrections of the head covariances in steady aquifer flow," *Water Resour. Res.*, **21**, 573-578, 1985.

[38] Dagan, G., *Flow and Transport in Porous Formations*, Springer-Verlag, 1989.

[39] Darcy, H.P.G., *Les fontaines publiques de la ville de Dijon*, V. Dalmont, Paris, 1856.

[40] Davies, B. and Martin, B., "Numerical inversion of Laplace transform: a survey and comparison of methods," *J. Comp. Phys.*, **33**, 1-32, 1979.

[41] Detournay, E. and Cheng, A.H.-D., "Fundamentals of poroelasticity," Chap. 5 in *Comprehensive Rock Engineering: Principles, Practice & Projects, II, Analysis and Design Method*, ed. C. Fairhurst, Pergamon, 113-171, 1993.

[42] Domenico, P.A., *Concepts and Models in Groundwater Hydrology*, McGraw-Hill, 1972.

[43] Domenico, P.A. and Schwartz, F.W., *Physical and Chemical Hydrogeology*, Wiley, 1990.

[44] Driscoll, F.G., *Groundwater and Wells*, 2nd ed., Johnson Division, 1986.

[45] Dupuit, A.J.E.J., *Etudes théoriques et pratiques sur le mouvement des eaux courantes*, 2nd ed., Carilian-Goeury, Paris, 1863.

[46] Earlougher, R.C. Jr., *Advances in Well Test Analysis*, Monograph Series, **5**, Society of Petroleum Engineers, 1977.

[47] EL Harrouni, K., Ouazar, D., Walters, G.A. and Cheng, A.H.-D., "Groundwater optimization and parameter estimation by genetic algorithm and dual reciprocity boundary element method," *Eng. Analy. Boundary Elements*, **18**, 287–296, 1996.

[48] Forchheimer, P., "Über die Ergiebigkeit von Brunnen-Anlagen und Sickerschlitzen," *Zeitschrift der Architekten und Ingenieur-Verein*, **32**, 539-564, 1886.

[49] Freeze, R.A., "A stochastic-conceptual analysis of one-dimensional groundwater flow in nonuniform homogeneous media," *Water Resour. Res.*, **11**, 725-741, 1975.

[50] Gambolati, G., "Transient free surface flow to a well: An analysis of theoretical solutions," *Water Resour. Res.*, **12**, 27-39, 1976.

[51] Gelhar, L.W., "Stochastic subsurface hydrology from theory to applications," *Water Resour. Res.*, **22**, 135S-145S, 1986.

[52] Greenberg, M.D., *Application of Green's Functions in Science and Engineering*, Prentice-Hall, 1971.

[53] Greenberg, M.D., *Advanced Engineering Mathematics*, Prentice-Hall, 1988.

[54] Gringarten, A.C., "Interpretation of tests in fissured and multilayered reservoirs with double-porosity behavior: Theory and practice," *J. Pet. Tech.*, April, 549-564, 1984.

[55] Gutjahr, A.L. and Gelhar, L.W., "Stochastic models of subsurface flow: Infinite versus finite domains and stationarity," *Water Resour. Res.*, **17**, 337-350, 1981.

[56] Hantush, M.S. and Jacob, C.E., "Plane potential flow of ground water with linear leakage," *Trans. Am. Geophys. Union*, **35**, 917-936, 1954.

[57] Hantush, M.S. and Jacob, C.E., "Non-steady radial flow in an infinite leaky aquifer," *Trans. Am. Geophys. Union*, **36**, 95-100, 1955.

[58] Hantush, M.S., "Analysis of data from pumping tests in leaky aquifers," *Trans. Am. Geophys. Union*, **37**, 702-714, 1956.

[59] Hantush, M.S., "Modification of the theory of leaky aquifers," *J. Geophys. Res.*, **65**, 3713-3726, 1960.

[60] Hantush, M.S., "Hydraulics of wells," in *Advances in Hydroscience, Vol. 1*, ed. V.T. Chow, Academic Press, 281-442, 1964.

[61] Hemker, C. J., and C. Maas, "Unsteady flow to wells in layered and fissured aquifer systems," *J. Hydrol.*, **90**, 231-279, 1987.

[62] Hennart, J.P., Yates, R. and Herrera, I., "Extension of the integrodifferential approach to inhomogeneous multiaquifer systems," *Water Resour. Res.*, **17**, 1044-1050, 1981.

[63] Herrera, I. and Figueroa, G.E., "A correspondence principle for the theory of leaky aquifers," *Water Resour. Res.*, **5**, 900-904, 1969.

[64] Herrera, I., "Theory of multiple leaky aquifers," *Water Resour. Res.*, **6**, 185-193, 1970.

[65] Herrera, I., "A review of the integrodifferential equations approach to leaky aquifer mechanics," in *Advances in Groundwater Hydrology*, ed. Z. Saleem, Am. Water Resour. Assoc., 29-47, 1976.

[66] Herrera, I. and Yates, R., "Integrodifferential equations for systems of leaky aquifers and applications: 3. A numerical method of unlimited applicability," *Water Resour. Res.*, **13**, 725-732, 1977.

[67] Herrera, I., Hennart, J.P. and Yates, R., "A critical discussion of numerical models for multiaquifer systems," *Adv. Water Resour.*, **3**, 159-163, 1980.

[68] Holland, J.H., *Adaptation in Natural and Artificial Systems, An Introductory Analysis with Applications to Biology, Control, and Artificial Intelligence*, MIT Press, 1992.

[69] Hörmander, H., *Linear Partial Differential Operators*, Springer-Verlag, 1963.

[70] Huyakorn, P.S. and Pinder, G.F., *Computational Methods in Subsurface Flow*, Academic Press, 1983.

[71] Indelman, P., Dagan, G., Cheng, A.H.-D. and Ouazar, D., "Sensitivity analysis of flow in multiple leaky aquifer systems," *J. Hyd. Eng.*, ASCE, **122**, 41–45, 1996.

[72] Jacob, C.E., "The flow of water in an elastic artesian aquifer," *Trans. Am. Geophys. Union*, **21**, 574-586, 1940.

[73] Jacob, C.E., "Notes on determining permeability by pumping tests under water table conditions," USGS Mimeo. Rept., 1944.

[74] Jacob, C.E., "Radial flow in leaky artesian aquifer," *Trans. Am. Geophys. Union*, **27**, 198-205, 1946.

[75] Jacob, C.E. and Lohman, S.W., "Nonsteady flow to a well of constant drawdown in an extensive aquifer," *Trans. Am. Geophys. Union*, **33**, 559-569, 1952.

[76] Khalili-Naghadeh, N. and Valliappan, S., "Flow through fissured porous media with deformable matrix: Implicit formulation," *Water Resour. Res.*, **27**, 1703-1709, 1991.

[77] Leake, S.A., Leahy, P.P. and Navoy, A.S., "Documentation of a computer program to simulate transient Leakage from confining units using the modular finite-difference ground-water flow model," USGS Open File Report 94-59, 1994.

[78] Leap, D.I., "Geological occurrence of groundwater," Chap. 1 in *The Handbook of Groundwater Engineering*, ed. J.W. Delleur, CRC Press, 1998.

[79] MacDonald, J.R., "Accelerated convergence, divergence, iteration, extrapolation, and curve fitting," *J. Appl. Phys.*, **10**, 3034-3041, 1964.

[80] McElwee, C.D. and Yukler, M.A., "Sensitivity of groudwater models with respect to variations in transmissivity and storage," *Water Resour. Res.*, **14**, 451-459, 1978.

[81] McDonald, M.G. and Harbaugh, A.W., "A modular three-dimensional finite-difference ground-water flow model," Techniques of Water-Resources Investigations, 06-A1, USGS, 1988.

[82] Meinzer, O.E., "Compressibility of elasticity of artesian aquifers," *Econ. Geol.*, **23**, 263-291, 1928.

[83] Mercer, J.W., Thomas, S.D. and Ross, B., *Parameters and Variables Appearing in Repository Models*, U.S. Regulatory Commission, NUREG/CR-3066, Washington, D.C., 1982.

[84] Moench, A.F., "Double-porosity models for a fissured groundwater reservoir with fracture skin," *Water Resour. Res.*, **20**, 831-846, 1984.

[85] Morohunfola, O.K., "Analytical and numerical solutions for multiple leaky aquifer systems," Ph.D. dissertation, Univ. Delaware, Newark, 1992.

[86] Morris, D.A. and Johnson, A.I., "Summary of hydrologic and physical properties of rock and soil materials as analyzed by the Hydrologic Laboratory of the U.S. Geological Survey," U.S. Geol. Survey Water Supply Paper 1839-D, 1967.

[87] Neuman, S.P., "Transient flow of groundwater to wells in multiple-aquifer systems," Doctoral dissertation, University of California, Berkeley, 1968.

[88] Neuman, S.P. and Witherspoon, P.A., "Theory of flow in a confined two aquifer system," *Water Resour. Res.*, **5**, 803-816, 1969.

[89] Neuman, S.P., "Theory of flow in unconfined aquifers considering delayed response of the water table," *Water Resour. Res.*, **8**, 1031-1045, 1972.

[90] Neuman, S.P., "Supplementary comments on 'Theory of flow in unconfined aquifers considering delayed response of the water table'," *Water Resour. Res.*, **9**, 1102-1103, 1973.

[91] Neuman, S.P., "Analysis of pumping test data from anisotropic unconfined aquifers considering delayed gravity response," *Water Resour. Res.*, **11**, 329-342, 1975.

[92] Neuman, S.P., Preller, C. and Narashiman, T.N., "Adaptive explicit-implicit quasi three-dimensional finite element model of flow and subsidence in multiaquifer systems," *Water Resour. Res.*, **18**, 1551-1561, 1982.

[93] Oberhettinger, F., *Fourier Transforms of Distributions and Their Inverses*, Academic Press, 1973.

[94] Ouazar, D. and Cheng, A.H.-D., "Application of genetic algorithms in water resources," Chap. 7 in *Groundwater Pollution Control*, ed. K. Katsifarakis, WIT Press, 293-316, 1999.

[95] Papadopulos, I.S. and Cooper, H.H., "Drawdown in a well of large diameter," *Water Resour. Res.*, **3**, 241-244, 1967.

[96] Piessens, R., "A bibliography on numerical inversion of the Laplace transform and applications," *J. Comp. Appl. Math.*, **1**, 115-126, 1975.

[97] Piessens, R. and Dang, N.D.P., "A bibliography on numerical inversion of the Laplace transform and applications: a supplement," *J. Comp. Appl. Math.*, **2**, 225-228, 1976.

[98] Pinder, G.F. and Gray, W.G., *Finite Element Simulation in Surface and Subsurface Hydrology*, Academic Press, 1977.

[99] Pirson, S.J., "Performance of fractured oil reservoirs," *Bull. Am. Assoc. Petrol. Geol.*, **37**, 232-244, 1953.

[100] Premchitt, J., "A technique in using integrodifferential equations for model simulation of multiaquifer systems," *Water Resour. Res.*, **17**, 162-168, 1981.

[101] Press, W.H., Teukolsky, S.A., Vetterling, W.T. and Flannery, B.P., *Numerical Recipes in Fortran, the Art of Scientific Computing*, 2nd ed., Cambridge Univ. Press, 1992.

[102] Prickett, T.A., "Type-curve solution to aquifer tests under water-table conditions," *Ground Water*, **3**, 5-14, 1965.

[103] Reed, J.E., "Type curves for selected problems of flow to wells in confined aquifers," Chap. B3 in *Techniques of Water Resources Investigations, Book 3*, USGS, 1980.

[104] Rouse, H. and Ince, S., *History of Hydraulics*, Dover, 1957.

[105] Rubin, Y., Hubbard, S.S., Wilson, A. and Cushey, M.A., "Aquifer characterization," Chap. 10 in *The Handbook of Groundwater Engineering*, ed. J.W. Delleur, CRC Press, 1998.

[106] Sauveplane, C.M., "Analytical modelling of transient flow to wells in complex aquifer systems: Characterization and productivity of these systems," Doctoral Dissertation, Univ. Alberta., 1987.

[107] Scheidegger, A.E., *The Physics of Flow Through Porous Media*, 3rd ed., Univ. Toronto Press, Toronto, 1974.

[108] Smith, B.T., et al., *Matrix Eigensystem Routines—EISPACK Guide*, 2nd ed., v. 6, *Lecture Notes in Computer Science*, Springer-Verlag, 1976.

[109] Smith, N., *Man and Water*, Charles Scribner's Sons, 1975.

[110] Stehfest, H., "Numerical inversion of Laplace transforms," *Comm. ACM*, **13**, 47-49 and 624, 1970.

[111] Streltsova, T.D., "Hydrodynamics of groundwater flow in a fractured formation," *Water Resour. Res.*, **12**, 405-414, 1976.

[112] Sun, N-Z., *Inverse Problems in Groundwater Modeling*, Kluwer, 1994.

[113] Terzaghi, K., *Theoretical Soil Mechanics*, Wiley, New York, 1943.

[114] Theis, C.V., "The relation between the lowering of the piezometric surface and the rate and duration of discharge of a well using groundwater storage," *Trans. Am. Geophys. Union*, **16**, 519-524, 1935.

[115] Thiem, A., "Die Ergiebigkeit artesischer Bohrlöcher, Schachtbrunnen, und Filtergallerien," *Journal für Gasbeleuchtung und Wasserversorgung*, **14**, 450-467, 1870.

[116] Todd, D., *Groundwater Hydrology, 2nd ed.*, Wiley & Sons, 1980.

[117] Vuković, M. and Soro, A., *Hydraulics of Water Wells, Theory and Application*, Water Resources Publ., 1992.

[118] Walton, W.C., *Groundwater Resources Evaluation*, McGraw-Hill, 1970.

[119] Walton, W.C., *Groundwater Pumping Tests*, Lewis Publ., 1987.

[120] Warren, J.E. and Root, P.J., "The behavior of naturally fractured reservoirs," *Soc. Pet. Eng. J., Trans. AIME*, **228**, 245-255, 1963.

[121] Wenzel, L.K., "Methods for determining permeability of water-bearing materials," *U.S. Geol. Survey, Water Supply Paper* 887, 1942.

[122] White, R.R. and Clebsch, A., "C.V. Theis, the man and his contributions to hydrogeology," In: *Selected Contributions to Ground-Water Hydrology by C.V. Theis, and a Review of His Life and Work,* (ed.) A. Clebsch, USGS Water-Supply Paper 2415, 1994.

[123] Wilson, C.R. and Witherspoon, P.A., "Steady state flow in rigid networks of fractures," *Water Resour. Res.,* **10**, 328-335, 1974.

[124] Witherspoon, P.A., Wang, J.S.Y., Iwai, K. and Gale, J.E., "Validity of cubic law for fluid flow in a deformable rock fracture," *Water Resour. Res.,* **16**, 1016-1024, 1980.

[125] Wolfram, S., *The Mathematica Book, 3rd ed.,* Wolfram Media/Cambridge Univ. Press, 1996.

Index